FIRST ECOLOGY

It is true: Man is the microcosm: I am my world.
Ludwig Wittgenstein

FIRST ECOLOGY

SECOND EDITION

Alan Beeby

Anne-Maria Brennan

London South Bank University

OXFORD

UNIVERSITY PRESS

OXFORD

UNIVERSITY PRESS

Great Clarendon Street, Oxford OX2 6DP

Oxford University Press is a department of the University of Oxford.
It furthers the University's objective of excellence in research, scholarship,
and education by publishing worldwide in

Oxford New York

Auckland Bangkok Buenos Aires Calcutta Cape Town Chennai
Dar es Salaam Delhi Hong Kong Istanbul Karachi Kolkata
Kuala Lumpur Madrid Melbourne Mexico City Mumbai Nairobi
São Paulo Shanghai Taipei Tokyo Toronto
and associated companies in Berlin Ibadan

Oxford is a registered trade mark of Oxford University Press
in the UK and in certain other countries

Published in the United States
by Oxford University Press Inc., New York

The moral rights of the authors have been asserted

© Alan Beeby and Anne-Maria Brennan

Database right Oxford University Press (maker)

First edition published 1997
Second edition published 2004

British Library Cataloguing in Publication Data
Data available

0-19-926124-5

Typeset by Newgen Imaging Systems (P) Ltd, Chennai, India
Printed by Ashford Colour Press Limited, Gosport, Hampshire

■ CONTENTS

SECOND THOUGHTS

We used Ludwig Wittgenstein's assertion about the nature of reality to set the theme for the first edition of this book, and continue to misinterpret its meaning in this new edition. Confident that he would have forgiven us then, we quickly dismissed any thought of replacing it here. We still wish to keep humankind at the centre stage and try again to introduce ecology from a human perspective, using our species as both example and as principal actor.

Human beings are, each of us, a microcosm illustrating some basic ecological principles. With the right perspective, we can learn much ecology simply by comparing ourselves with other species. We are the products of natural selection and environmental pressures shaped us. Even the most technologically advanced peoples are subject to the ecological processes of the world as we find it. Its ecological organization places checks on what we might do and we are beginning to realize that our understanding of these is critical to the future of our species.

Ecology helps us to understand our species, offering new perspectives on our biology and our position in the world. In this book, the human context is the starting point for each chapter. Many of the basic ecological ideas are introduced using common experience and are illustrated with examples with which most people are familiar, or that touch upon an aspect of human evolution or cultural development. Once the ideas have been explored in purely ecological terms, each chapter ends with an application or an environmental issue in which human beings have been a central player.

The prologue (*First Words*) introduces the reader to the general organization of the book and the approach it adopts. It uses a current pollution problem from which our species may itself be at threat, but using a very different ecological scenario. This encourages the reader to adopt a questioning approach, warning in advance that there are few simple or clear-cut answers. It also demonstrates the range of factors that need to be considered to address some of the questions. The prologue ends with a brief set of directions, describing some of the features of the book and its associated web site.

This second edition retains several of the themes of the first edition, although the sequence of chapters has been changed. We continue to use Mediterranean-type ecosystems as a recurring example, though the chapter dealing with their organization is now earlier in the sequence (Chapter 5). Not only are these ecosystems an interesting and well-worked test of whether ecological communities repeat themselves, they allow us also to look in detail at the organization of one community, one which we can re-visit throughout the text. It is also perhaps the one community type relatively close to most peoples of the world. Within the Mediterranean basin itself, environmental degradation is becoming a major issue, as it is in equivalent communities around the globe, but here we can match the history of this change to our well-documented past.

Our primary aim has been to produce a text which is accessible and easy to read. Rather than a textbook which is highly sectionalized, we have sought to produce a narrative that encourages the reader to complete a chapter. The connections between subject areas are given by references to related sections and on the companion web site we provide routes through the text linking various topics. This allows the reader to pick out all the relevant sections, and thereby build a more complete story of subjects we re-visit.

To maintain the flow of the text, we have decided not to cite supporting references in the text; this has been a difficult choice (which some of our reviewers have repeatedly warned us against!). To keep our text clear, we name key authors (but omit dates) and provide the references for each section at the end of the book, so our sources are easy to trace. Nevertheless, we are aware of our obligations to the reader and the prologue gives an example and encourages students to cite references properly in their own work.

Boxes are used throughout the book to provide illustrative examples in some detail or extension material that students may find useful. Several new boxes have been added and others brought up to date. New exercises have been included at the end of each chapter. Additional exercises and the answers (including some of those used in the first edition) are now also available at the web site. This also contains up-to-date information on the web sites cited at the end of each chapter.

First Ecology companion web site

First Ecology companion web site

First Ecology is supported by a range of electronic resources, available through the book's companion web site at **www.oup.com/uk/booksites/biosciences**. The web site is divided into two basic sections:

(i) Student support materials

This includes

- **Answers to exercises** posed in the book, to allow students to self-check their answers as they attempt the exercises
- A section entitled '**Routes**' that provides a detailed map of how some key themes are developed through the book and includes some **case study titles**, relating to each theme, that could direct student-based research into these key areas. Students may find this useful if they wish to research or improve their understanding of these themes, and we hope it will also demonstrate how the various themes in ecology are linked.
- A **web link library** including all the URLs listed at the end of chapters in the book. Additional web links to other sites for specific topics are also listed on a sectionalised basis. Please feel free to suggest links that should be added to this site, preferably offering a commentary on the content (see below for contact details). We hope this will serve to keep the web links up-to-date, though of course neither we, nor the publishers, can accept responsibility for any of the links posted here. We will endeavour to review these links periodically and offer our own commentary.

(ii) Tutor support materials

Access to the tutor support materials is **free**, but is password-protected to enable tutor-only access. To obtain a password simply register with us by visiting the *First Ecology* companion web site, clicking on the 'lecturer resource' button, and following the 'Not yet registered?' instructions.

The tutor-only part of the site includes:

- **Additional exercises** to supplement those in the book
- Answers to the additional exercises
- **Downloadable figures** — those for which we hold copyright in the book
- **Course outlines** — bullet-pointed PowerPoint slides covering the main topics in each chapter, kindly devised by Erik Scully, Towson University, USA.
- A feedback site where corrections or suggestions can be posted. There are better questions, and undoubtedly better answers to the questions, than those we have posed and we would welcome suggestions on both counts. Similarly, any comments on the text *per se* would also be very useful, especially those which are attributed to which we can respond.

Lecturers might also consider some of the suggested topics for further study using the 'Routes' function that picks out themes and connections across the different chapters. Again we would welcome suggestions for any additional '**Routes**' covering other topics and themes that tutors might develop or, indeed, resources on their own or other web-sites to which links could be posted.

Contact us

To contact us with comments and suggestions, simply click through to our companion web site from **www.oup.com/uk/booksites/biosciences** and click on the 'contact us' button.

Grateful thanks

Our debt to the team of reviewers who persevered through the manuscript is immense and the final version has been improved immeasurably by their insights and perceptive suggestions. We hope this final version is a close approximation to what they collectively judged to be a better text. With their help we hope that most of our errors have been corrected.

Our grateful thanks go to Institutio Veneto Scienze Lettre ed Arti (VSLA), Cathy Bach, Eastern Michigan University (USA), J. Franklin, Guy Beaufoy (WWF), Tony Bradshaw (UK), Andy Bright (UK), Paddy Coker (UK), Eleanor Cohn, Wolverhampton University (UK), Marty Condon, Cornell University (USA), Mike Dobson, Manchester Metropolitan University (UK), Terence D. Fitzgerald, Cortland College (USA), Kevin Fort, University of California Davis(USA), Alastair Grant, University of East Anglia (UK), Rex Lowe, Bowling Green State University (USA), Alex Meinesz (University of Nice), Jonathan Mitchley Imperial College (UK), Margaret Ramsey (Royal Botanic Gardens Kew), Vinnie Jones (USA) Mark Seaward (University of Bradford), Walker Smith, Virginia Institute of Marine Science (USA), Walter J Tabachnick, University of Florida (USA), and Catherine Toft, University of California (USA). We particularly wish to thank Walker Smith for providing Box 8.6 (iron fertilization of marine phytoplankton) and Erik Scully, Towson University, for compiling the bulleted slides of each chapter. Our thanks also to Richard Dawkins and especially to Jonathan Crowe and Laura Hodgson of OUP who have been immensely supportive throughout the work.

In addition, we also wish to thank again all those whose help and cooperation in the first edition endures in the form of the second, including Ambio, Martin Angel, Blackwell Scientific Publications, the Broads Authority, Bob Carling, Cambridge University Press, Luca Cavalli-Sforza, John Currey, John Dodd, John Feltwell, French Ministry of Culture, Ken Giller, Elliot Gingold, Lee Hannah and Conservation International, Mike Harding, Gary Haynes, Alan Hopkins, John Hopkins, T. Jaffre, Steve Jansen, Unversity of Leuven (Belgium), B. Jedrzejewska, Susan Jenks, Joint Nature Conservation Committee, Tim Johns, Hefin Jones, Nancy Laurenson, Rachel Leech, George Lees, the Linnean Society of London, Joe Lopez-Real, Lynn Margulis, Lord May of Oxford, the late Chris Mead, Patrick Morgan, NASA, Natural History Museum, Kate Neale, Mike Newman, Mike Nicholls, Simon Parfitt, Panos Institute, Arthur Penhally, Val Porter, George Potts, Dominic Recaldin, Roger Reeves, John Rodwell, the Royal Entomological Society of London, Peter Saville, Candy D'Sa, Helen Sharples, Liz Sestito, Roger Tidman, Martin Tribe, Will Wadell, and Robert Wayne.

Finally, we have to thank our families, friends and colleagues who have supported us, which included moral support (Larry Richmond and Jackie Beeby), holding the bicycle (Ralph Beeby), or by suggesting ways in which the book could be improved, mainly by not writing about ecology (Kate Beeby).

ACKNOWLEDGMENTS

Cover

Courtesy of NASA Visible Earth/Earth Observatory. Image compiled by Reto Stockli, Robert Simmon and the MODIS Earth Imaging Teams using data from the MODIS Terra Satellite at the Goddard Space Flight Center.

First words

Frontispiece

Photograph courtesy of Dr Mette Mauritzen, Norwegian Polar Institute

Four- or five-year-old female on drift ice in the Northwen Barents Sea, August 1999.

Figure 1. Photograph courtesy of Visible Earth, NASA, Jacques Descloitre and the MODIS Land Rapid Response Team. From the satellite Terra, taken June 7, 2001.

Figure 3. Drawn from data in Carlsen *et al.* (1992).

Chapter 1

Frontispiece—Alan Beeby

Figure 1.5. Modified, from Cavalli-Sforza, L. L. 2001. *Genes, Peoples and Languages*, Penguin.

Chapter 2

Frontispiece—Anne-Maria Brennan

Plates

2.1—(a) Andy Bright; (b) Anne-Maria Brennan; (c) Roger Tidman; (d) John Feltwell
2.2—Natural History Museum London
2.3—Robert Wayne
2.4—Anne-Maria Brennan
2.5—Alan Beeby

Figures

2.1—The Linnaean Society of London
2.5—Anne-Maria Brennan
2.6—(a) and (b) Anne-Maria Brennan

2.11—Based on findings of Schluter, D. 1994. Experimental evidence that competition promotes divergence in adaptive radiation *Science 266*, 798–801.
2.12—Anne-Maria Brennan
2.13—Anne-Maria Brennan
2.14—Redrawn from Ammerman, A. and Cavalli-Sforza, L. L. 1971. Measuring the rate of spread of early farming in Europe. *Man 6*, 674–688.

Chapter 3

Frontispiece—Alan Beeby

Plates

3.1—Bialowieza forest—Alan Beeby
3.2—Regenerating forest—Anne-Maria Brennan
3.3—Elephant grazing—Nancy Laurenson
3.4—Elephant damage—Nancy Laurenson
3.5—Attaching polar bear collars—Photograph courtesy of Dr Mette Mauritzen, Norwegian Polar Institute

Figures

3.6—Data redrawn with kind permission from Jedrzejewska, B., Okarma, H., Jedrezejewska, W., and Milkowski, L. 1994. Effects of exploitation and protection on forest structure, ungulate density and wolf predation in Bialowieza Primeval Forest, Poland. *Journal of Applied Ecology 31*, 664–676.
3.8—Redrawn from Mauritzen, M., Derocher, A. E., Wiig, O., Belikov, S. E., Boltunov, A. N., Hansen, E., and Garner, G. W. 2002. Using satellite telemetry to define spatial population structure in polar bears in the Norwegian and western Russian Arctic. *Journal of Applied Ecology 39*, 79–90.
3.9—Micropropagation unit, Royal Botanic Gardens Kew.
3.11—Rhino (Martin B Withers/FLPA.)
3.12—Nancy Laurenson
3.13—Diagram compiled from data from Ashley, M. V., Melnick, D. J., and Western, D. 1990. Conservation genetics of the Black Rhinocerus (*Diceros bicornis*).

1. Evidence from the mitochondrial DNA of three populations. *Conservation Biology 4*, 71–77 and Worldwide Fund for Nature.

Table 3.1—Table compiled from data supplied by the Worldwide Fund for Nature (http://www.world-wildlife.org/spec) and the International Union for Nature Conservation (http://www.iucn.org).

Chapter 4

Plates

4.1—Roger Tidman
4.2—Anne-Maria Brennan

Figures

4.3—Anne-Maria Brennan
4.5—Christoph Sheidegger
4.6—Anne-Maria Brennan
4.7—Alan Beeby
4.8—Redrawn from Crombie, A. C. 1946. Further experiments on insect competition. *Proceedings of the Royal Society of London, Series B 133*, 76–109.
4.14—Anne-Maria Brennan
4.15—Professor Alex Meimesz, University of Nice.

Chapter 5

Frontispiece—Alan Beeby and Ralph Beeby

Plates

5.1—Sclerophyllous leaves—Alan Beeby
5.2—Kalymnos phrygana—Alan Beeby
5.3—Mallee—Kate Neale
5.4—Languedoc maquis—Alan Beeby
5.5—Chaparral—John Feltwell
5.6—Gulley erosion on Kalymnos—Alan Beeby
5.7—Colonization of volcanic lava—John Feltwell
5.8—Photo montage of sand dune succession—Alan Beeby and Anne-Maria Brennan
5.9—*Carpobrotus*—Anne-Maria Brennan
5.10—Pilat—Alan Beeby
5.11—Eucalpytus—Alan Beeby

Figures

5.3 and 5.4—Modified after Trabaud, L. 1981. Man and fire: impacts on Mediterranean vegetation. In: F. Di Castri *et al.* (eds), *Mediterranean-type Shrublands*. Elsevier, Amsterdam.
5.6—Modified after Di Castri, F. 1981. Mediterranean-type shrublands of the world. In: F. Di Castri *et al.* (eds), *Mediterranean-type Shrublands*. Elsevier, Amsterdam, pp. 1–52.
5.7—Drawn using data from Cody, M. L. and Mooney, H. A. 1978. Convergence versus nonconvergence in Mediterranean-climate ecosystems. *Annual Review of Ecology and Systematics 9*, 265–351.

Chapter 6

Frontispiece—Anne-Maria Brennan

Plates

6.1—Anne-Maria Brennan
6.2—John Feltwell

Figures

6.1—NOAA
6.2—Redrawn from Zscheile, F. P. and Comar, C. L. 1941. Influence of preparative procedure on the purity of chlorophyll components as shown by absorption spectra. *Botanical Gazette 102*, 463–481.
6.3—Rachel Leech
6.4—Derived from data from Schultz, E. D. 1970. De CO_2—Gaswechsel de Buche (*Fagus sylvatica* L.) in Abhangigkeit von den Klimafaktoren in Feiland. *Flora Jena 159*, 177–232; Schultz, E. D., Fuchs, M., and Fuchs, M. I. 1977a. Spatial distribution of photosynthetic capacity and performance in a mountain spruce forest in Northern Germany. I. Biomass distribution and daily CO_2 uptake in different crown layers. *Oecologia 29*, 43–61; and Schultz, E. D., Fuchs, M., and Fuchs, M. I. 1977b. Spatial distribution of photosynthetic capacity and performance in a mountain spruce forest in northern Germany. III. The significance of the evergreen habit. *Oecologia 30*, 239–248.
6.5—After Orshan, G. 1963. Seasonal dimorphism of desert and mediterranean chamaephytes and its significance as a factor in their water economy. In: A. J. Rutter and F. H. Whitehead (eds), *The Water Relations of Plants*. Wiley, New York.
6.7 and 6.8—After Swift, M. J., Heal, O. W., and Anderson, J. M. 1979. *Decomposition in Terrestrial Eosystems*. Blackwell, Oxford.

6.11—Anne-Maria Brennan

6.12 and 6.13—Drawn from data from Golley, F. B. 1960. Energy dynamic of an old-field community. *Ecological Monographs 30*, 187–200.

6.14 (a) and (b)—Data from Whittaker, R. H. 1975. *Communities and Ecosystems*, 2nd edn. Macmillan, New York.

6.14 (c) and 6.15—Data from Varley, G. C. 1970. The concept of energy flow applied to a woodland community. In: A. Watson (ed.), *Animal Populations in Relation to their Food Resource*. Blackwell, Oxford.

6.16—Drawn using data from Polischuk, S. C., Nortstrom, R. J., and Ramsay, M. A. 2002. Body burdens and tissue concentrations of organochlorines in polar bears (*Ursus maritimus*) vary during seasonal fasts. *Environmental Pollution 118*, 29–39.

6.17—Redrawn with modifications from Duckham, A. N. (1976). Environmental constraints. In: A. N. Duckham, J. G. W. Jones, and E. H. Roberts (eds), *Food Production and Nutrient Cycles*. North Holland Publishing Company, Amsterdam.

6.18—After Tivy, J. 1990. *Agricultural Ecology*. Longman, Harlow.

6.19—After Balch, C. C. and Reid, J. T. 1976. The effiicinecy of conversion of animal feed and protein into animal products. In: A. N. Duckham, J. G. W. Jones, and E. H. Roberts (eds), *Food Production and Nutrient Cycles*. North Holland Publishing Company, Amsterdam.

6.20—After Tivy, J. 1990. *Agricultural Ecology*. Longman, Harlow.

6.21—Based on data from Slesser, M. 1975. Energy requirements of agriculture. In: J. Lenihan and W. W. Fletcher (eds), *Food Agriculture and the Environment*. Blackie, Galsgow & London; and Tivy, J. 1990. *Agricultural Ecology*. Longman, Harlow.

6.22—Guy Beaufoy

6.23—Derived from Smith, D. F. and Hill, D. M. 1975. Natural agricultural ecosystems. *Journal of Environmental Quality 4*, 143–145

Chapter 7

Frontispiece—Morgan/A. Pengally

Plates

7.1—John Feltwell

7.2—Anne-Maria Brennan

7.3—Venice Institute of Science, Letters and Arts

7.4—NASA

7.5—George Lees

7.6—NASA

7.7—Provided by Jonathan Mitchley; photographed by Mike Griggs; copyright TML.

7.8—Provided by Roger Reeves and reproduced by permission of T. Jaffre.

7.9—Anne-Maria Brennan

Figures

7.5—From data after Reinoso, J. C. M. 2001. Vegetation changes and groundwater abstraction in SW Donana, Spain. *Journal of Hydrology 242*, 197–209.

7.6—Modified from Machita (1973) in *Carbon in the Biosphere*. United States, NTIS, Washington DC.

7.8—Redrawn from Odum, E. P. 1989. *Ecology and our Endangered Life-support Systems*. Sinauer Associates, Sunderland, MA.

7.9—John Dodd

7.11—(a) Ken Giller; (b) Joe Lopez-Real

7.13—Broads Authority

7.14—Drawn from data after Bettinetti, A., Pypaet, P., and Sweerts, J.-P. 1996. Application of an integrated management approach to the restoration project of the lagoon of Venice. *Journal of Environmental Management 46*, 207–227.

7.15—After Bradshaw, A. D. 1984. Ecological Principlesand land reclamation practice. *Landscape Planning, 11*, 35–48.

7.18—From Pan, J.-X. 1591. *A review of river flooding control* (in Chinese).

Chapter 8

Frontispiece—Alan Beeby

Plates

8.1—Alan Beeby

8.2a—Panos Picture

8.2b—Photograph courtesy of Earth Observatory, NASA GSFC, Jacques Descloitre and the MODIS Land Rapid Response Team. From the satellite Acqua, taken April 18, 2003.

8.3—Photograph taken by the crew of the space shuttle (STS037-152-091), taken April 1991, courtesy of NASA/Johnson Space Center.

8.4—Alan Beeby

8.5—John Feltwell

8.6—Nancy Laurenson

8.7—Anne-Maria Brennan

8.8—Candy D'Sa

8.9—John Feltwell

8.10—Paddy Coker

8.11—Paddy Coker

8.12—Alan Beeby

8.13—Photograph courtesy of Earth Observatory, NASA GSFC, Jacques Descloitre and the MODIS Land Rapid Response Team. From the satellite Acqua, taken April 20, 2003.

8.14—Redrawn with the kind permission of Stott, P. A. *et al.* and which appeared in Stott, P. A., Tett, S. F. B., Jones, G. S., Allen, M. R., Mitchell, J. F. B., and Jenkins, G. J. 2000. External control of 20th century temperature by natural and anthropogenic forcings. *Science 290*, 2133–2137. We are grateful to Dr Tim Johns in helping to supply the figures from the Hadley Research Centre of the UK Meteorological Office.

Figures

8.2—John Feltwell

8.3—Alan Beeby

8.8—After Whittaker, R. H. 1975. *Communities and Ecosystems*, 2nd edn. MacMillan, New York.

8.13—John Feltwell

8.17 and 8.19—Redrawn using data from Houghton, J. T., Jenkins, G. J., and Ephraums, J. J. (eds). 1990. *Climate Change: The IPCC Scientific Assessment*. Cambridge University Press, Cambridge.

8.18—Redrawn using data principally from the UK Natural Research Council 1989. *Our Future World: Global Environmental Research*.

8.20—Data from Whittaker, R. H. and Likens, G. E. 1973., cited by Paul Colinvaux in *Ecology 2*, John Wiley, New York.

8.21—Drawn from data in Behling, H. 2002. Carbon storage increases by major forest ecosystems in tropical South America since the Last Glacial Maximum and the early Holocene. *Global and Planetary Change 33*, 107–116.

Table 8.1—Based on Forman, R. T. T. 1995. *Land Mosaics. The Ecology of Landscapes and Regions.* Cambridge University Press, Cambridge.

Table 8.2—Data from Houghton, J. T., Jenkins, G. J., and Ephraums, J. J. (eds). 1990. *Climate Change: The IPCC Scientific Assessment.* Cambridge University Press, Cambridge.

Chapter 9

Frontispiece—Mike Harding

Plate 9.1—Photograph courtesy of Earth observatory, NASA, GSFC, Jacques Descloitve, and the MODIS Land Rapid Response Team. Image M.: ev23565-Nepal. A20023000505.

Figures

9.2—After MacArthur, R. H. and Wilson, E. O. 1967. *The Theory of Island Biogeography.* Princeton University Press, Princeton.

9.3—Drawn from data compiled by the World Resources Institute for 2000–2001 using various sources including the UN and the International Union for the Conservation of Nature (IUCN).

9.7—Redrawn from Tilman, D. and Downing, J. A. 1994. Biodiversity and stability in grasslands. *Nature 367*, 3633–3635.

9.8—Used with kind permission of Dr M. V. Angel of the Southampton Oceanographic Centre, Empress Dock, Southampton SO14 3ZH, UK.

9.9—Used with kind permission of Dr D. J. Currie and the Editor of the *American Naturalist.* University of Chicago Press.

9.11—Used with kind permission of the Editor of the *American Naturalist.* University of Chicago Press.

9.12—After Schwartz, M. W., Brigham, C. A., Hoeksema, J. D., Lyons, K. G., Mills, M. H., and van Mantgem, P. J. 2000. Linking biodiversity to ecosystem function: implications for conservation ecology. *Oecologia 122*, 297–305.

9.13—Redrawn from Tuckwell, H. C. and Koziol, J. A. 1992. World population. *Nature 359*, 200.

FIRST WORDS

Four or five year old female polar bear on drift ice in the north-west Barents Sea, August 1999.

FIRST WORDS

What the polar bear tells us

Polar bears are the largest land carnivores alive today yet they live on a part of the globe where there seems to be little for any animal to eat. It is, of course, the seas beneath their feet that provide their food. Curiously, these are terrestrial animals living in a marine habitat.

In fact, the bears are only forced onto land when the ice begins to melt and life on the floes becomes too precarious. In the summer they can no longer hunt seals from the ice and they have to live off the fat reserves accumulated during the winter. Half the body weight of a well-fed polar bear may be fat; by the end of the summer this will fall to just 10 per cent.

Despite their isolation in the far north, well away from most humans, it seems that the bears may be affected by our industrial activity. A survey around Svalbard, a group of islands 800 km off the northern edge of Norway, found a small percentage of the female bears with malformed genitalia. Their external reproductive organs have partially developed male characteristics. And it seems this may be our fault: researchers from the Norwegian Polar Institute suggest that industrial pollutants in the bear's tissues could be the cause of these changes—compared to those in Alaska, the bears captured in Svalbard have 20 times the concentration of some industrial chemicals.

The case is far from proven, but if nothing else, this research shows that the pristine whiteness of the northern ice sheet belies the pollutants in the waters below. We do not know if the fertility of the bears is affected by their abnormalities. The Institute report catching fewer older bears in recent years, but this only indicates a shortening of their lifespan and says nothing about the number of cubs born. However, other aquatic mammals and a variety of reptiles also show abnormal sexual characteristics and elevated levels of these pollutants. Many are from waters much closer to major pollutant sources and have a greater exposure, so they may be more indicative of our own pollutant burden. Indeed, some scientists suggest that we too are showing the effects: compared to 50 years ago, human sperm counts have been halved and, on average, men today produce less semen.

We could go on, but this example may have already raised some questions in your mind. Such as—could this low level of deformity be 'normal' in polar bears? That is, could it be a genetic defect, a variation that has arisen spontaneously in the Svalbard population, quite independently of its pollutant exposure? Can we demonstrate that the pollutants in its tissues are indeed responsible?

And why are the pollution levels so much higher in the Svalbard population anyway? If it is something in the water, are all mammals affected in the same way? And how did it get to the Arctic? Are human beings at greater risk living further south, closer to the pollution sources?

We could go on (and do . . . see Box 1). We could ask questions about the ecology of the polar bear and whether this makes them more susceptible to the pollutants. Does the malformation have any effect—can the females still give birth and if so, is there any reduction in the fertility of the Svalbard bears? Is the population likely to decline when these changes affect only a small number of females? And finally, if the polar bear has a poor reproductive future, do we also?

There are few answers and the questions, as you will have noticed, have a habit of multiplying. Some of the answers come from understanding the adaptations of polar bears to their environment and their relationship to other species. The Svalbard bears have higher concentrations because of the location of these islands (Figure 2) but all polar bears will accumulate organic pollutants in their fat. Part of their adaptation to survive the arctic cold is to maintain a layer of fat beneath the skin, to act as insulation. It is this feature of

BOX 1   Nature or nurture . . . or neuter

What is it to be a man? In mammals, maleness means having a Y-chromosome which carries a gene that codes for a protein that triggers the development of particular cells which regulate sperm production and produce a hormone that interacts with other genes to drive the development of secondary sexual characters, such as a penis and a low voice. This sequence of switches needs to be set to on, early in the life of the embryo to stop it becoming female.

All mammals share the X–Y chromosome mechanism for sex determination, but the switches can be different in other animals and a variety of chromosome combinations govern sex in insects, birds, and reptiles. For alligators, it is the prevailing temperature of incubation that determines the eventual sex of the embryo—at relatively low temperatures females are produced. In our case, gender is determined by the embryo's *nature*—the characters it has inherited with the sperm—but in alligators it is *nurture*—the temperature at which it is raised.

As with much of the rest of the plant and animal kingdoms, being male or female says much about the individual's ecology. However, being a man at the beginning of the twenty-first century is proving increasingly difficult. Sperm counts have fallen by half amongst United States males since 1940 (Figure 1), a pattern that seems to be reflected in most other parts of the world. Males show increasing genital variability, with more cases of undescended testicles and lower testicular weights. This has been described as 'feminization', a tendency observed in some other mammals and thought by many scientists to be associated with the pollution burden many people carry.

The link with high levels of polychlorinated biphenyls (PCBs) and other persistent organic chemicals (the so-called 'oestrogen-mimics') is, however, far from proven. Neither is the response consistent: amongst the Svalbard bears the females are showing masculinization. We do not know whether the males have any signs of feminization. It may be that these pollutants act generally as 'endocrine disrupters', since the feminizing oestrogens and the masculine androgens share very similar chemical structures. Endocrine disrupters have been blamed for the increased incidence of reproductive disorders in humans—prostrate and testicular cancers in the United Kingdom, the proportion of ectopic pregnancies in the United States, and a more general increase in breast cancer. Again, this is matching trends to rises in environmental levels of the pollutants, but not establishing a

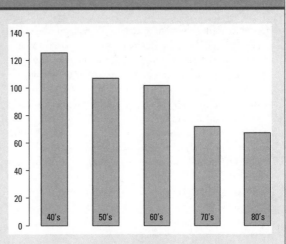

FIG. 1 Average sperm counts (millions per millilitre) in US males by decade, 1940–90.

connection. Some surveys of mammals from contaminated habitats have failed to show any such effects.

We need to establish that the Svalbard bears are responding to the pollutant rather than a spontaneous (non-induced) genetic change that occurred in one female several generations ago. Describing the relationship between a dose of the poison and the response of the bears would require a series of controlled experiments. Fortunately, perhaps, the bears are neither sufficiently abundant nor cooperative to allow us to do this properly. We can, however, draw on experimental evidence from other mammals—grey seals show disorders of their immune system, skeletal and reproductive disorders when fed fish contaminated with PCBs.

Some of the best evidence for a link comes from alligators. PCBs are passed from mother to offspring in the eggs. Lake Apopka in Florida has high levels of persistent organic pesticides and in one survey, over 80 per cent of the eggs of its alligators failed to hatch (compared to an average of 20–30 per cent in nearby lakes). Apopka's hatchling survival rate was only one-tenth that of other lakes. Female juveniles had oestrogen levels twice those considered normal and produced twice as many eggs. In contrast, males had low levels of testosterone and adult males show signs of ovary development. They also have penises around one-third their normal size. In experiments, painting alligator eggs with the two pesticides contaminating the lake (DDT and DDE) caused the hatchlings to develop abnormalities seen in the local population. Rather like the polar bears, alligator populations distant

BOX 1   Continued

from any significant source acquire PCBs from their diet, albeit at much lower levels, and pass them on to their eggs.

There are enough examples of human beings changing their gender using hormones for us to know that what is written in the genes, even those defining sex, only represent possibilities. The bears and the alligators emphasize a very important general point—our genetic nature defines our potentialities and not our outcomes. The success or otherwise of an individual to reproduce depends on the interaction of these genes with the environment in which they find themselves, and for the alligators of Apopka, such nurture may lead to neutering.

Certainly, the more species we find with reproductive abnormalities that correlate with high pollutant levels, the more likely it is that these compounds are responsible for our own falling sperm counts. This would be a sharp reminder that humans are mammals sharing physiologies that respond in the same way to these toxins. We cannot take ourselves out of the food chain or absent ourselves from the ecological processes which sustain the environment. We are poisoned for the same reasons that the bears are poisoned and we rely on an ice sheet at the poles just as much as they do.

FIG. 2 Svalbard from space. A satellite photograph of the islands with the wind direction picked out by the cloud formations passing over the archipelago. As a measure of the scale, the long thin island off the west coast, Forlandet, is 85 km long.

their biology and ecology that puts the polar bears at risk.

Ecology is the science that seeks to describe and explain the relationship between living organisms and their environment. In this case, everything that determines the distribution and numbers of the polar bear, from what they eat to their position in the global traffic of wind and water (Box 1). Their anatomy and physiology reflect the polar conditions where they live, the waters in which they swim, and the long periods they have to survive without food. Continually subject to natural selection, the bears are adapted to their habitat and the prey they chase.

To understand why the Svalbard bears are affected by the pollutants, we need to describe the movement of pollutants through oceans and food chains, in the currents and winds that move across the planet, and the plants and animals that move with them. Why some pollutants move along food chains—and others do not—is partly explained by the pollutant's chemistry but also by the habits, physiology, and biochemistry of the organisms at each step. To explain the exposure of the Svalbard bears requires an appreciation of everything from the global climate to cell metabolism.

Starting with the cell. At the moment we do not know whether the bears are showing a genetic abnormality, or simply the natural variation found between individuals. However, the presence of similar malformations in a range of aquatic animals suggests some common factor. It is unlikely that seals, bears, alligators, and whelks all share the same genetic factor, but it is probable that some basic feature of cellular structure or function—fat metabolism, perhaps—has been impaired by the contaminants. Some groups are clearly more susceptible than others, possibly because of their genetic constitution, but probably because of their physiology or feeding habits.

The pollutants most often blamed are the polychorinated biphenyls (PCBs), which are thought to mimic the mammalian sex hormones, the androgens and oestrogens (Box 1). At key stages in the growth of a young mammal the levels of these hormones govern the development of the gonads (ovaries and testes) and the genitalia. Like the natural hormones, PCBs are fat-soluble but as pollutants they accumulate in the fat of animals. Marine mammals need this fat (blubber) as both insulation and as an energy reserve. Seals from the waters around Svalbard are known to have high concentrations of PCBs in their blubber and these pollutants are readily transferred to predators consuming them. Indeed, very often polar bears will preferentially consume the blubber of a seal, selecting this tissue for its high energy and high vitamin content.

At the global level, the affected Svalbard bears are on a principal route of pollution into the Arctic—major air and water movements around the islands bring pollutants from the rivers of northern Europe. Contaminated air arrives from the industrialized regions of Europe and North America (Figure 3) but most importantly, these pollutants arrive in the tissues of fish and seals migrating from the south. The isolation of the bears on these islands close to the North Pole is not as great as it seems and the pollutants found in its tissues testifies to the ease by which some contaminants can move around the globe. Thus, the ecology of the polar bear, where it lives and what it eats, exposes it to higher concentrations of PCBs.

The story of the Svalbard bears increases the likelihood that our wastes are also affecting us. We are also mammals, with essentially the same sex hormones and reproductive physiology, so that the polar bears may be an important warning signal for us. There are, of course, major differences between the cold-adapted polar bears and ourselves: we are, after all, a tropical or subtropical species, only recently (in the last few thousand years) colonizing the higher latitudes. But the biology we share with the bears, and the possible effects of contaminants in their tissues, should prompt further study for both our sakes (Box 1).

Even so, the case against the pollutants is not proven and we may be on thin ice. Indeed, we have chosen this example to begin our survey of ecology because the data are not clear-cut—the connections have not been confirmed. We wish to encourage you to adopt a critical and evaluative approach to all of the ideas we describe in this book. Effective science proceeds by ruling out possibilities and generating alternative explanations. This is how we design experiments and collect data. Identifying and evaluating alternative explanations is crucial to the scientific process.

This book is a primer in the science of ecology and the major environmental issues. The polar bears of Svalbard have already introduced some of the key concepts. Ecology seeks to describe and quantify the connections between an organism and its environment. The biology and behaviour of the polar bear can only be fully understood by seeing how it is adapted to its living and non-living environment—how it survives the cold, finds a mate, catches its prey. Ecology describes how its population changes with time and the effect this has on the species around it. The bear's success in catching seals means these prey are being selected to evade capture, and the bears, in turn, must adapt to remain effective hunters. Ultimately, ecologists seek to explain all the features of polar bear ecology as a product of evolution through natural selection.

Through a sequence of such interactions, the fortunes of the bears can be linked to those of the fish on which the seals feed, and the invertebrates on which the fish feed, and so on. By these interactions, collections of species are bound together to form an ecological community. Many of the connections within a community can be tenuous and we have to decide which species and which connections are critical to the nature and performance of an ecological community. How much would the community change if the polar bears were lost? Is the collection of species within a community predictable for a given set of environmental conditions—and if so, why are polar bears only found in the Arctic?

Populations and communities grow and shrink in space and time, communities change across landscapes and between seasons, whilst life on the planet has a 4 billion year history. Understanding the mechanisms by which species evolve, flourish, and disappear will give us some insight into our own future and our impact on the Earth.

FIG. 3 The major movements of wind and water currents into the seas around Svalbard in the Arctic circle. Migratory fish and mammals following these currents and are a major source of some organic pollutants. The figure also shows the extent of the pack ice in April and September, from which Polar bears hunt their main prey, seals.

How to use this book

First Ecology provides a sequential development of ecology, running from the chromosomal basis of inheritance and evolution, through populations, communities, and ecosystems, to landscape and planetary ecology. The value of this hierarchy in describing some key ecological patterns becomes apparent when we start to explore high-level processes in Chapter 8.

Our starting point throughout the book is to introduce ecology from a human perspective. Whilst we assume little background biology, we hope to draw upon your everyday experience and understanding of natural processes. Each chapter begins with familiar ecological processes and ends with some example of how our ecological understanding can be used to manage or safeguard our environment.

Placing human beings in the foreground should help us to see humanity as part of the ecology of the planet. Over the last 100 years we have learnt much about our own origins and the changes in the environment that prompted our evolution. We can also trace the environmental changes that have followed in our wake. Mediterranean communities are as old as we are and, for the most part, each region has been greatly influenced by our activity, so throughout the book we use these as an exemplar ecological community. Because these ecosystems are found in five isolated areas of the plant, they provide an insight into how ecological communities are put together (Chapter 5). They are also a community type most likely to change with significant global warming over the next 100 years.

Some of the basic terms of ecology are given in Box 2 and a glossary is also provided at the end of the book.

Ecology is a relatively young science and formal experiments on communities and ecosystems are confined to the last 100 years. This is a small sample of the Earth's history, important to remember when ecologists are asked to predict global changes over the next 100 years. The fossil record helps us to appreciate the current scale of change, most especially the current rate of species extinctions (Chapter 9), comparable in some respects to the drastic change which has followed major environmental change in the planet's history. Today, we know only a small

fraction of the species that have ever inhabited the Earth and we are losing them faster than we can count them.

Ecological processes that take place over large areas tend to happen very slowly. For the most part, regional change is slow and local change is rapid. Ecology has to encompass processes operating from the microscopic to the planetary level, sometimes exploring the relationship between the two (Chapters 8 and 9). Between microbe and biome are highly complex communities of species that divert and delay the transfer of energy and nutrients (Chapters 6 and 7).

In a book of this kind some topics are not covered neatly in one chapter. Themes recur and concepts are developed in several places. This book does not offer a convenient set of notes—it is not a reference book—but rather a narrative that will visit a topic from several perspectives. Seeing the connections and achieving the overview is essential to fully grasping the ideas and adopting an evaluative approach. To aid your navigation, we provide a series of chapter summaries and references in the text to related sections. The companion web site for the book also has a section termed 'Routes', which allows you to follow certain themes through the text, with suggestions for

---

BOX 2   A few definitions

The formidable collection of names in biology can be a major barrier to understanding. A glossary is provided at the end of the book, but a few important definitions are needed to help clarify your study from the outset.

A *species* is any group of individuals that can actually or potentially breed with each other to produce viable and fertile offspring. Because there are degrees of genetic difference between all individuals, the demarcation of where one species ends and another begins can be very difficult to draw. *Hybrids* derived from two different (if closely related) species are possible, but these are often infertile and therefore a genetic dead-end. This definition of a species is fine for sexually reproducing organisms, but becomes problematical in others. We look at the concept of the species in greater detail in Chapter 2.

Within a species there may be several *populations*— individuals of a single species occupying a particular

location at a particular time. Notice that both place and time need to be defined carefully (Chapter 3).

Populations of different species are collected together into *communities* (Chapters 4–7). These interact, feeding on each other, competing for resources or cooperating through special relationships. Various types of community, consisting of regular associations of plant species, *biomes*, are found in different parts of the globe (Chapter 8).

A community together with its physical environment is an *ecosystem*. Very often, physical features, such as light, moisture and so on, define the sort of community that can be supported. Within the system, nutrients and energy move between the living (*biotic*) components and the non-living environment (its *abiotic* components) (Chapters 6 and 7).

We have just described a hierarchy moving from the species to the ecosystem. This is the sequence we use in this book to explore the science of ecology.

First Ecology companion web site

First Ecology doesn't end with this printed book. There is also a companion web site at www.oup.com/uk/booksites/biosciences.

The site includes:

- **Answers to exercises** that we've provided in the book so that you can check your own answers to the exercises as you work through them. We encourage you to attempt many of the exercises, to recap on the factual material and to test your understanding of the concepts.

 There are also additional exercises (with answers) that we have provided in a separate section of the site for your tutors to use.

- A **'Routes'** section, providing a map of how some key themes are developed through the book and including some case study titles, relating to each theme, that could direct your own research into these key areas. You may find this useful if you wish to research or improve your understanding of these themes, and we hope it will also demonstrate how

the various principles in ecology need to be considered as a whole to see the complete picture.

- A **web link library**, including all the URLs listed at the end of chapters in the book. Additional web links to other sites for specific topics are also listed on a sectionalised basis. Please feel free to suggest links that should be added to this site, preferably offering a commentary on the content (see below for contact details). We hope this will serve to keep the web links up-to-date, but cannot accept responsibility for any of the links posted here. We will endeavour to review these links periodically and offer our own commentary.

 Finally, we would welcome your general comments, especially concerning those topics that need to be clarified in the text.

 To contact us with suggestions for additional web links, or with any other comments, click through to our companion web site from **www.oup.com/uk/booksites/biosciences** and click on the **'contact us'** link.

---

case studies that you might complete by visiting the linked sections. The web site also provides additional web-addresses for specific topics.

At the end of each chapter there are also directions for further reading, with a brief commentary of their content or particular merits. We have also used boxes to extend material (with some examples described in detail) or to provide quick overviews of the key concepts. Finally, there are exercises at the end of each chapter to test your understanding. The answers and additional exercises appear on the web site. Many of the tutorial and seminar topics given at the end of each chapter can be researched using the 'Routes' facility of the web site.

Citing of references

Although we have minimized our citing of references in the text, you should cite references properly in your own work. Consider the example of Box 3.5 ('Gene flow on the ice floes'). This draws on several papers, but principally

Mauritzen, M., Derocher, A. E., Wiig, O., Belikov, S. E., Boltunov, A. N., Hansen, E., and Garner, G. W. 2002. Using satellite telemetry to define spatial population structure in polar bears in the Norwegian and western Russian Arctic. *Journal of Applied Ecology 39*, 79–90.

which would have been cited as Mauritzen *et al.* (2002) in the text. Formal scientific writing demands that statements are supported in this manner and you have to develop this routine in your own writing. To keep our text clear, we name key authors (but omit dates) and provide the references for each section at the end of the book, so our sources should be easy to trace. Note that the complete reference given above (and in the bibliography) is the model that you should follow in your own work.

You can find out more about Svalbard at:

www.unis.no

and the Norwegian Polar Institute at:

http://npiweb.npolar.no

1 ORIGINS

'Nature . . . or survival of the fittest, cares nothing for
appearances, except in so far as they are useful to any being.'

Charles Darwin: *The Origin of Species*

Human ancestry as represented by the skulls of *Australopithecus robustus, Homo sapiens neanderthalensis, Homo erectus,* and *Homo sapiens sapiens.*

ORIGINS

The beautifully streamlined profile of a shark immediately suggests its way of life. The finely shaped head, lifted on the wings of its pectoral fins and driven by the powerful motor of its long tail—are the lines of an open water predator built for speed and manoeuvrability. As sleek as a jet plane, the shark contrasts with the heavy armour and forward-facing weaponry of the lobster. Lobsters are the tanks of the sea-bed, protected and armed to stand and fight, rather than for speed.

Human beings are not so obviously shaped for one way of life. We do not have obvious defensive structures or adaptations to a particular diet. We are clearly terrestrial animals, primarily a tropical species needing a warm habitat with freshwater. Yet our species has ranged far more widely. We have burrowed into the ocean floor, marked our territory on the moon, and sniffed around active volcanoes. Even without the armour plate of the beetles, we have visited more extreme environments than any other animal. And we can do this because of a single oversized organ—the large programmable machine in our head. It is our brain that enables us to survive deep oceans or lunar landscapes. The brain is our most adaptable feature and its capacity to store and process information, to learn, to accept new instructions, or develop new solutions, is the key to our success.

We are products of our environment in the same way that sharks or lobsters are, but our adaptability comes from a capacity to modify environments to suit our needs. We have freed ourselves from the constraints of a single environment and one way of life. Given the advantage our species enjoys, we must inevitably ask why other animals have not developed their mental capacity in the same way. Or to put the question the other way around, why should a thinking machine become so significant for one group of animals, at some particular location, at some point in the past? What was so different about our ancestor's ecology that made a powerful mental capacity so advantageous? Or indeed, possible?

We may be able to protect ourselves in submarines or spaceships but it was the pressures of an environment selecting our ancestors that made us what we are today. In this chapter we start our exploration of ecology by looking at our own evolutionary history, as products of an African grassland. Later we look in detail at the theory of evolution and the mechanism of natural selection. We also consider the detail of genetic change, how genetic information is coded and stored, selected and inherited. We also see how the information written in our genes can help us reconstruct our past.

1.1 Origins of humanity

New species are produced when environments change and indeed, we know that early humans evolved at a time of considerable climate change in tropical Africa. This almost certainly created the genetic distance between gorillas, chimpanzees and ourselves: our ancestors were those apes that became adapted to life beyond the forest, where the trees gave way to grass.

The earliest human fossils are found in Eastern and Southern Africa in deposits dating from around 5 million years ago. Six million years ago we shared an ancestor with chimpanzees, but already the climate had started to become cooler and drier, and the forests in East Africa were retreating. In the West, the forests persisted, and a variety of apes could maintain a life in the trees. It was here, in the last million years or so that gorillas and chimpanzees appeared, adapted to a variety of forest habitats. Other apes, those most able to survive the open grassland, came to dominate in the drier East.

Bipedalism, standing and walking on two legs, is the feature of the apes we now designate as human. All hominoid apes (gibbons, orangutan as well as the great apes of Africa—Figure 1.1) will stand and walk upright on the ground, but it was *Australopithecus*—southern ape-men—which first walked persistently on two legs. Bipedal walking is a more efficient way of moving over the ground, at least when we compare modern humans with a chimpanzee using four limbs (Table 1.1). Standing tall and being able to see above the grass of the plain also helps in finding food and avoiding predators. It

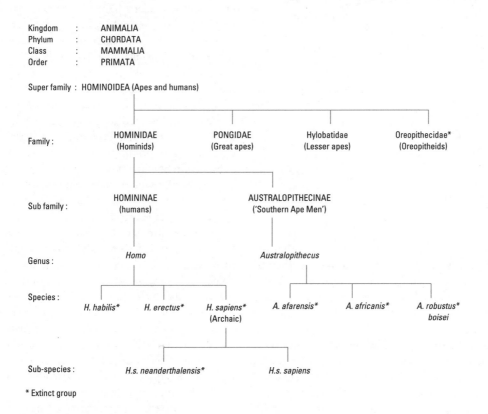

FIG. 1.1 A simplified classification of Hominids. This sets out the phylogenetic relationships of our species—species which are closer together on the phylogenetic tree share more of their ancestry than those far apart.

TABLE 1.1 In comparison to chimpanzees, human development is slower, extending the time the infant is dependent on the mother. Humans also have a significant post-reproductive lifespan possibly because of the role adults of all ages play in educating the young. In contrast to the small growth in the chimp brain, the human brain quadruples its size after birth. Notice also the greater efficiency of bipedal walking in terms of energy usage, compared to four-limb walking by chimps on the ground.

	Chimpanzee	Modern human
Gestation time	34	38 weeks
Age when		
Head is held erect	2	20 weeks
Walk on all fours	20	40 weeks
Bipedal standing	40	56 weeks
Bones of pelvis fuse	1	7–14 years
Age at adulthood	9	13 years
Years surviving after reproduction has ceased	0	20 years
Brain size		
At birth	300	350 cm^3
At adulthood	400	1450 cm^3
As a ratio of adult body weight	0.001	0.025
Relative energy use		
Slow walk	1.7	1
Fast walk	1.5	1

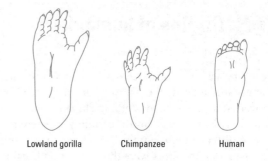

Lowland gorilla   Chimpanzee   Human

FIG. 1.2 The feet of three living hominoidea. Two retain the capacity to grasp, the other is adapted entirely to walking on flat ground.

line of the spine and are swung as a balancing aid in walking. In its vertical position, the rib cage has to move up and down, and in modern humans becomes barrel-shaped. Not surprisingly, the changes are also clearly seen by comparing the feet of modern hominids: those still associated with the forest retain the capacity to grasp whilst modern humans have a propulsive pad (Figure 1.2). Interestingly, *Australopithecus* had not entirely relinquished the safety of the trees because their feet retained the ability to grip branches.

Around 3.5 million years ago, *Australopithecus* had a brain the size of a chimpanzee's, but stood upright. Notably the fossil record suggests that bipedalism pre-dated any increase in brain size by over a million years. Perhaps, as Charles Darwin suggested, an upright posture might have promoted subsequent human mental development by freeing the hands, allowing them to manipulate objects.

We know from their teeth that early *Australopithecus* species were herbivores, but around 2 million years ago a new genus of humans appeared, with a much larger brain (Figure 1.3). *Homo habilis* was probably a flesh-eater, suggested by the crude stone tools dating from the time of its fossils, and found alongside the butchered remains of other animals. It seems that *Homo* had developed new skills. Using the hands to manufacture tools, or to manipulate objects would require a high degree of coordination, which Darwin argued would promote the expansion of their mental capacity. So too would the forethought required to envisage an end-use for the tool.

Whether these new humans were primarily hunters or scavengers remains a matter of some debate, though like many other predators, they probably consumed the remains of kills by other species. Perhaps these

frees the hands to carry food and infants so that the whole family group or band can move together, to feed or escape danger. An upright posture will also have helped cool the body in the open savanna since this reduces the area directly exposed to the sun, and increases heat loss.

Unlike other apes, all human species spent most of their time upright, a change in posture that requires significant anatomical changes. The pelvis changes to locate the legs directly beneath the body and the spine becomes S-shaped to bring the centre of gravity over this line. The muscles in the buttocks are used to pull the spine upright. The forelimbs now fall along the

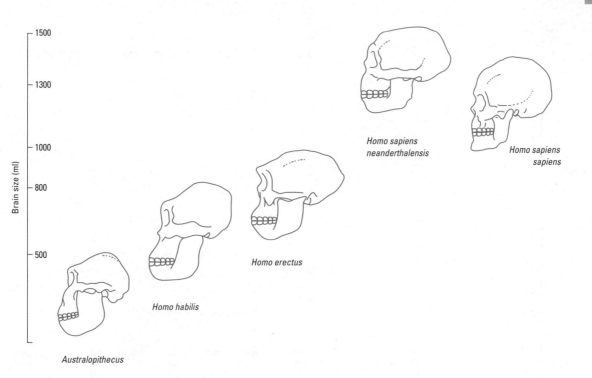

FIG. 1.3 Changes in skull shape and the increase in brain size in the hominids. The finer jaws and reduced teeth are possible because of the use of tools to aid feeding. Much of the increase in its size is in areas of the brain responsible for higher mental processes (thinking and memory) and speech.

first tools, simple stone clubs and stone flakes, were used to break open the bones for their marrow or to cut meat from a carcass. For nearly a million years *Australopithecus robustus* and *Homo habilis* lived together in East and Southern Africa, one primarily a fruit-eater and the other a herbivore and an opportunistic scavenger. The two species probably shared the same ancestor (*Australopithecus africanus*), but *H. habilis* had a larger brain and flatter, less protruding face (Figure 1.3).

The continued reduction in the size of the jaw and teeth we see in *Homo* follows the advent of tools as well as this shift in diet. As Darwin speculated, our smaller teeth would have been no disadvantage if tools were used to butcher and cut meat. The evolution of *Homo* goes hand in hand with the evolution of their tools: over hundreds of thousands of years, tool development, from simple cutting edges through hand-axes to arrow heads, mark the advent of different human species (Figure 1.4).

Within half a million years, the climate had begun to dry again and the first human predator, *Homo erectus*, appeared. With its arrival *Homo habilis* and

Australopithecus robustus disappeared from the fossil record. Again, *Homo erectus* had a larger brain and with a wider range of tools. Indeed, the arrival of all subsequent *Homo* species—generally with a larger brain and flatter face—are marked by their artefacts and the new skills they were practising.

Along with the switch to hunting and gathering, new technologies opened up new habitats: fire, clothing and the building of shelters provided protection and warmth, allowing early humans to expand northwards into less hospitable environments.

Equally important, perhaps, was the need to communicate when hunting. Cooperating to hunt, or to ward off predators, would have been essential for these not very fast apes. The social cohesion of the group and the induction of the young into its traditions and skills benefit from a precise language. Using words and putting them together in a meaningful way requires conventions accepted by all. These rules impose a discipline on those trying to make themselves understood. With a developed syntax and vocabulary, teachers can convey complex and

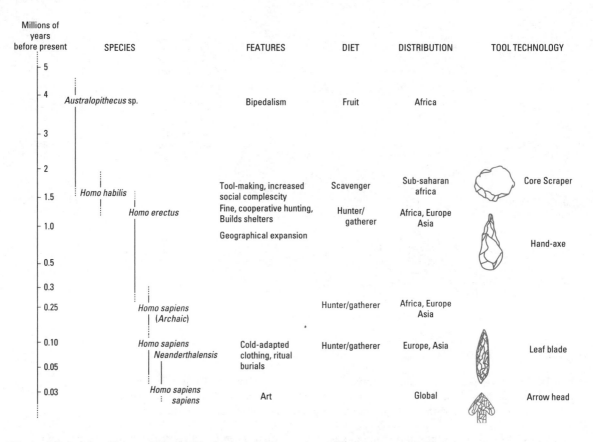

Millions of years before present	SPECIES	FEATURES	DIET	DISTRIBUTION	TOOL TECHNOLOGY
5					
4	*Australopithecus* sp.	Bipedalism	Fruit	Africa	
3					
2					
1.5	*Homo habilis*	Tool-making, increased social complescity	Scavenger	Sub-saharan africa	Core Scraper
1.0	*Homo erectus*	Fine, cooperative hunting, Builds shelters	Hunter/ gatherer	Africa, Europe Asia	Hand-axe
		Geographical expansion			
0.5					
0.3					
0.25	*Homo sapiens (Archaic)*		Hunter/gatherer	Africa, Europe Asia	
0.10	*Homo sapiens Neanderthalensis*	Cold-adapted clothing, ritual burials	Hunter/gatherer	Europe, Asia	Leaf blade
0.05					
0.03	*Homo sapiens sapiens*	Art		Global	Arrow head

FIG. 1.4 The origin of human beings. This diagram shows our immediate ancestors, which are distinct from the other African great apes (chimpanzees and gorillas). The hominids began with *Australopithecus* and its capacity to walk upright, and whose fossils are found at a time when the forests in East Africa were shrinking. The next major development was a change of diet and way of life, from being fruit-eater to a scavenger/carnivore. Tools used to obtain their food gradually improved and became more finely fashioned. Whether *Homo habilis* or *Homo erectus* was on a direct line leading to modern man is still debated. Equally, there is controversy whether *Homo sapiens sapiens* arose from a single stock of archaic *Homo sapiens* in Africa, or developed simultaneously in several locations in Africa, Europe, and Asia.

abstract ideas. Tool-making and other skills can then be learnt by instruction, not simply by copying the behaviour of others.

Individuals can rehearse and structure their thoughts using these same rules. Our verbal messages are usually what we wish the listener to hear and we anticipate the reaction our words might provoke. However, like many other species, we also send messages to deceive. While the obligations to share food and to cooperate hold a social group together, there are circumstances, say in competition for a mate, when the advantage is with those who think ahead,

who predict outcomes and make trade-offs. Social intrigues and duplicities are known in groups of chimpanzees and some anthropologists have suggested that this was key to our own mental evolution, the second-guessing of social interactions for individual advantage and reproductive success.

Language demands memory and a capacity for imagination and abstract thought, and its increasing importance is recorded in the later fossils. Changes in the braincase of these skulls indicate an expansion of the areas of the brain associated with language, and the anatomy of the mouth and the position of the larynx, or voice box also shift. A larger space developed above the larynx allowing a wider range of sounds to be produced. Unfortunately, it also increases the chance of choking on food, so that articulating speech must have been advantageous indeed to risk this hazard.

The significance of language and mental skills can be seen from looking at other aspects of our biology. Compared to other primates we are born at an early stage of development, before our head grows too large to pass through the mother's pelvic girdle. Relative to our body mass, our brain grows considerably after birth. Childhood itself is extended so that many of the anatomical changes associated with an upright posture and mobility occur relatively late in our development (Table 1.1). Even then, the child remains with the mother and the family for a considerable time. Along with mother's milk we pass on the tricks of survival, feeding both the body and the mind, so that offspring are inducted into the experience and knowledge held collectively by the adults.

Unlike most wild chimpanzees, we also live on for some years after we have ceased reproducing. Drawing on their longer experience, grandparents often play a crucial role in raising the young, and in many cultures they are seen as repositories of a society's wisdom. It is perhaps the lessons we pass on to the young that give our post-reproductive years some adaptive significance.

The complexity of our language, the relative precision of its terms and structure, puts the greatest distance between ourselves and the apes (Box 1.1). Other animals are able to walk upright, some use tools and most communicate with each other. Some, like chimpanzees, do all these things and live in highly integrated and duplicitous social groups, often with distinctive cultures of behaviour or even technology. But it is our spoken language, and the culture it creates, which distinguishes the hominids. So central is language to human evolution some linguists suggest it is less of an acquired skill and more of an inherited instinct (Box 1.2).

Their success allowed *Homo erectus* to escape from their cradle in the East African Rift Valley, moving through the Middle East into Europe and Asia. Surviving in these different environments would have demanded adaptability and ingenuity, and the ice ages that followed would have further tested human resourcefulness. Yet the hominids continued to spread.

Our species, *Homo sapiens sapiens*, appeared around 100 000 years ago in East Africa, and moved out through the Middle East. In step with the glaciations, they migrated northwards to join their near relatives, the Neanderthals, in Europe and Asia. Neanderthals had been in the north before the evolution of modern humans and had the anatomy of a cold-adapted people with large chunky bodies. They also had larger brains.

Genetic evidence suggests Neanderthals and modern humans shared an ancestor about half a million years ago, so *Homo sapiens sapiens* were not a descendent of *Homo sapiens neanderthalensis*. Modern humans arrived in Europe around 40 000 years ago; by 30 000 years ago Neanderthals had disappeared. For a period the two sub-species lived side by side, but it may have been competition for resources with our species that led to the demise of the Neanderthals. Some have claimed that the Neanderthals were 'absorbed' by interbreeding with modern humans (pointing to an ancient skeleton found in Portugal that shows characteristics of both sub-species) but there is no genetic evidence for this. Another suggestion was that there was a more direct conflict, of which our species were the victors. Today we can only imagine what it would have been like to look into Neanderthal eyes, perhaps even to speak with them and learn directly of another species' view of the world.

Hominids appeared because of a series of environmental changes, to which their biology adapted. These apes evolved into humans by developing their mental abilities. It is our mental agility that makes us different—it provides us with a more flexible range of responses to a changing environment, more rapid than any physiological or genetic adaptation. It is the mind that has made us so successful and allowed us to become manipulators of our environment. However, this has consequences we are about to explore in the rest of this book.

We have been describing how one species has developed from another, the sequence of changes that describe human evolutionary history. Compared to chimpanzees and gorillas, we have a relatively rich collection of fossils from our recent ancestry. We continue to search for remains but now we can also look inside our own cells at the history written in our genes. It is changes in this code, slowly selected over

BOX 1.1 Different from the rest

For some thinkers, it is consciousness that makes human beings distinct from the other animals. Although we are often impressed by the computational skills of some animals—such as the navigational skills of migratory birds or butterflies, our species has always revelled in manifesting its own mental powers. Building pyramids or sending men to the moon is as much a celebration of our mental prowess as our desire to know more. Indeed, our mental power may be the reason we now find ourselves alone, a single species in a single genus.

Our mental abilities allowed us to expand out of Africa and were the reason we could survive the ice ages of the North, moderating those severe conditions with fire and shelter. We may also have won our past competitive battles because our species could use limited resources more efficiently or because we were more aggressive. It is unlikely that we displaced the cold-adapted Neanderthals on the basis of our biology alone.

Whilst it is hedged around with many philosophical problems, our mental capacity and particularly our sense of self-consciousness, is one of the important ways in which we define our species. Consciousness is the capacity to think about thinking. It gives our mental machinery the capacity to learn and to adapt, to improve from experience. Daniel Dennett argues that humans are uniquely conscious because, unlike other animals, we can reflect on our thinking processes. That is, we can isolate concepts, the principles we use to define something, and then treat them as objects. Objects that can be mentally manipulated and refined, much as we might improve a tool through repeated handling. Being reflective, we can review how we think about things, refine our ideas by exploring their properties and their relations. For example, the concept 'bear' may include all the indicative features by which we define bears—their fur, their ears, their snout and so on—but our idea of *bear-ness* can encompass both a cuddly toy and a threatening predator.

We can also combine concepts and play with them—bears roaring, bears sleeping . . . bears planting a flag on the moon—creating images that we may have never seen. Novel combinations can produce novel solutions, some of which may be adaptive in the same way that novel gene combinations can be. The evidence is that even the brightest of other animals cannot manipulate concepts to create new ideas in this way. Without a fully verbalized language, chimpanzees do not come close. With no prompt to conceptualize—deriving the principle behind a word that it represents—thoughts are not easily manipulated or reviewed. And words, as symbols of these ideas, are easily combined. As Dennett notes, without words, such manipulations are not easy:

A polar bear is competent vis-à-vis snow in many ways that a lion is not, so in one sense a polar bear has a concept that a lion does not—a concept of snow. But no languageless mammal can have the concept of snow in the way we can, because a languageless mammal has no way of considering snow 'in general' or 'in itself'.

Humans go much, much further. Not only do we use language to extract the principles, we play with the ideas. We revel in our lingual skills, toying with the words and enjoying the clash of symbols. Dennett demonstrates his own playfulness a few lines later:

We can speak of the polar bear's implicit . . . knowledge of snow (the polar bear's *snow-how*) . . . but then bear in mind that this is not a wieldable concept for the polar bear.

Not only can we manipulate concepts, and occasionally come up with novel solutions, our language provides us with one other key advantage over other animals. We can pass these concepts on to our offspring. Certainly other animals learn from their parents, but they do not deal in concepts and they do not learn to manipulate them. We famously stand on the shoulders of the great thinkers who have gone before us. Mercifully, we do not have to invent the wheel or the word-processor anew with each generation.

BOX 1.2 An ecology of language

A language, like a species, when extinct, never . . . reappears
Charles Darwin, *The Descent of Man*

A language is the embodiment of a people's culture and, as Charles Darwin observed, has the characteristics of a living species. Languages evolve (and merge, unlike most species) and can also become extinct. Languages carry ways of thinking and of seeing the world, passing ideas from one generation to the next.

Indeed, the cultural and genetic history of peoples is closely associated and, in recent years, there have been attempts to trace the origins of our languages, to describe ancestral stock languages, from which the roots of all languages can be traced. The most widely accepted is Proto-Indo-European which provides a root for languages as dispersed as Sanskrit and Gaelic. This can be traced back to a single tribe, which some locate originally in Europe and Southern Russia and others put in Anatolia. The latter is favoured by genetic evidence from populations in Turkey today. One suggestion is that the spread of their language was possible because they had a successful agricultural technology which was quickly adopted by other peoples. If this is true, human genes and a language spread on the back of a set of ideas.

Genetic evidence also supports the suggestion that there are three basic families of languages associated with the Americas, associated with three waves of colonization from Asia. This scheme groups languages into families, by similarities in names and the way words are used. The largest and oldest family is Amerindian which includes most of the languages of the two continents (Figure 1.5). This is followed by Na-Dene, largely confined to the North and North West (but including Navajo and Apache) whilst the extreme North has the Eskimo-Aleut family. Although this grouping is highly contentious, the distribution does match three migrations detected from the archaeological record and also three distinct peoples based on genetic differences. The first colonizers, associated with Amerindian family of languages, reached North America from Asia about 30 000 years ago.

When cultures clash, languages often suffer, so subsequent invasions often mean that a language disappears. As Steven Pinker says, languages go extinct for the same reason that species disappear, by loss

Eskimo-Aleut
NA-Dene
Amerindian

FIG. 1.5 The distribution of three families of languages throughout the Americas. The oldest, associated with the first colonizers from Asia, is Amerindian, to which most languages have been allocated. Although these groupings are to some extent subjective, this division does match archaeological evidence and an increasing body of genetic evidence.

of habitat. Its speakers are assimilated into other cultures, or in some cases, laws are passed to suppress a language and its culture. Currently, there are around 6500 languages globally, of which up to 90% are threatened with extinction. There may be no more than 300 extant (living) languages in a hundred years time. Michael Kraus notes that 150 (out of 250) North American languages are under threat. California alone had 50 native languages and those still used are spoken by only older citizens. The last speaker of the Northern Pomo language died in 1995 and with her that living tongue. Michael Kraus estimates that a language needs a minimum viable population of several hundred thousand people to persist.

Languages can be re-born from a written record. Cornish died out 200 years ago but today there are 150 speakers. Electronic media and the speed of global communications have contributed to this demise, but it also provides the means to preserve languages. Rather like frozen embryos or stored DNA, we have the prospect of preserving something of their diversity on recorded media.

millennia, which have produced modern human beings. We now need to describe the mechanism by which changes at this molecular level could give rise to a new species.

1.2 Evolution by natural selection

The **theory of evolution by natural selection** proposes that species change over generations because of the selection of inherited characteristics. Chance and the environment determine which individuals go on to reproduce and their success means that their traits are represented in the next generation. Over many generations, a trait that confers some selective advantage will become dominant in the population. So it was, we believe, that bipedalism became dominant amongst early humans, allowing them to be better fitted to life in open grassland.

Over many generations, the accumulation of a series of small, gradual changes may lead to a population becoming distinct from its neighbours. When these differences prevent successful matings between populations a new species has arisen (Section 2.6). We can explain this process most easily with an example.

Imagine a field of annual plants, plants that die at the end of the year, but which leave behind seeds to germinate the following spring. The seeds are viable for 1 year only, so each generation is derived solely from the parents of the previous year. When space or nutrients are scarce the newly germinated seedlings compete for the available resources, and those best able to survive will grow and contribute the most seed to the next generation. Others fail to reproduce because they lose these competitive battles.

Now, although they belong to the same species, we can see differences between the plants. Some are larger, some are dull in colour, and some produce more flowers. One reason for their variation lies in the soil. Perhaps the larger plants are found where nitrogen is abundant. All individuals would be larger if they had the same supply of nitrogen. This is a physiological response to the nutrient and such differences are termed **phenotypic variation**.

On the other hand, some plants produce more flowers than their neighbours, irrespective of soil quality. We shall say this is a trait they have inherited. This difference between the plants reflects information stored in their genes, in this case either the code for single or for multiple blooms. Such inherited differences are known as **genotypic variation**.

Thus, there are two possible sources of variation between individuals: genetic differences inherited from the parents, and therefore coded in the genes, and phenotypic differences which are not. Phenotypic characteristics may be either reversible, such as wilting at times of water shortage, or irreversible. Irreversible differences are fixed during the development of the organism—for example, once a plant has grown tall on a nitrogen-rich soil it will not shrink again when the nitrogen supply is depleted.

Our main concern here is with genetic differences. The instructions for building a new plant are stored in the **chromosomes** of the fertilized egg, and the code written into the **DNA**. DNA—**deoxyribose nucleic acid**—is a long molecule, the spine of the chromosome, whose chemical structure carries the information of the genes (Box 1.3).

We shall assume that our flowers are unable to fertilize themselves. Let us say that having multiple blooms gives a better chance of producing seed in very wet years, primarily because their pollinating insects only fly when there is no rain. A flower lasts a short time in wet weather and few are pollinated before they deteriorate. A plant with a series of flowers is available for pollination for longer and is more likely to produce seed.

In a wet year, the single-flowering plants are unsuccessful. Most die unpollinated and few produce any seed. Those with a sequence of flowers were pollinated in the early months and now have a higher proportion of seed in the soil. Next year the field is a carpet of blue for the whole season, since most plants carry the gene for multiple blooms, and so, in turn, will their seed. This gene has increased its frequency within the population and our plant population has changed (Figure 1.8).

Despite being hypothetical, this example demonstrates the inevitability of evolutionary change in a

BOX 1.3 Storing the information to make a living organism

A chromosome is a long strand of protein and **DNA—deoxyribose nucleic acid**—held in the **nucleus** of a cell. All higher organisms have several **chromosomes** held within a nuclear membrane, though bacteria and their relatives have a single chromosome not confined to a discrete nucleus. Most higher plants and animals are termed **diploid** because they have two sets of chromosomes ($2n$). Human beings have 23 pairs of chromosomes ($n = 23$), one of each pair derived from either parent.

The information is actually encoded by a series of **genes**, built up from subunits of the DNA called **nucleotides** (Figure 1.6). There are four chemical forms of nucleotides, distinguished by the type of nitrogenous base in their structure. There are four bases (C, G, T, A) and it is the sequence of these bases along the chromosome which encode the instructions. This is rather like a Morse code message, except it has four possible characters rather than two. The code is read in groups of three bases called triplets. As with any language, much of the meaning is derived from the position of each unit of information. Note however that genes are defined not by the length of their code but by their function. For this reason the number of nucleotides will vary from one gene to another.

The position of a gene on a chromosome is called its **locus**. A gene is 'read' in a process that leads to the production of a **polypeptide**, a chain of many amino acids joined together (Figure 1.7). Either on its own, or in combination with other polypeptides this can be folded to produce the three-dimensional structure of a finished protein. The activity of the protein is determined both by its chemical behaviour and its structure, according to the folding of the polypeptide chain.

Paired chromosomes carry instructions for the same functions and are termed **homologous**. Each gene from the father is matched by an equivalent gene from the mother. When the code is the same on both chromosomes the individual is described as **homozygous** for that character. Otherwise it is **heterozygous**. Alternate forms of the same gene are called **alleles**. Within a population, there may be many different alleles for a character. The totality of genes in a population is termed its **gene pool**.

A key distinction is drawn between the genetic information carried by an individual, its **genotype**, and the trait it shows, its **phenotype**. The two will differ because some of the genetic information is not expressed in the phenotype. For example, a child with brown eyes may actually have the code for both brown

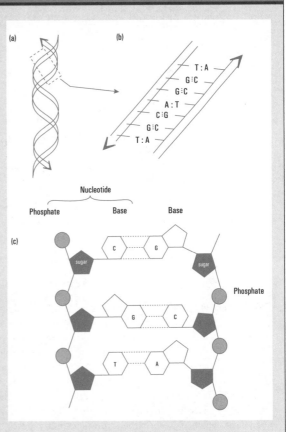

FIG. 1.6 The storage of the genetic code in the chromosome. DNA consists of two backbones of sugar-phosphate groups, coiling around each other in a double helix. The information is actually stored in the sequence of bases (A, G, T, C) that bind the two strands together. The direction, sequence and position of these bases give the code its meaning, and is read in blocks of three bases. The bases come in two forms—purines (**adenine** and **guanine**) and pyrimidines (**thymine** and **cytosine**), distinguished by their chemical structure. Notice that only two combinations of bonds will form, A with T and G with C. This means that one strand of the DNA is a complement or 'mirror-image' of the other. The two strands separate and the sequence of bases is read on the strand carrying the required code. The vast amount of information needed is packed into such a small space by winding the double helix around protein bundles, and then packing these 'beads' or nucleosomes into a sequence of loops that eventually comprise one chromosome.

BOX 1.3 Continued

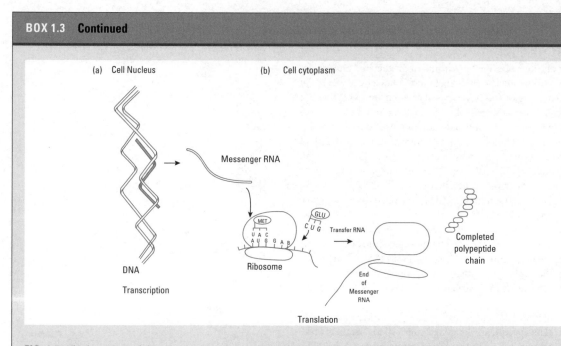

FIG. 1.7 The language of the chromosome is translated into the language of proteins. The linear code of nucleotides on the DNA strands has to be translated into the linear sequence of amino acids used to build a polypeptide. There are two stages in this process: **Transcription**, where the two strands of the DNA are parted so that free nucleotides can pair with their corresponding bases on the strand carrying the code. Note that uracil replaces thymine as the complement of adenine in all RNA. The read sequence is joined together with RNA to form a length of messenger RNA (mRNA) and the completed mRNA then enters the cytoplasm. **Translation** itself takes place in the ribosome. Here the mRNA is read sequentially, with three base sequences matched against their complementary bases attached to a transfer RNA (tRNA). Each tRNA molecule has an amino acid attached that corresponds to a particular triplet code. This makes the translation. As the appropriate tRNA molecule is selected by the complementary base-pairing, that amino acid is added to the line being assembled. Particular codes also signal the start and finish of the linear sequence needed to synthesize each protein.

and blue eyes, but the action of one gene masks any activity by the other. The brown gene is dominant over its allele. We cannot tell from looking at the phenotype (the brown eyes) whether the child is heterozygous or homozygous. Heterozygotes in the population thus carry some hidden genetic variation with the two alleles at this locus.

changing world. If a selective pressure favours some individuals over others, and if the selected trait is inherited, then a change in the population must follow. Individuals best adapted to a particular habitat will produce more offspring than their competitors. Traits that improve reproductive success will proliferate within the population.

This is genotypic adaptation and its essential elements are inheritance, variation, and selection.

Consider the example again. It makes several important points about the nature of genotypic adaptation:

• *The character selected must be inherited*: to become part of future generations a trait has to be written in the genetic code. Phenotypic adaptations are lost with each generation—and must develop anew each year.

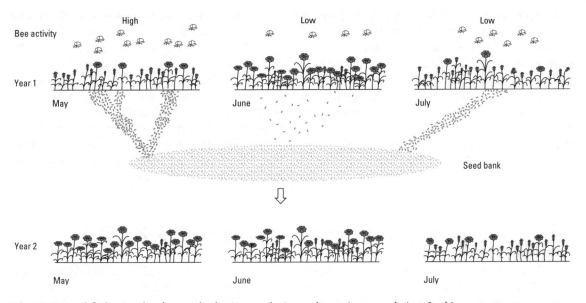

FIG. 1.8 A simplified example of natural selection producing a change in a population. In this case, the selective pressure is exerted by bees, responsible for pollinating a flower, and thereby ensuring its seed production. In this hypothetical case, the bees only fly during fine weather. In year one this happens only in May, so those plants which flower over a long period, producing several blooms during the season, are pollinated. Those plants which only flower once, perhaps in June or July miss out when the weather is poor and the bees are not active. In this particular year the majority of the seeds in the seed bank contain the gene for multiple blooms. Next year this variety dominates, and the field is covered in flowers in May as well as June.

- *An individual carries the traits that were of selective advantage in its parent's environment*: the new generation is dominated by multiple flowers because such plants were more frequently pollinated in the wet conditions prevailing last year. Yet this year conditions may be dry. It was the selection pressures operating upon the parents that determine the nature of the current population. An individual is adapted to its environment only as far as it matches the conditions under which its parents lived.

- *One type dominates only while environmental conditions favour it, or there is no selective pressure against it*: if there was no selective pressure against multiple blooms in the drier years their proportion in the population would not change. Only if they were at some disadvantage would their numbers decline with a series of dry years, for example, if their seed were viable for shorter periods or they germinated less readily under dry conditions. Then a series of dry years might shift the balance in favour of single blooms.

- *Individuals need to differ for there to be selection*: without variation, all plants would have the same chance of producing seed. With variation, those less able produce fewer seed. Success—the relative fitness of each type—is measured as the proportion of offspring that each contributes to the next generation.

- *The gene responsible for the adaptation was not induced by the environment*: chance threw up the successful genotype; it was not produced in *response* to a wetter summer. Genetic change is blind to selective pressures operating in the environment. Most major changes in the genetic code are disadvantageous, or confer no advantage. We discuss this further below.

- *The effects of natural selection are seen as changes in the proportions in the population*: a consistent shift to multiple blooms from one year to the next might lead us to suspect this variety was being selected. This directional change in the population, over a number of generations, might lead to complete dominance or **fixation** of the favoured gene.

We have observed only a change in the preponderance of one variety of the plant, reflecting a change in the frequency of just one gene. But now imagine that the climate turns wetter for an extended period, so that over the generations, fewer and fewer single-flowered plants survive. Eventually, the gene for single blooms is lost. If the population remains isolated, not swapping genes with other populations, other genetic changes may follow. An accumulation of differences may lead to reproductive isolation (Section 2.6).

The important difference between the plants is the availability of the flowers because this is the phenotypic character selected by the environment, in this case by the bees. Yet future generations will only reflect this selection as far as the character is written into the genes. Only the selection of genetic code can produce inherited change, and thus new species derive from genotypic variation, not phenotypic variation.

This distinction between the phenotype and genotype is important. To understand its significance we need to recall the facts of life.

1.3 The advantages of sex: the blooms and the bees

The curious might reasonably ask why the plants should flower at all. Rather than hoping for a dry year and the attention of some rather fussy bees, why do they not simply produce some form of runner or tuber that could survive the winter? Then they could do without flowers, gametes and seeds.

With the picture we have painted so far, this might indeed be a good strategy. This would be fine if life in the field never varied. The plant need not change and an over-wintering stage may cost less to produce and also have a greater chance of producing new plants. The species could then avoid all the tribulations and risks of sexual reproduction and a successful genotype would be passed, unaltered, into the next generation.

But if the environment were to change, perhaps with the climate getting wetter, such plants, despite all their runners and tubers, would find it increasingly difficult to survive. It may be that in a wetter soil the tuber is more likely to rot and not produce a new plant next year. Relying solely on asexual reproduction, the capacity of a species to adapt genetically is very limited; once its phenotypic flexibility has been exceeded there is little prospect of accommodating further change. Sexual reproduction may be risky, but it does at least produce new combinations of genes and a possibility of increasing the range of tolerance.

Enter the bees. Inadvertently acting as go-betweens, they carry the male genes in the pollen grains to the egg held in the flower. The flower is simply a device to attract the bee, a feeding station that advertises its presence. The cost to the plant is worth bearing given all the advantages of sexual reproduction.

Not only are the costs of sexual reproduction high, so too is the risk of failure. Each individual has to meet a partner and two **gametes**, the sperm and the egg, must fuse to form a single cell, the **zygote**. For the female, who may protect and nourish the egg, this means sacrificing half of her genetic code whilst supporting the male genes. The zygote must have a viable combination of genes, always with the risk that the new genotype may be unable to support the development of a new individual. In multicellular organisms, the zygote has to undergo a series of divisions and transformations to develop all the tissues of the adult. In the process it must, in turn, produce specialist organs for sexual reproduction, such as flowers.

To survive to reproductive age, each offspring, and the new combination of genes it represents, needs to be well matched to the environment in which it finds itself. Producing large numbers of offspring, all slightly different from each other, improves the chance that some will survive, but there is wastage and this is a major cost to the parents. Invariably, some combinations, some new genotypes will be less fit than their parents. Asexual reproduction avoids all these risks and costs, and so is a good bet when the environment is not changing.

The merits of sex are not obvious to biologists and there is a long-running debate about its selective advantages. A wide range of plants and animals have a few species that are **parthenogenic** (Section 2.3), where the female produces viable eggs that develop without fertilization. In several cases, the mother's genotype passes unaltered into the next generation, with no role for a male. In others the sperm is used only to stimulate the egg to begin its development but its chromosomes play no part again in the next generation.

However, the evidence is that such species are short-lived in evolutionary time. Clearly, the swapping of genetic information has some major advantages. Sexual reproduction allows for genotypic adaptation by throwing together novel gene combinations and more potential solutions when selective pressures are continually changing. Another view is that sexual reproduction is favoured because uncommon genotypes may allow escape from parasites seeking to adapt to the internal environment of the host (Box 4.5). Notice, however, that these benefits are indirect, arising from the chance combinations of genes in the individuals of the next generation. This variation cannot benefit the individuals producing gametes and reproducing in this generation. And

then, only the genes whose code is passed on to the successful offspring benefit. It must be the genes that promote sexual reproduction—which prompts this shuffling of the combinations—which are favoured in a variable world.

For most higher plants and animals, sexual reproduction is the only viable strategy. A changeable environment favours different traits at different times. The variation that comes from shuffling the genes in sexual reproduction means new genotypes that will, occasionally, translate into new phenotypes better fitted to the environment. Sex also provides the means by which deleterious variations might be lost, at meiosis and at fertilization (Figure 1.8), and via those individuals that fail to reproduce. Variation itself may confer some advantage, primarily for heterozygous individuals that have alternative alleles, perhaps providing them with a greater range of adaptive responses.

1.4 Sources of variation

The infinite variety possible from sexual reproduction provides for change in a changing world. These adaptations follow from alterations in the molecular sequences carrying the genetic code. In Chapter 2, we shall see how the gradual accumulation of such changes over many generations can produce new species. Now we look at how variation between individuals and across generations is created.

Traits may be inherited from either parent, but what is actually transferred by the sperm to the egg? By what mechanism is genetic information passed on? Charles Darwin could only speculate about the nature of inheritance and never knew that Gregor Mendel had begun to describe the mechanism just 7 years after he had published *The Origin of Species*. Much later, the **chromosomes** within the **nucleus** of the cell were identified as the site of its genetic information. The code is written as a series of discrete units, the genes, down the length of the chromosome (Figure 1.6).

The code is read sequentially, bearing the instructions either for reading the code (its 'punctuation'), or for building a protein (Figure 1.7). The protein is

actually constructed at the **ribosome** from a series of smaller molecules, **amino acids**. This product is then used in a cellular structure or in cellular metabolism. Genotypic change occurs when part of the code reads differently. We detect this as a phenotypic change when the sequence of amino acids produces a protein with different structural and chemical properties.

In most higher plants and animals, chromosomes come in matching pairs, one from each parent (Box 1.3). A chromosome has to be copied every time the cell divides so that each daughter cell has a copy of the full genetic code (Figure 1.9). Each chromosome is replicated, producing an identical copy, which then separate, one to each cell. Thus, in this process of **mitosis**, the chromosome number is maintained and copies of chromosomes pass unaltered into the daughter cells. This is not true of the gametes, where each cell has only a single set of chromosomes. Sperm and eggs have undergone **meiosis**, a process that firstly duplicates and then separates the paired chromosomes (Figure 1.9). When the gametes fuse at fertilization, the full chromosome number is restored.

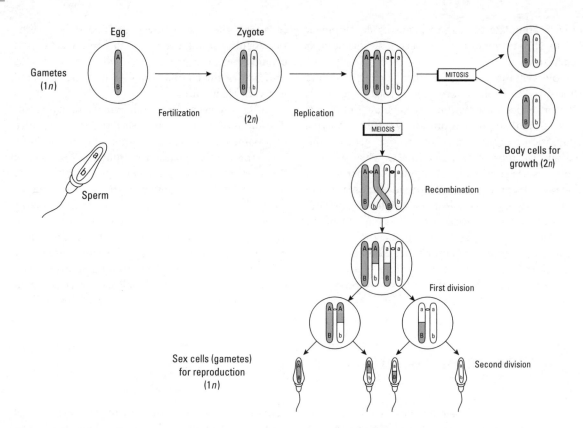

FIG. 1.9 Chromosome replication and division. In most higher plants and animals, fertilization produces an individual with matching pairs of chromosomes, one from each parent. To simplify matters, here there is only one chromosome ($n = 1$). Thereafter, chromosomes are replicated for either growth or the production of gametes.

Mitosis is the replication of the chromosomes to produce two identical daughter cells, each with a full complement of chromosomes ($2n$). The chromosome number remains the same and the chromosomes themselves remain intact, so that the genetic code in each cell is the same. Mitosis precedes cell division within the non-reproductive cells of an organism, and has no significance for evolution. However, notice that the $2n$ cell may contain recessive genes (represented by the lower case letters) which are not expressed in the phenotype. This is a store of variation, retained in the population since it is never exposed to selection.

Meiosis produces haploid ($1n$) gametes carrying just one set of chromosomes. When these fuse at fertilization the diploid condition is restored ($2n$). Meiosis also involves replication, but now there are two division stages. The first division allows a chromosome and its replicate to remain together (termed sister chromatids). Homologous pairs of chromosomes separate into each cell at this division. A second division parts the two chromatids prior to their separation into individual cells. This is the haploid state. Notice that when homologous chromosomes form pairs, they become attached at various points. In some cases, pieces of the chromosome may cross over, and become transposed between non-sister chromatids. This is **recombination**, an important way in which genetic information can become shuffled, producing new combinations and variation. Meiosis does not change gene frequencies in the gene pool, but it does produce new combinations of genes. Variation will also occur between individuals because gametes do not contain equivalent combinations of chromosomes. This is termed **independent assortment**. Gametes differ from each other because of both independent assortment and recombination.

You look a lot like your parents, but you are not a perfect blend of your mother and your father. Genes carry discrete quanta of information and the proportion of traits you have inherited from each parent depends on the shuffling at meiosis and the chance unions at fertilization. Each parent supplies two of each chromosome to meiosis, but any egg or sperm will only have one of each pair within it.

The chromosomes can also be altered in the process of meiosis. A **crossing over** of sections of chromosomes may occur between pairs of chromosomes during meiosis (Figure 1.9) producing new gene combinations. Each pair of chromosomes are **segregated**, one to each gamete, so the union of sperm and egg gives a vast number of possible combinations of traits. Thus, any individual inherits only some of each parent's genetic code.

Most of the variation within a generation thus comes from the segregation and the shuffling of existing code during sexual reproduction. Entirely novel code requires that the sequence of bases along the chromosome be changed—for mistakes to be made in their replication. Genes are copied and read chemically in a highly reliable process, and cells have mechanisms to try to preserve the fidelity of the copying procedure. But mistakes do occur and these may be passed on with the genes in gametes that produce the next generation.

A **mutation** is either a change in the code at a single point, or deletions or insertions of lengths of code. A mistake that affects the code for a protein may change the sequence of its amino acids, or the way in which the molecule is folded. Sometimes this has little effect on its function and causes no significant change in the phenotype or its performance. Mutations that cause major change are often deleterious and some are lethal. Very rarely, a mutation improves performance or enables a protein to carry out a new function.

Chromosomal aberrations also result from the duplication and separation processes in meiosis. Large pieces of code can be swapped between non-homologous (non-matching) chromosomes (**translocation**), or chromosomes are duplicated or lost entirely to a gamete. Again, most chromosomal changes are likely to disrupt the development of the organism after fertilization and so do not survive into the next generation.

Genetic and biochemical analyses show that a large amount of variation is normal in most populations. Much of the variation appears to have little selective advantage, so minor code changes have no apparent impact on reproductive success. This is termed **neutral variation**. Some biologists believe, however, that a high level of variation may, in itself, contribute to the vigour of a species.

Variation also resides in the code hidden or not expressed in the phenotype. At a heterozygote locus, where one allele dominates another, the recessive gene is not subject to selection (Box 1.3). It is only ever exposed to selection in a homozygous individual, that is when it is expressed in the phenotype. Recessive genes not lost from the population can be an important source of additional variation at recombination.

In some situations 'ancient' code can re-enter the population. With our hypothetical flowers, we carefully arranged for all seeds from one generation to be derived from plants alive the previous year. If, instead, some seeds were several years old, their code would come from a slightly different gene pool, reflecting the environmental conditions prevailing when their parents were alive. Germination of very old seed (and some seeds can remain dormant for centuries) can re-introduce variation into the gene pool.

Variation can also be lost. Over the generations, a directional change in the code may occur as the most fit alleles are selected for a particular trait. Rapid change in the population can lead to a gene becoming 'fixed' at its locus. Then there is little or no variation in this characteristic. This may happen in small and localized populations where chance and circumstance means a small gene pool and breeding between a small number of individuals, relatives perhaps, who share much of the same code (Section 3.6). Then they are said to have undergone **genetic drift**. Isolated from other populations, with no genetic exchange between them, these changes may become sufficient to create a new species.

1.5 **Rates of evolution**

The idea of species changing over time was not particularly revolutionary amongst the scientists of Darwin's generation. They knew that the fossil record included unknown and strange animals that died out millions of years previously. Remains of many familiar species appeared in much younger sediments. Giant mastodonts and mammoths recalled living elephants, with similarities suggesting that modern species might be derived from ancient relatives. The experience of breeding domesticated plants and animals demonstrated the range of variation that can exist within a species and how readily traits could be selected. What was needed was a description of the mechanism in nature by which inherited traits were selected and new species could arise.

The same answer suggested itself to two British naturalists, both of whom had experienced the diversity of species in tropical forests. Within a few years of each other, Charles Darwin and Alfred Russel Wallace independently originated the idea that the selective force was imposed by the demands of the environment. Given competition for limited resources, and the capacity of all organisms to produce large numbers of offspring, only the fittest, the best adapted, would survive to reproduce. Traits selected over many generations, and the accumulation of small-scale changes, would eventually produce the differences by which a new species would become distinguished.

In the middle of the nineteenth century, the commonly held belief was that species were ordained by a supernatural power, following a fixed design. Darwin and Wallace showed how generations would change naturally under a selective pressure. Not only were species not from a fixed template, the disturbing implication was that humans had evolved from an ancestor shared with the great apes. The human design, once deemed to be perfect, even divine, was now seen as a compromise born out of chance.

When it was published, there was much that the theory did not explain. Species were not obviously changing from generation to generation so it was thought the process was too slow and the Earth too young to produce the great diversity of plants and animals. We now know much more about the real timescale of life on Earth. Today we can trace life back perhaps 3.5 billion years and have evidence of several bursts of speciation in the fossils embedded in the rocks of different eras (Section 9.2).

Several times slates have been wiped clean and new species have evolved to exploit the opportunities as competitors have disappeared. Darwin knew his theory discounted a single day of creation but would have been surprised how frequently and how rapidly species have arisen in the geological past. Massive upheavals in the ecology of the planet have led to both the extinction and the evolution of species. In-between times, the fossil record shows long periods with few discernible evolutionary developments. Even today it is difficult to know how rapidly species were changing at different times, given the imperfect record in the rocks.

The study of **speciation** is a branch of evolutionary biology in its own right. A species forms when populations become isolated and genes cease to pass between them. This reproductive isolation can develop for reasons other than just the physical separation of the populations, a subject we take up in Chapter 2. Rapid speciation will follow rapid environmental change, but the mechanism remains the same—small, gradual changes, building over many generations to produce the differences that will separate populations into new species.

At the cellular level, there is great uniformity in biological systems: all living organisms share similar mechanisms of metabolism and reproduction. These essential features are very conservative in their detail, so that even the flowers and the bees share much of their cellular machinery. The differences we see between species result from their interaction with the environment, from their adaptations to their way of life. Today we can map differences in the code of individuals and populations and describe in great detail how they change with time. In this way we have been able to reconstruct how early humans colonized the earth. We can compare geographical distances with the genetic differences between peoples to reconstruct both our evolutionary and cultural histories (Box 1.2).

It is the information in the chemical code, often minor differences, that distinguish one species from another (Box 2.2). Life became possible when molecules able copy themselves were large enough to carry information that aided their own replication. Chromosomes are immensely complex macromolecules with elaborate mechanisms for reproducing themselves and for reading the signal they carry. Replication and transcription are chemical processes, requiring energy to drive the bonding of the molecules.

If there is an inevitability to life, it is that it continues to improve itself, to better match itself, chemically, biologically, to the environment in which it finds itself, to ensure that this code is replicated. It is the chemical imperative by which genes replicate themselves that drives the whole process (Box 1.4). The inevitability of change follows from the variation we see at the level of the molecule—simply the sequence of bases we observe down the strand of DNA and the different chances of survival these variations in the code have. By such differences are individuals are separated, and over many more loci, so are species separated. Famously, we share 98.5 per cent of our genetic information with chimpanzees, but with a billion base pairs in the human genotype, that odd 1 per cent is enough to create a very

---

BOX 1.4 Units of selection

What does natural selection operate upon? The DNA of the chromosome is not subject to the selective pressures of the environment directly, but rather it is the success of different phenotypes that determines who contributes their genes to the next generation. Each individual represents the expression of thousands of genes and the phenotype is their combined effect. Yet it is some part of the genotype, the instructions written in the genes, which survives into the next generation, not the phenotype. Genes replicate themselves, individuals do not.

Some biologists thus emphasize the role of genetic code in the process of evolutionary change. Richard Dawkins distinguishes between the genes as 'replicators' and the 'vehicle' in which they travel, the individual. Individuals are communal vehicles for a collection of replicators. A gene improves its chances of survival by endowing the vehicle, the phenotype, with a trait that better fits it to the environment. But since the code alone only survives into the next generation, the selection, however indirectly, is only between replicators.

The instruction coded in a gene is, in effect, in competition with its alleles. Actually, Dawkins does not want to identify the 'gene' as the sole unit of selection but rather any length of DNA, any part of a chromosome, which survives intact into the next generation. Replicators can thus be whole sequences of genes. The length of the replicator depends upon the frequency at which they are fragmented when the gametes are produced after meiosis (Figure 1.8). While individuals and populations change over the generations, the replicator remains the same.

The alternative, and more traditional view is put by Ernst Mayr, who emphasizes that it is individual phenotypes which are adapted to their environment. Genes, however defined, are according to Mayr, not the determinants of their own fate. It is the phenotype, the collective expression of the genotype, which succeeds or fails to reproduce. Mayr contends that the effect of a gene and its value to the phenotype can only be understood in the context of that individual in that environment. Under some conditions a gene may be lethal; under others it is advantageous.

This is largely undisputed by the other side. However, they contend that the phenotype, as a collection of genes, is part of the environment in which a replicator finds itself. It succeeds or fails in this context. Combinations of genes that work, mutually compatible replicators, are more likely to reproduce and remain together.

For many replicators it is enough not to be selected against—a gene may not be expressed in the phenotype and yet still find its way into the next generation—it has not been selected, but it has been inherited. Thus far it has been a success, simply by not lowering fitness. The traditional view argues that the unit of selection has to be that which determines fitness (the phenotype or individual) but phenotypes are not inherited and genes are, even the quiet ones.

different species. It may be that the real difference is the way some genes are switched on and off, especially in the process of building a brain. Some feature of our past ecology selected this trait but in providing another adaptable system—our mental processing—we gave ourselves a headstart on the rest of nature.

SUMMARY

The origin of humanity from an ancestral stock shared with the great apes followed a series of changes in the environment of Africa around 5 million years ago. The advent of bipedalism in these apes, prompted by the advancing grasslands, freed the hands for carrying and for tool-making. The techniques of making tools and of working within groups would benefit from a precise language, all of which fostered mental development. In time, the adaptability of mental skills made for a more adaptable species.

Such developments had their origins in shifts in the genetic code stored in the chromosomes. Natural selection determines which individuals are able to pass their genes onto the next generation. A gradual accumulation of genetic changes can lead to a population becoming reproductively isolated from its relatives and, over the generations, evolving into a new species.

FURTHER READING

Dawkins, R. 1986. *The Blind Watchmaker*. Penguin. (A superbly written account of how evolution by natural selection can produce complexity.)

Leakey, R. 1994. *The Origin of Humankind*. Weidenfeld & Nicolson. (The story of human evolution and the various theories surrounding our origins by one of its foremost investigators.)

Cavalli-Sforza, L. L. 2001. *Genes, Peoples and Languages*. Penguin. (How genes can be used to retrace the history of our species and our culture. A readable and accessible overview.)

WEB PAGES

General evolution pages:
www.ucmp.berkeley.edu/history/evolution.html
www.bbc.co.uk/education/darwin/index.shtml
www.nhm.ac.uk

Human evolution pages:
www.anth.ucsb.edu/projects/human/ (Includes many fossils as 3D images).
www.bbc.co.uk/science/cavemen/chronology/

A useful directory of web resources on cell and molecular biology:
www.cellbio.com/

EXERCISES

1 (a) Which genus of humans moved out of Africa?

  (b) Which species was the first to leave Africa?

  (c) The arrival of which species of *Homo* coincided with the disappearance of the last *Australopithecus* species?

  (d) Which species of human first achieved a global distribution?

  (e) Which species was probably the ancestor of *Australopithecus robustus* and *Homo habilis*?

2 If the code for two codons on the chromosome is CCGAGT

  (a) What is the code on the messenger RNA that carries this information to the ribosome?

  (b) What is the code on the two transfer RNA molecules that match the code in the ribosome?

3 Explain why a recessive gene is only subject to selection in a homozygous individual.

4 State the role that meiosis and mitosis play in both sexual and asexual reproduction. At this level, why is sexual reproduction more risky?

5 The current view is that *Homo sapiens sapiens* are not direct descendents of *Homo sapiens neanderthalensis*. Give two pieces of evidence that supports this conclusion.

6 Using the terms given below, complete the paragraph by inserting the appropriate word in the gaps.

The ____ information of an individual is stored within the ____ present in each cell. A ____ individual has ____ sets of chromosomes ____ set from the mother and ____ set from the father. Each chromosome consists of ____ strands, one the ____ of the other. In ____ the chromosomes are duplicated before separating into daughter cells. The chromosome number is therefore maintained. In contrast, ____ includes a cell division which splits the ____ chromosomes, so that each daughter cell is ____. These will produce gametes. When the ____ and ____ fuse at fertilization the ____ condition is restored.

diploid	haploid	one	mitosis	meiosis
gamete	strand	number	genetic	four
crossover	complement	fusion	chromosome	nucleus
mitochondria	three	sperm	two	egg

7 Consider the example illustrated in Figure 1.5. Speculate on how the phenotype of the flower population might change if it came to rely on bees that only flew in May, and then when it was dry.

Tutorial/seminar topics

8 Examine the ways in which the language of a people carries its culture.
Are ways of speaking ways of thinking?
Has the discipline of language promoted our ascendency as a species? And if so, how?
Does our language change or evolve with our understanding of the world? If so, does this have an adaptive significance?

9 'Human beings are successful because they are not specialized'. Consider Table 1.1 and review which features are advantageous in maintaining our adaptability. Is our adaptability a phenotypic or genotypic trait?

10 Sexual reproduction is a genetically determined trait. What does this suggest about the unit of selection?

2 SPECIES

'*That which we call a rose
By any other name would smell as sweet.*'

William Shakespeare: *Romeo and Juliet*

Mirror orchid *(Ophrys speculum)*.

SPECIES

Classification comes naturally to us. We try to compare the unfamiliar with objects or events which are already part of our experience. Recognizing similarities allows us to place something into a category and assume it shares the properties of that group. Establishing categories and describing their characteristic properties is central to the way in which we learn.

This is a good adaptive strategy. It means that we can make predictions and perhaps anticipate situations that we have never encountered before. Indeed, pattern recognition is common to all animals with enough neural machinery to store the information. Identifying key patterns is so central to the behaviour of many higher animals that some plants have developed means of exploiting it. The male bee looking for a mate is readily fooled by the orchid displaying the signals associated with the category 'female'. The flower may have no wings, no legs, no genitalia, but it does, however, have what a male looks for in a female bee, or at least enough for him to change his behaviour.

One difference between us and the animals is that we use our language to name the categories. Names are abbreviations that imply the properties of a category, a term that serves to summarize a complex description. The general term 'dog' implies all that we take to be characteristic of this category from being hairy with a leg at each corner to the fine detail that distinguishes dog from cat. Even then we are sometimes surprised to learn that our category 'domestic dogs' contains everything from a chihuahua to a great dane.

The danger, as Romeo reminds us, is to mistake the name for the category. We need to draw a distinction between the agreed term for an object, its name, and its essential properties. Sometimes the distinctions we draw are not a true reflection of the real world. At other times, we are not sufficiently discerning to pick out the fine detail of nature. This is certainly true of biology, where we have had to repeatedly refine our ideas about the variety and diversity of life.

All sciences begin by creating categories and classifications to help us understand the relations between their important elements. This is invariably a hierarchy, ordered into levels of increasing similarity. In biology, this is termed **systematics** and it groups species according to their shared properties. In the process, biologists can pick out the evolutionary lines linking species, more formally called their **phylogenetic relationships**.

In this chapter, we introduce systematics and also the conventions used in describing, naming and classifying organisms—**taxonomy**. The basic unit in biology is the species, which we defined earlier as individuals sharing a common gene pool and able to produce viable fertile offspring (Box 2). Although this is a good functional definition, we shall see that the boundaries between some species are poorly defined, sometimes because we are witnessing the process of speciation in action, sometimes because we are unable to measure the distinguishing traits or at other times because no real distinctions exist. We go on to look at how new species arise and the increasingly fine lines we have to draw to accurately represent the variety of life.

2.1   What's in a name?

The closest we get to a universal language is in the sciences. Even if nations do not use the same names or even the same alphabet we invariably share the same symbols. Chemists speak to each other using one or two letter symbols as short-hand for the elements of the periodic table. Biologists have the convention of referring to species with two Latin names, the **binomial system** (Box 2.1). This was developed by Linnaeus in the eighteenth century, who actually streamlined earlier systems that used several names.

The most important property of the binomial system is that it establishes a unique combination of two names for each species. When Europeans and Americans use the term 'blackbird' they are invariably referring to two different species and these have different Latin names. *Turdus merula* is the European blackbird and this belongs to an entirely different genus to the North American blackbird (*Agelaius phoeniceus*). The same is true for plants: the British bluebell is *Hyacinthoides non-scripta*, of

---

BOX 2.1   Keys to the Kingdom

Modern classification is a hierarchical system based on Aristotle's concept of the Kingdom as being the highest category in which living organisms can be grouped. This was further developed into the binomial system by the Swedish botanist Carl Linne, who latinized his own name to Carolus Linneaus (Figure 2.1). His system consisted of seven nested categories:

- Kingdom (originally either plant or animal)
- Phylum/Division (suffix –phyta for plants)
- Class (suffix –phyceae for plants)
- Order (suffix –ales for plants)
- Family (suffix –aceae for plants and –idae for animals)
- Genus
- Species.

In Linnaeus's time, it was not always easy to identify the precise relationships between species in the intermediate groups (Class, Order, and Family) and so genus and species tended to be the most reliable and precise way of naming an organism. The hierarchy still works well, but biologists have found it necessary to extend it considerably, dividing and sub-dividing these main categories. Nevertheless, it retains its important property—that species grouped together at the lowest levels have more in common than those only grouped together further up. Plants or animals belonging to the same genus or family share more characters than others in the same order or class. In effect, the hierarchy helps us pick out the evolutionary history of a species.

By the nineteenth century, problems were starting to emerge with the Two-Kingdom system. The development of the microscope revealed an unexpected world populated by microorganisms. These could not be easily classified as either plants or animals and so in 1866, Ernst Haeckel proposed a third Kingdom, the **Protista**, to include the protozoa, primitive algae and fungi. These groups were lumped together with the prokaryotic bacteria so this Kingdom was little more than a biological dustbin for groups that did not fit anywhere else. Eventually a further Kingdom, the **Monera**, was created to contain the bacteria and the blue-green algae (more properly known as the cyanobacteria). Finally, in 1969, Robert H. Whittaker separated the Fungi from the Protista into a Kingdom of their own, so

BOX 2.1   Continued

FIG. 2.1 Carolus Linnaeus (1707–78).

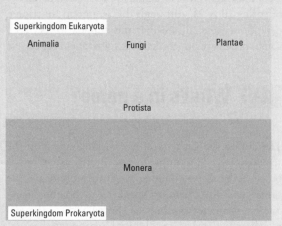

Superkingdom Eukaryota

Animalia   Fungi   Plantae

Protista

Monera

Superkingdom Prokaryota

FIG. 2.2 The five Kingdoms can be divided between two Superkingdoms. All organisms without a discrete nucleus—bacteria, cyanobacteria (blue-green algae) comprise the Monera and are the sole members of the Prokaryota. Protozoans, algae, and slime moulds together form the Protista. The rest—higher plants, animals, and fungi—each have their own Kingdom within the Superkingdom Eukaryota.

producing the Five-Kingdom system generally accepted today (Figure 2.2). On top of these, Lynn Margulis has introduced the concept of **Superkingdoms**, the **Prokaryota** and **Eukaryota**, to acknowledge the fundamental difference in cellular structure between these groups.

There are problems, however, as some life forms do not seem to fit into any group. Viruses are so varied and so different from anything else that Lynn Margulis once half-joked that they were more closely related to their hosts than each other. To this day, viruses remain outside mainstream classification as little more than replicating molecules that hijack the replication machinery of living cells.

Taxonomists have an altogether different problem when it comes to lichens. Although these may look like individual organisms, they are in fact a close association between algal and fungal cells. Lichens are the only true naturally occurring **chimeras**—organisms that have two genotypes (Section 4.2). There are relatively few algal species involved in these associations (and in some cases) several different algae occur in a single lichen). However, each 'species' of lichen has a different fungal species and so they are classified on the basis of their fungal partners.

the lily family, whilst the plant called a bluebell in California is *Phacelia tanacetifolia* (Plate 2.1) is from an entirely unrelated family, the Hydrophyllaceae.

Notice the detail of using a Latin (or scientific) name:

- Each consists of two names (hence the **binomial** system).

- The first name is the **generic name**, the name of the genus (this always begins with a capital letter).

- The second name is the **specific name**, the name of the species (this always begins with a lower case letter).

- A scientific name is always printed in italics or if handwritten, underlined.

- It is the combination of the generic and specific names that is unique. So two species may belong to the same genus, and therefore share the same generic name, but will share different specific names—for example, *Mustela erminea*—the stoat and *Mustela nivalis*—the weasel. Two species belonging to different genera may share the same specific name—for example, *Primula vulgaris* the primrose and *Calluna vulgaris*, heather.

- After first being cited in full, the name may subsequently be abbreviated if there is no possibility of confusion, for example, *P. vulgaris* and *C. vulgaris*.

- Often the name is descriptive, even for those with just a little knowledge of Latin or Greek. For example, 'vulgaris' means 'common'.

Sometimes the citation also gives the name of the **authority** immediately afterwards, for example, *Helix aspersa* Müller (the garden snail). The authority is the first person to describe that species in its currently accepted classification and who gave it the name. Scientific papers usually cite the name of the authority when the Latin name is used for the first time. There are strict rules about naming and giving authorities for species, with well-known authors having their names abbreviated. In *Homo sapiens* L., the L refers to Linnaeus, the authority behind our own species name.

These rigorous rules make the binomial system universal. Species names are agreed by an international committee which operates the International Codes of Biological Nomenclature. Animals are dealt with by the International Code of Zoological Nomenclature (ICZN), which recognizes organisms down to sub-species level, whilst the International Code of Botanical Nomenclature (ICBN) goes further, recognizing not only sub-species, but also varieties, sub-varieties, forms and sub-forms (Box 2.1). Classification of most groups is under constant review, not only as new species are found and named, but also as we learn more about the evolutionary history of known groups.

Even within a well-defined species there can be considerable variation between individuals and this can make it difficult to decide where one species ends and another begins. We can observe this simply by looking around a room of people, noting the range of sizes and shapes. For many species, the variation we observe will be incidental, possibly phenotypic variation acquired during the life of the individual that cannot be passed on to the next generation. Our problem is to decide whether the characters we observe are important in a species' evolutionary past or, indeed, whether their variation could be important for its future.

2.2 Origins of the major groups

In its original form, the binomial classification system used mainly morphological characters to classify around 10 000 plants and animals. Grouping species into this hierarchy, according to their shared characters, made it obvious which species were closely related. Darwin and other naturalists recognized that this gives us important clues to the evolutionary history or **phylogeny** of a species, helping us to pick out the ancestry of multicellular organisms (Box 2.1). Today, we no longer rely solely on external morphology, but now draw upon the evidence written in the cells, in sequences of genes and the structures of key proteins.

Our understanding of the most fundamental divisions between the major groups of living organisms has advanced considerably since we have been able to compare their cellular and biochemical organisation. This has led some molecular biologists to attempt to reformulate the whole classification of life, in some cases based on a single, albeit ancient molecule. The genetic sequence in one type of ribosomal RNA (Figure 1.7) has been used to divide all organisms into just three major 'domains': the bacteria, the eukaryotes and a group called the **archaeobacteria**. The archaeobacteria share an ancestry with the eukaryotes (the group that includes higher plants and animals) but they are largely confined to very extreme environments (Box 6.1). Perhaps not surprisingly, there is considerable debate about the merits of using just a few phenotypic characters, but other DNA-based approaches are increasingly confirming this three-way division.

The traditional classification, based principally on morphology, recognizes five kingdoms: Monera, Protista, Fungi, Animalia and Plantae. At a higher level, these are grouped between two superkingdoms—the **prokaryotes** and the **eukaryotes** (Figure 2.2). Prokaryotes, including the bacteria and blue-green algae, have no nucleus and a simple looped chromosome. Eukaryotes have cells in which their DNA is wound around proteins to form long chromosomes that are held within a nuclear membrane (Figure 1.6).

Other organelles in the eukaryote cell are equally well packaged, including the mitochondria, (whose enzymes drive much of the cellular metabolism) and the chloroplasts of photosynthetic plants. According to one view, this compartmentalization evolved as a series of infoldings of the cell membrane, isolating the organelles in their own membranes. An alternative theory, now widely accepted, suggests the eukaryote cell formed from the assimilation and association of a series of prokaryote cells. All the cells in this partnership cells benefit from this **symbiosis** so that now, after many generations evolving together, they can only reproduce and survive in association. In her **endosymbiont theory**, Lynn Margulis argues that chloroplasts and mitochondria, amongst others, were derived from prokaryotic cells that combined symbiotically with a host cell. This would explain why the chloroplast has its own DNA, and why it is so similar in size and shape to blue-green algae (cyanobacteria) cells. In eukaryote cells, the chloroplasts and mitochondria are synthesized using the DNA of the nucleus, so their code had to find its way into the host chromosomes. Although this is a complication, there are known mechanisms by which this could happen.

2.3   The species

Ideally, our classification system would accurately reflect the important differences between organisms. Then, each species, as the basic element in the phylogeny, would represent a unique collection of traits, the current endpoints of different evolutionary histories. However, the boundaries may be blurred where two species have only recently diverged or if the trait which distinguishes them is not obvious.

Distinctions between orders, families or genera are usually easily made. For example, it is easy to differentiate between a grasshopper and a butterfly. Though they are both insects, the differences in their anatomy (the enlarged hind legs of the grasshopper and the broad, dusty wings of the butterfly) readily place them in different orders. As we descend down the hierarchy, deciding where a family or genus begins or ends requires closer and closer observation (Figure 2.3). At the level of the species, the difficulty is deciding whether the differences observed between individuals are distinguishing traits. Are the variations in the colouration of the butterfly's wings part of the natural variation within a species, or a character that helps to define a distinct species? Often, variations in colour mark only a different race, a variety confined to a particular area, but whose members are still capable of interbreeding with the rest of the species.

Shared morphology remains the most common means of classification and the basis of most identification keys for higher plants and animals (Figure 2.4). In a key, a series of descriptions are offered at each level in the hierarchy. An unknown specimen is compared against each description and placed in the category with the closest match. This continues, running down a sequence of alternatives, each giving directions to the next stage. By choosing the most appropriate description at each step, the key arrives at an identification and provides a sequence of descriptions that could only match the named species.

This works best, of course, with species that have been well described, where we know the most reliable characters to observe. Without these, classifications based on appearances alone can be misleading. For

Classification

European Blackbird	American Blackbird	Category	European Bluebell	Californian Bluebell
Animalia (animals)	Animalia (animals)	**Kingdom**	Plantae (plants)	Plantae (plants)
Chordata (chordates)	Chordata (chordates)	**Phylum**	Trachaeophyta (vascular plants)	Trachaeophyta (vascular plants)
Aves (birds)	Aves (birds)	**Class**	Angiospermae (flowering plants)	Angiospermae (flowering plants)
Passeriformes (songbirds)	Passeriformes (songbirds)	**Order**	Liliales (lilies)	Polemoniales (Jacob's ladders)
Muscicipidae (warblers)	Icteridae (icterids)	**Family**	Liliaceae (lily family)	Hydrophyllaceae (phacelia family)
Turdus	*Agelaius*	**Genus**	*Hyacinthoides*	*Phacelia*
merula	*phoenicius*	**Species**	*non-scripta*	*tanacetifolia*

FIG. 2.3 The full hierarchical classification of European and North American blackbirds and bluebells.

example, morphological variation caused Edward Poulton considerable difficulty when he tried to classify the African mocker swallowtail butterfly (*Papilio dardanus*). The males were obvious enough, but there seemed to be no females of the species. Females of the genus were known, but there were at least three different types and these were so variable that they neither resembled the males nor each other. Each group of females was therefore classified in a species of its own. In fact, the females come in three forms, each of which mimics one of three species of an unpalatable butterfly (*Amauris* species) (Box 4.3). Despite being **polymorphic** in appearance the different female forms shared the same internal anatomy. Variation on this scale in the morphology of the males is disadvantageous because they need to be recognized by the female to be allowed to mate. However, there is also a risk for females which vary too much—those whose appearance does not closely match the unpalatable species are more likely to be preyed upon.

Another complication in classifying from morphology alone is that natural selection can produce very similar structures from very different starting points. This is known as **convergent evolution**. One of the most spectacular examples is the thylacine or marsupial wolf. Probably now extinct, this was, to all outward appearances, a member of a dog family, but, as a marsupial, it was more closely related to possums and kangaroos (Plate 2.2).

Such structural resemblance is termed **homoplasy** and is particularly common amongst plants. Again, similar selective pressures will tend to produce similar results. Mediterranean-type plant communities are found in five distant regions of the world, but although they are composed of very different plant groups, many species have evolved thick, tough leaves and a short shrubby stature. These are adaptations against a prolonged summer drought and periodic fires (Section 5.1, Plate 5.1). A more general example of homoplasy are tendrils. Plants that have evolved tendrils are able to raise themselves toward the light, avoiding the costs of substantial stems and trunks. Instead, they exploit the supporting structures of their neighbours. Various families have evolved tendrils, but as modifications of different structures (Figure 2.5) including leaves, stems, petioles (leaf stalks), and stipules (scales).

(1)

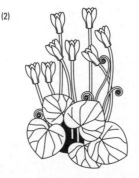

(2)

Cyclamen purpurascens
(Cyclamen)

(3)

Primula vulgaris
(Primrose)

Hottonia palustris
(Water violet)

(4)

Primula veris
(Cowslip)

(5)

Primula elatior
(Oxlip)

1. Leaves in basal rosette—go to 2
   Leaves in rosette and along stem—water violet (*Hottonia palutris*)
2. Petals flat/turned forwards (Primula species)—go to 3
   Petals turned backwards—cyclamen (*Cyclamen purpurascens*)
3. Flowers in cluster—go to 4
   Flowers solitary—primrose (*Primula vulgaris*)
4. Flowers fragrant—cowslip (*Primula veris*)
   Flowers not fragrant—oxlip (*Primula elatior*)

FIG. 2.4 A simple key for some members of the primrose famil (Primulaceae).

The biological species

The complications of morphological variation and convergent evolution have led biologists to treat morphological species with a good measure of caution. Today, most biologists use a more functional definition of a species, termed the **biological species concept**. A biological species comprises those individuals which can interbreed successfully to produce fertile and viable offspring forming a gene pool and sharing a common set of adaptations. This definition recognizes that species are variable and changeable, but by referring to an interbreeding group, it does not rule out the possibility of change in the future, perhaps with some individuals becoming excluded from the breeding population. Over time, species evolve and new breeding populations form.

FIG. 2.5 *Clematis* uses its petioles as tendrils.

FIG. 2.6 An apomictic species: dandelion (*Taraxacum officinalis* agg).

However, the biological species concept cannot be applied universally. It does not readily encompass the many species that reproduce asexually. More complicated still are groups which reproduce sexually and asexually, and do so within defined groupings.

Although it is given a single binomial, *Taraxacum officinale*, the dandelion (Figure 2.6) is more properly described as an aggregate, (*Taraxacum officinale* agg.) because it comprises 10 sexual types, and a further 2000 types that reproduce asexually. These latter types are called **apomictic**—they produce seed without pollination—and are consequently genetically isolated. Since the flow of genes from one individual to another has stopped, different apomictic species (also known as **micro-species**) are often found growing very close together, but remaining distinct from each other: John Richards discovered that as many as 100 micro-species of dandelion could be found within a single hectare. Micro-species are notoriously difficult to differentiate and we simply avoid the complication by adding the abbreviation agg. after the specific name.

A further complication is the freedom with which some organisms exchange genes across species boundaries and thus fail to conform to the definition of the biological species. Sometimes, perhaps under special circumstances (such as cultivation or captivity), the barriers that divide species may be breached and interbreeding occurs. Any offspring from such a union is called a **hybrid**. Interspecific hybrids are rare amongst most animal groups and these are rarely fertile, though even the normally sterile mule (the hybrid from a horse and donkey) may, very occasionally, parent its own offspring. Other hybrids are consistently sterile, so that ligers and tigons (crosses between tigers and lions sometimes produced in zoos) are inevitably genetic dead-ends.

More interestingly, there are examples of fertile hybrid animals which have been classified as different species. Indeed, we have only learnt about their true ancestry after looking at their genetic signature (Box 2.2). An example was the discovery that the red wolf (*Canis rufus*) was actually a fully fertile hybrid between two distinct species, the coyote and the grey wolf (Plate 2.3).

The red wolf underwent dramatic decline after 1900 due to hunting, loss of habitat and increasing hybridization with the coyote, which had been extending its range. Robert Wayne and Susan Jenks compared the mitochondrial DNA (mtDNA) of the few red wolves collected together in a captive breeding programme with that of coyotes and grey wolves. They also collected mtDNA from six pelts of red wolves that dated before hybridization was considered to have begun on a large scale. The mtDNA and its cytochrome b sequences revealed that the red wolf is a hybrid of coyotes and grey wolves, and that their range represented a zone of hybridization between the two separate species. Indeed, Wayne and Jenks suggest that some of the wild animals they sampled in the wild would have been classified morphologically as grey wolves but actually approximate more closely to the red wolf genotype.

Fertile hybrids are quite common in the plant Kingdom. Many species of orchid belonging to the genus *Dactylorhiza* will readily cross with each other. Some hybrids appear to be a perfect blend of the two parent species, while others resemble one or other parent. This makes them particularly difficult to identify (Plate 2.4).

The results of crosses between fertile hybrids also make it difficult to trace phylogenies. For these groups, the idea of a species as a discrete entity is misleading and we have to accept a more fluid classification, of a highly variable group of individuals passing genetic code readily between themselves. Nowadays, molecular biology and genetics allows us to explore these differences in increasing detail (Box 2.2).

BOX 2.2   Reading the molecular code

The genetic code, the blueprint for a species, may seem the obvious place to look for the real differences between one species and another, but it is only recently that we have been able to read gene sequences. Before that, molecular biologists had to look at the products transcribed from the code (Figure 1.7). This meant comparing the composition of proteins, especially those responsible for key functions, to reconstruct the phylogenetic relationships of a species. The greater the genetic distance between individuals, the less likely they are to share proteins with exactly the same structure.

Even in the absence of a selective pressure, genotypic differences and chance variations will change the amino acid composition or sequence and we can quantify this

BOX 2.2   Continued

relatively easily. **Immunological methods** use the capacity of mammalian immune systems to detect 'non-self' proteins. Antibodies produced against the foreign protein are highly specific and the degree of binding between an antigen (the protein under test) and its antibody can give us a gross measure of the structural similarities of proteins from different species.

Gel electrophoresis separates proteins according to their electrical charge and molecular weight. The greater the charge across a protein molecule and the smaller its size, the faster it will migrate toward one pole of an electrical field. Two protein samples migrating the same distance through a gel are likely to have the same configuration and amino acid composition. This technique is widely used because it can pick up subtle differences in protein structure and is particularly good at identifying hybrids that otherwise look the same.

Sequencing the amino acids in a protein, using a series of enzymic scissors and assays, have paved the way for more detailed analysis of proteins. One protein, **cytochrome**, is of particular interest because it is found in all living organisms and is therefore very ancient. Cytochrome plays an important role in the energy-generating activity of mitochondria and chloroplasts. Analysis of its 104 amino acids has allowed biologists to group species according these sequences. All share a sequence of 33 amino acids in the same position along the protein chain, but the remainder fall into groups—a hierarchy of sequences that mirrors the phylogeny of living systems. All vertebrates, for example, fall into a group of their own, as do the invertebrates, plants, and fungi. The cytochrome of human beings and chimpanzees is identical.

The most direct way of deciding the genetic differences between individuals and species is to read the genetic code itself. Part of the double strand of DNA is unzipped when the genetic code is being read (Figure 1.7). Strands can also be separated by heat treatment and will re-associate as they cool. In **DNA hybridization**, DNA from two sources is heat treated, mixed together and allowed to cool. Some strands from each source will associate to form hybrid strands. The hybridized DNA is then separated again by heat treatment, but the closer their match, the higher the temperature required to dissociate the strands. Using the temperature scale over which complete dissociation takes place between hybridized and non-hybridized

strands, we can calculate the a percentage similarity for an entire genome.

More precise information can be gained from **gene sequencing**. Central to the technique is the **polymerase chain reaction** (PCR), which allows minute samples of DNA to be copied (amplified) and then sequenced by a process known as the Sanger method. This technique uses a series of dyes to tag each of the bases that make up DNA. PCR is used to build copies of the DNA fragment with the tagged bases which are then separated out using gel electrophoresis and read off using a laser scanner linked to a computer. The result is a series of bases corresponding to the gene sequences within each gene fragment. The more fragments shared between two individuals, the greater their relatedness and the longer their common evolutionary history. These techniques can be used wherever DNA occurs. PCR allows us to amplify fragments of DNA from the remains of long-dead organisms (including Neanderthal human remains) or identify living individuals through DNA 'fingerprinting'.

We can also distinguish the DNA of the nucleus from that of the mitochondria or the chloroplast. **Mitochondrial DNA** (mtDNA) is only inherited from the mother with the cytoplasm supplied in the egg cell, making it particularly useful for working out relatedness within parthenogenic species where females produce young without having to mate.

It also helps us sort out the parents of hybrids. Using this technique, Steven Carr was able to unravel the complex mating system of white tailed deer (*Odocoileus virginianus*) and mule deer (*Odocoileus hemionus*) in Texas. He found that mtDNA of the hybrids was the same *as O. virginanus*, indicating that crosses between the two species invariably had a mule deer male and a white tail female as their parents.

Classifying species and working out their origins by genes alone can be problematic. Changes in DNA may not accurately reflect evolutionary change. DNA molecules are more changeable in some groups than others. This makes it difficult to construct accurately timed phylogenies on the basis of molecular evolution alone. For example, plants within the sunflower family, the Asteraceae, show little variation compared with other flowering plants. This may mean that sunflowers are a recent evolutionary development or simply that the DNA within their chloroplasts is slow to change.

Defining species by their chemistry

We could also take Romeo's advice and identify a rose by its scent. Flowers are recognized for their ability to produce complex chemical signals to attract insect pollinators. Some scents even mimic an animal's sex pheromone, the chemical signal used to attract a potential mate. For centuries, a group of Panamanian orchids had defied all attempts at classification. Defeated, taxonomists simply bundled the whole lot into a single species, *Cycnoches egertonianum* and referred to them as the 'Egertonianum complex'. Unable to classify the plants by traditional means, Katherine Gregg looked at the sex lives of the bees that pollinate them. After they have attempted to mate with the orchid, male bees collect the delicate lemon scent and use it as a pheromone to mark out their mating sites. Several species of bees use this strategy and field observations showed that each were extremely selective about their choice of plant. A chemical analysis of the scents revealed that they were composed of up to 18 separate compounds and Gregg found their differences reflected links with different pollinators. By reference to the species of bee and type of scent she was able to make sense of the Egertonium complex, dividing it up into four 'chemotypes', which are now recognized as four separate species.

Although the bees might have got there first, our use of chemistry to classify organisms also has a long history. In fact, the standard test for lichen identification was developed by Nylander back in 1866 and serves to group lichens into six chemical races. The molecular classification of higher plants is also well developed because of our detailed knowledge of the secondary plant metabolites (Section 4.4) which have medicinal or economic value. Genera like *Eucalyptus* hybridize freely and it is not always apparent which species were the parents, especially when one of the parents may itself have been a hybrid. A particular group of these compounds—the terpenoids—have proved invaluable in sorting out such plant paternity cases and these have also been used to work out the origins of hybrids in other aromatic species such as junipers (*Juniperus* species) and pines (*Pinus* species).

Although morphological characteristics remain the basis of most keys to multicellular plants and animals, chemical and biochemical techniques enable us to make finer distinctions. Our goal is to arrive at a classification that reflects the real phylogeny of a species, but the complication is the great variability from one individual to another.

2.4 Variation within a population

Variation between individuals is usually fairly obvious. We use these differences to distinguish one person from another and to recall their name, at least most of the time. Under other circumstances, we might wish to know what significance these differences have for an individual's survival and reproductive success.

If we plot some measure of fitness (say growth rate or abundance) against some key environmental factor (such as temperature or water availability) we invariably find a species has an optimum range (Figure 2.7). Usually this produces a characteristic curve, termed a normal distribution. Most individuals are found close to the mean value and numbers rapidly tail off with distance away from the mean.

This tells us something about the species' adaptation to that parameter. Away from the optimum, conditions would be less favourable and individuals will incur physiological costs to maintain themselves. For a warm-blooded animal (or endotherm), for example, this may mean using energy to keep warm or to keep cool. If the costs are high, or have to be sustained for a long time, an individual will have fewer resources to devote to growth or reproduction (Section 4.1). Those forced to live at the margins of their physiological range may thus fail to reproduce or grow.

However, individuals may arise which are better adapted to living outside the optimum range of the larger population. Then some change in their genotype allows them to flourish where most struggle.

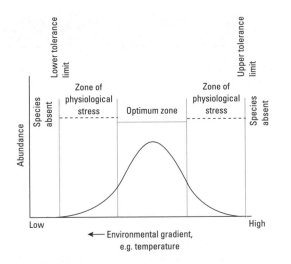

FIG. 2.7 A plot of abundance of a species against some environmental parameter, in this case, temperature. This shows the range to which a species is best adapted and the same sort of pattern would be found for any important parameter, such as water availability, nutrient level, salinity, and so on. Outside the optimum range, numbers decline rapidly because individuals have to expend energy or other resources to adapt physiologically to the poor conditions. At extreme temperatures, conditions are too harsh and the species is not found. We could make the same sort of plot for any other measure of a species' performance, such as reproductive success or growth rate.

This is the nature of variation within a species: not every individual responds in the same way to a selective pressure and there will be some who favour conditions away from the population optimum.

A dramatic example of the outcome of genetic variation within human populations is seen in disease resistance. When challenged by unfamiliar diseases, those able to resist the pathogen, either through variation in their immune system or some other adaptation, stand a greater chance of survival. Nowhere is this more starkly seen than in the responses of native peoples to the diseases of in-comers. The landing of Columbus and his successors sealed the fate of many millions in the Americas. It is estimated that the indigenous population of Mexico fell from around 30 million to just 3 million between 1519 and 1568 following their first exposure to smallpox and measles. The native peoples simply had no immunity to the diseases carried by the conquistadors. The genetic variability within the native population meant that some were able develop disease resistance, and it is their descendants we see today. Isolated Amazonian tribes have more recently succumbed to the diseases of invaders, though on this occasion carried by lumbermen and miners (Box 2.3).

Much of the variation we observe between individuals does not have an adaptive significance and some differ simply because of the chance combination of genes they were dealt at fertilization. This is termed neutral variation (Section 1.3): much of the variation in proteins (Box 2.2) appears not to be adaptive and impart no selective advantage or incur any selective cost. Even so, in a new or a changing environment, such traits may, again by chance, equip some individuals to survive and reproduce with a higher frequency than their neighbours. Then the amount of variation in some critical trait allows a species to adapt and evolve quickly. A population lacking such variation has fewer genotypes from which to select, and less chance of adapting to significant change in its habitat.

BOX 2.3   The shock of the new

The native Americans arrived in the New World around 13 000 years ago when they crossed the Bering Strait between Alaska and Siberia. Their genes record the history of their walk through the continent. The early colonists had to survive the harsh and unpredictable conditions of the Arctic Circle, to pass through what has been called the 'arctic filter'. It seems that the severe conditions in the north favoured particular genotypes, most especially those who could survive periods of famine.

Life must have been easier as they moved south into the temperate regions, where a thriving and expanding population quickly established itself. However, what started as a relatively rapid spread then seems to have slowed. Archaeologists and anthropologists have repeatedly found evidence that colonization was checked further south—some new challenge must have faced those moving towards Central and South America. One suggestion is that these peoples were confronted with a range of diseases for which they had no immunity, diseases prevalent in the tropical forest and resident in the mammals of these regions. Human populations had to adapt and acquire immunity before they could move on. Again, only those genotypes able to pass through the tropical filter would survive to enter South America.

That this was the cause of a second genetic bottleneck is suggested by the genetic make-up of native South American peoples to this day. However, their geographical and genetic isolation also become obvious when they were exposed to the diseases of later European colonists. Many millions perished due to the diseases of the Conquistadors and those who followed.

Ironically, their geographical and cultural isolation saved some native peoples from the worst of these epidemics, at least in the first waves of European colonization. The Yanomamo number 20 000 individuals spread across 150–200 villages in remote forest areas on the border between Brazil and Venezuela. They survive through a regime of hunting, gathering and cropping, but this requires large areas of forest so their villages tend to be widely dispersed. Because there is little contact between the villages, individuals tend to marry within their own community and between-village pairings occur only rarely. The small size of communities and their reproductive behaviour had important consequences for the disease resistance of the Yanomamo.

For a viral disease like measles or influenza to remain within a population there has to be a sufficient number of individuals to act as a reservoir of infection, depending on the infection pattern of the disease. Measles, for example, is only virulent within an individual for 14 days after which it needs to move on to someone else. Epidemiologists estimate that for measles to remain

active within a population there needs to be a minimum population size of 200 000 individuals. In small, isolated communities such a disease is easily lost, and with it, over a number of years, the immune response of the population. When the disease returns, it can devastate a population with no immune protection.

This is exactly what happened when the Yanomamo people were 'discovered' by the outside world in the latter half of the twentieth century. Their first encounters with anthropologists, explorers and missionaries in the 1960s led to a series of epidemics amongst the natives in which the 'diseases of civilization' such as measles, mumps and influenza swept through their villages. These are diseases which much of the world's population take in their stride because they are frequently exposed through the regular contacts between peoples, helping to build and maintain the immune response of the larger population.

The Yanamamo have not had this frequency of exchanges and were thus further devastated when these and other diseases arrived with later colonists. The building of the Northern Circumferential Highway in the 1970s brought with it construction workers and more recently an influx of loggers, ranchers and miners seeking to exploit the Yanomamo lands. Now tuberculosis and HIV/AIDS are the major killers of the Indians and a whole way life is under threat. The paradox here is whilst genotypic variation is important for disease resistance, so is exposure to low levels of infection from childhood. This is a phenotypic response, so that without exposure, the immune system cannot provide a responsive protection. Globally, the human population is threatened in the same way by a new pathogen or a variation on the influenza virus at regular intervals. Today it is the SARS virus, another pathogen thought to have crossed the barrier between species.

Isolation only offers protection as long as it lasts. Thereafter, the shock of the new can be fatal.

2.5 Ecological niche

If a species is defined in nature by its collection of adaptations, which selective pressures have been most important in shaping it? And what selective pressures have led to it becoming separated from its close relatives?

Major morphological adaptations, such as those needed for flight, have a long evolutionary history and will separate animals high up in the hierarchy, at the Class level (Class Aves, the birds) or Order level (Order Chiroptera, the bats). Smaller adaptive

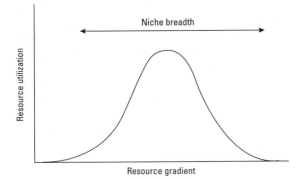

FIG. 2.8 Niche, defined by one environmental gradient—in this case, the resource being used. For every environmental parameter we can define an optimum (Figure 2.7), but the full range exploited by a species is termed its niche breadth. For a plant, such a gradient might be light levels of different intensities, whereas the niche breadth for an insectivore would be the size-range of insects on which it feeds.

changes—variations in colour, for example—may not require major physiological or structural changes. They distinguish closely related species which have diverged relatively recently. Even more recent are the distinctions between varieties of plants or sub-species of animals that perhaps are only separated geographically rather than genetically.

But what about closely related species that share the same geographical range, but apparently remain distinct? Often, it is the details of their life history, or the habitat they use, which keeps them separate. For example, two micro-moths, *Lampronia rubiella* and *Lampronia praeletella* share a similar range, and as adults, have similar markings. Their caterpillars, however, are leaf miners, and one attacks strawberry leaves (*L. praeletella*), the other raspberry fruit stalks (*L. rubiella*). They have evolved to use different resources and, indeed, live their larval stage in different habitats. This separation at one stage in their life cycle means they are not in direct competition with each other. The larval food plant is one of a large range of factors to which each species has to adapt, any one of which can serve to distinguish them from other species in the same genus.

For any environmental factor, whether biotic or abiotic, we can define the optimum range for a species and measure its performance over this range (Figure 2.8). The distribution of the species will reflect its adaptation to that factor and, more generally, to the other myriad factors to which it responds. This totality of factors to which a species has adapted is called its **ecological niche**.

Niche is a complex idea that tries to describe the fit between a species and its environment. An ecological niche is not a place in the normal sense of the word and is much more than a species' habitat. It represents the interaction between a species and its habitat, describing where a species lives and how it lives there. Niche is sometimes described as a species' role in its community. That role depends on its interaction with other members of the community and so can include everything from being, say, a strawberry leaf miner to a prey item for a bird or bat, or a host to a parasite.

Although niche is difficult to define, it does, however, have a very real ecological meaning. We can see this in convergent evolution, where species filling equivalent niches in different areas develop the same adaptive solutions. The thylacine and the dog is one example (Plate 2.2). Anteaters in the three southern continents have elongated snouts and long probing tongues—equivalent adaptations to the same selective pressure. Like the thylacine, one of these (*Myrmecobius*) is a marsupial, another (the echidna) is an egg-laying mammal, while the pangolins of Africa or the tamandua of South America are placental mammals. These fundamental differences in reproductive physiology place their separation high up in the hierarchy (at the level of the sub-class), indicating a large phylogenetic distance between them. Yet each has evolved equivalent adaptations to collect ants.

The match is not perfect and differences between most convergent species are obvious on even the most casual inspection. This emphasizes that a niche is defined by the organism and its interactions with its living and non-living environment. Rather than simply being a role that might be played by a number of species, a full description of a niche includes the detail of the species which occupies it. So, part of the niche occupied by *L. rubiella* is defined by the raspberry stalk it uses when a larva.

Clearly, there are practical difficulties in measuring every significant factor in a species' niche. Usually, studies have concentrated on those which seem to be important in separating species and which are also easily measured. **Niche breadth** is the range of a factor over which a species is found (Figure 2.8). This could be abiotic factors like temperature, or (more often) the range of a resource the species uses. An insect that feeds on the leaves of a variety of plants is said to have a larger niche breadth than one which feeds only on a single species. Species with a large niche breadth are **generalist species**, able to exploit a wide range of resources. **Specialists**, however, are adapted to a narrow range of a resource, but attempt to out-compete other less well adapted species by being more efficient in its use. Being a generalist is a good strategy in an unpredictable habitat, when switching to a different part of the **resource spectrum** may be necessary at times of shortage. Being a specialist is only viable when the resource supply is very reliable, in constant or predictable habitats (Section 3.5). We would therefore expect to find frequently or recently disturbed habitats dominated by species with broad niches (Section 5.3).

Niche overlap is that part of a resource spectrum which two species occupy (Figure 2.9a). Sometimes, overlap indicates that two species are competing with each other for the resource. At other times it can mean the opposite.

Niche and competition

Different species invariably concentrate on different parts of the range of a resource—they are spread out along the resource spectrum (Figure 2.9b). Two species sharing a large zone of overlap may indicate a region where the two species are competing for the same prey or the same nutrients in the soil. It can also indicate the opposite—an abundant resource, not limiting the growth of either species; if an overlap persists over many generations, we may conclude there is little competition, at least along that resource gradient.

By the same token, a lack of overlap between two species only indicates there is no current competition for this particular resource. It tells us little about their past battles or their ongoing fight for other resources. We may be observing the outcome of competitive battles fought long ago, after each has become adapted to a different range.

Or we may be witness to a competitive battle in progress. Where a resource checks the growth or reproduction of both species, utilization by one species depletes the supply for the other. **Interspecific competition** (competition between species—Section 4.3) means either species would perform better in the absence of its competitor. Then niche overlap does indeed denote competition.

Under these circumstances, ecologists distinguish two types of niche. The range which a species could occupy in the absence of interference from other species is its **fundamental niche**. The range to which it is confined by competitors or predators is its **realized niche**. Under severe competition, a species may only use a very narrow part of a resource spectrum and have a small realized niche. Then natural selection will be very intense, favouring those individuals able to make best use of what is available. These will be the most successful reproducers and will soon dominate the population and gene pool. In this way, a species becomes highly specialised, often showing distinct morphological or other changes that adapt them to use a resource most effectively.

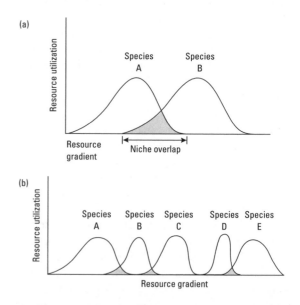

FIG. 2.9 (a) Two species that exploit the same part of a resource gradient (or 'spectrum') are said to shown niche overlap. In some cases, this may indicate competition between them for that resource. (b) A collection of species exploiting the same resource can only coexist if they exploit different parts of the gradient.

This is known as **character displacement** and is seen when two closely related species or races begin to diverge.

Competition between individuals of the same species is termed **intraspecific competition** (Section 4.3) and is one of the prime forces driving natural selection. Those most able to secure resources—food, space or shelter—will be more able to reproduce and ensure their genetic code enters the gene pool of the next generation. Ordinarily, we would expect those individuals close to their optimum range to have the greatest reproductive success. However, genetic change might adapt some individuals to a different part of the environmental gradient. Perhaps a change at one locus means a key enzyme is more effective at lower temperatures. This new genotype may then be able to thrive at a new optimum and its numbers increase. Mutation has produced a variety that eventually might lead to a new species, though at the moment, it still can still breed with the rest of the population. Over time, the entire population may become dominated by the new genotype and its optimum range shifts (Figure 2.10).

Dolph Schluter has shown how rapidly intraspecific competition can lead to significant character displacement. Schluter and his colleague Don McPhail work on the three-spined stickleback (*Gasterosteus aculeatus* complex), a variable group of species whose taxonomy has yet to be fully worked out. These fish have been isolated in a series of small coastal lakes in British Columbia since the retreat of the ice sheet 10 000–13 000 years ago. In some lakes, two different species are found—a larger and rounder fish that feeds on the invertebrates on the lake bottom ('benthic' species) and a smaller slender species that feeds on the phytoplankton closer to the surface ('limnetic' species). In lakes where there is only one species an intermediate form is found, which is able to exploit both food sources. These species will readily interbreed with each other.

Schluter set up an experiment to examine the pressure for character displacement when two species share similar morphologies and the same diet. He bred three intermediate forms which were either close to the limnetic form, close to the benthic form or true intermediates. These were then grown in experimental ponds in the presence or absence of wholly limnetic species. By measuring the performance of each species (in this case by measuring their growth rate over a season), Schluter was able to work out the response of different intermediate forms to the presence of a limnetic species (Figure 2.11).

As expected, the intermediates closest to the limnetic species grew less well in the presence of limnetics. In contrast, the same intermediates in control experiments showed no depression in their growth rate. The greater the similarity between an intermediate and the limnetic form, the greater the niche overlap and the greater the reduction in growth rate. Eventually this might result in the loss of intermediates as only those using the different niche would survive. Over a number of generations, under

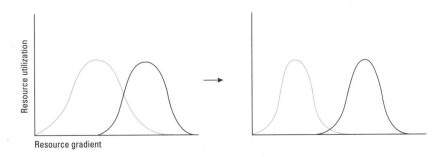

FIG 2.10 Where overlap represents a high competitive pressure on each species, one or both of them will, through natural selection, gradually change to reduce the overlap, shifting their niche. This is termed character displacement and may be represented by some change in physiology or morphology.

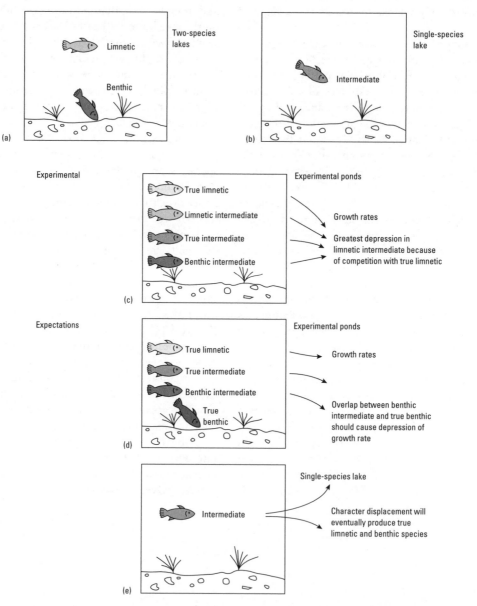

FIG. 2.11 Schluter's experiments with three-spined sticklebacks (*Gasterosteus aculeatus* complex) demonstrating character displacement. Lakes in British Columbia have either two species of stickleback with different morphologies or one which is intermediate between the two. (a) One form (the 'benthic' species) feeds on bottom-dwelling invertebrates and has a large mouth and rounded body; the other feeds on phytoplankton in open water and has a smaller mouth and slender body (the 'limnetic' species). (b) Intermediates feed in both categories. (c) Schluter has shown that intermediates with forms closest to the limnetic species have the greatest growth depression. (d) Similarly, we would expect intermediates in closest competition with benthic species to grow less well. (e) Over many generations, these intermediates would be expected to show further character displacement, feeding more and more in the opposite category of the resident specialist. Lakes now populated by intermediates are likely to differentiate into the two forms given sufficient time.

interspecific competition, the intermediates would be expected to show character displacement, presumably approximating to the larger, rounder form of benthic sticklebacks.

It seems that the niche separations we see in some lakes today are the result of past displacements. Perhaps, with time, the intermediate forms in single species lakes would begin to separate along the resource spectrum. Indeed, Schluter reports evidence that this process is already under way, with intermediates taking most of their food either from the phytoplankton or concentrating on the invertebrates. Again, it is the bigger fish that are predominantly bottom feeders. It may be that the same niche separation has occurred independently in five lakes over the last 13 000 years. Where single species intermediates are found, it seems to be happening again.

The alternative to character displacement may be oblivion. If the competition for a resource between two species is very intense, one species may lose out completely. Given sufficient time, any difference in their capacity to use a resource, or to interfere with the other's exploitation of it, will mean one species flourishes at the other's expense. This is the **competitive exclusion principle**. Simply stated, it says that two species may not occupy the same niche at the same time in the same place. One niche can only be occupied by one species (Section 4.3). Competitive exclusion and the partitioning of resources between species means there are probably limits to the number of species that a community can hold. The number of species that could be packed along a resource spectrum depends on how different they need to be to avoid excessive competition. This has proved very difficult to measure, partly because a separation along one environmental gradient is often tempered by the interactions between the species on other gradients

However, not every niche has to be filled. Sometimes an alien species manages to establish itself in a new habitat and find an unoccupied niche. Sometimes an invader may squeeze between two resident species, by being better adapted to a resource space neither fully occupies. For any newcomer to persist, it needs to be sufficiently different from its new neighbours to carve a new niche for itself (Box 5.3). There are also lots of examples of the alternative outcome, where an introduced species out-competes the native niche-holder, and the resident is lost (Section 4.3).

Niche theory is powerful because it links the evolution of species to their use of resources and the number of species that a community can contain (Section 5.2). It also links the range of resources used by a species to the constancy of their habitat, and explains why we might expect more species and more specialist species in constant or predictable habitats, a theme that we explore in some detail in Chapter 9.

2.6 Speciation

Whilst Dolph Schluter's sticklebacks are in the process of becoming new species, they are not yet perfectly separated because they can interbreed and produce intermediate hybrid forms. The reproductive barriers that would define them as entirely separate species have not yet formed. If two populations remain isolated for long enough, genetic differences emerge which eventually prevent them from mating at all. Their gene pools then become separate and any changes that develop in one pool are not transmitted to the other.

Consider the allele we described earlier, coding for an enzyme that had a different temperature optimum. Under character displacement, individuals with different alleles flourish in different places where the temperature regime is most suitable for each type. This means they are physically separated—perhaps in space, say by living on different sides of a hill, or perhaps they are active at different times of day. Either way, the effect may be the same: the two populations rarely meet or interbreed with each other and genes are only exchanged with individuals active at the same time or in the same place. Differences between two separated populations are thus never diluted by gene flow. With no exchange of genetic code the populations begin to diverge. Given a long period of isolation, their accumulated genetic differences may preclude mating, even if they are subsequently re-united.

Such divergence following a separation in space is termed **allopatric speciation**. With **sympatric speciation**, populations may live together, but gene flow is initially restricted because of some genetic change.

Allopatric speciation

Spatial separation can occur when a population becomes divided into isolated fragments, perhaps as some individuals become adapted to more marginal environments. Only a small number of individuals may show an adaptive change, but if this proves successful their numbers quickly build. If they can survive where others cannot, they benefit from a lack of competition, and may have part of a resource spectrum to themselves. Their separation also means they mate primarily with each other, rather than with the main population. As a result, their adaptive traits soon become fixed within their sub-population.

Sometimes, a small section of a population is cut off from the main population by a major disturbance, such as a flood or a forest fire, so that gene flow with the parent population is again restricted. In the same way, individuals invading a new habitat (perhaps an island) may only breed with those that have colonized the new area. Notice that in all these cases the new population is breeding from a small collection of individuals, either because they are colonists (or survivors) or because they are the first to show the new adaptive trait. These few individuals will have only some fraction of the genetic variation of the parent population. This has two important consequences for future genetic change.

First, traits that might occur at a low frequency in the main population may be more common here. The few individuals establishing the sub-population can be unrepresentative of the genotype of the larger population, if only by chance. From then on, their future generations may maintain and exaggerate these differences. This is known as the **founder effect**, and it means that small gene pools can rapidly diverge from the parent population. Given the different genetic stock each has to draw on, the two populations may quickly differentiate.

Second, in a small population rare alleles are easily lost by chance. With large numbers rare alleles will be found in a few individuals and may therefore persist down the generations. However, the same proportion in a small population implies very few individuals carry the rare allele and it has a small chance of passing into the next generation. The frequencies of some alleles can thus change very rapidly in a small population, a process called **genetic drift**. Isolated, a small gene pool becomes more and more distinct with time. Again, with no exchange with the parent population, the two gene pools quickly diverge, perhaps to reach a point when they can no longer interbreed.

Reproductive isolation and the genetic divergence that often follows underpins the concept of the **evolutionary significant unit** (ESU). The term, coined by Ollie Ryder, is applied to populations that are considered to be on a different evolutionary trajectory from their counterparts elsewhere. The aim is to identify populations requiring particular management regimes or protection to identify an appropriate conservation strategy. The ESU draws upon the **cohesion species concept**, which unites our understanding of a population's genetics, ecology and behaviour to determine distinct ESUs worthy of protection.

ESUs have been used extensively in management decisions concerning fishery stocks, in particular the Pacific salmon (*Oncorhynchus* spp.). In this case subspecies have been given the same level of legal protection as species in order to preserve their genetic diversity. For example one species of salmon, the Chinook (*Onchorhynchus tshawytscha*) has no less than 6 ESUs, some confined to a single river system in which they are reproductively isolated from the rest of their species. However, there are problems. First, maintaining variation of this nature is difficult, involving the identification of ESUs and then protecting them with an appropriate management regime. Second, there are concerns that overzealous attempts to preserve variation in these species might only serve to negate the evolutionary process itself, perhaps, deflecting or halting the very trajectory that makes the population evolutionarily significant.

Reproductive barriers

Of course, populations that have become physically separated can be re-united again, and gene flow may be resumed. For physical separation to produce a new species, it must last long enough for the populations to diverge. Then any subsequent mating between them is either impossible or unproductive.

Reproductive barriers take one of two forms according to whether they operate before or after fertilization—pre- and **post-zygotic barriers** (Table 2.1). Pre-zygotic barriers are mechanical, physiological, or behavioural. Many insects have intricately sculptured genitalia, so that the male fits the female rather like a key inside a lock. This is one way in which the female insect ensures that she is fertilized by the correct male, especially between species of the same genus.

An equivalent situation exists in apple (*Malus* × *domestica*) but which prevents self-fertilization, using a biochemically based pre-zygotic barrier (Plate 2.5). Apple-growers know that isolated trees seldom self-pollinate and it is necessary to have another variety nearby to ensure a good harvest. The reason is an 'anti-selfing mechanism' known as genetic

self-incompatibility (GSI)—a process controlled by a single gene at the S locus (referred to as the S allele). The S gene is highly polymorphic, so each variety of apple produces a slightly different S protein. The stigma (the female part of the flower) is able to differentiate between its own S proteins and those from other varieties. Fertilization proceeds only when non-self S proteins are detected.

Even when pollen is introduced to the ova or sperm are introduced to the egg, fertilization may still not happen. Often there are physiological barriers between the gametes of different species, and at the cellular level, the two cells have to recognize each other. A failure of an egg cell to recognize a sperm (or vice versa) prevents fertilization between species, and sometimes within species too.

Compared to the intricacies of cellular recognition, behavioural barriers to reproduction are much more easily observed. Often the female is highly selective, choosing a mate according to the signs and signals he uses to demonstrate his identity and worth. Within a species, females will often select between males according to some indication of the quality of

---

TABLE 2.1 Reproductive barriers

Pre-fertilization (pre-zygotic barriers)	**Post-fertilization** (post-zygotic barriers)
Ecological isolation Populations are separated by distance or barriers (such as mountains or water bodies).	**Hybrid inviability** Embryonic development may be impaired so a hybrid never reaches the adult stage
Temporal isolation Populations may be reproductively active at different times; they may flower at different times or have different breeding seasons	**Hybrid sterility** Offspring are produced but they are infertile, producing either dysfunctional gametes or no gametes at all
Behavioural isolation Without the correct signals to initiate reproductive activity, males and females of different populations may never interbreed	**Hybrid breakdown** Although the offspring are fertile and may reproduce, their young fail to develop properly, cannot reproduce or are poorly adapted to new habitat
Mechanical isolation Reproductive organs need to complement each other for the exchange of gametes. Anatomical differences can thus prevent fertilization	
Gametic isolation Unless the sperm and the egg recognize each other fertilization may be prevented by a failure of them to fuse	

their genotype. Males boast of their prowess by their plumage, or by their song or dance routine. This is termed **sexual selection** and explains why male birds of paradise or peacocks sport such elaborate plumage. More importantly, if the signals are not recognized by the female, the male may never get to pass his genes on. Part of the massive variety of fruit-flies (*Drosophila*) in Hawai'i is due to sexual selection. Males of some species are required to dance for the right to mate with a female. By their dance so they are known, and getting the steps right signals to the female that the male belongs to the correct species. Occasionally, the role is reversed and the male selects the female, especially if he makes a large contribution to the reproductive effort, perhaps by helping to feed or protect the young.

Post-zygotic barriers may operate at various stages in the development of the zygote. Sometimes a fertilized egg will not develop. A mismatch between the number of chromosomes in the sperm and the egg means that the development of the offspring is not likely to proceed very far. Even amongst closely related species, matching (homologous) chromosomes may differ according to their gene sequence and such incompatibilities usually cause meiosis to fail.

Sterile hybrids unable to produce effective and viable gametes are often the result of interspecific crosses. Even where a hybrid develops to full maturity, there are a number of ways in which it may be prevented from breeding, collectively called **hybrid breakdown**. Many hybrids are poorly adapted to the habitat. One example is the intermediate forms of *Papilio dardanus*. When females are a poor match to any of the distasteful butterflies mimicked by the rest of the species they are readily taken by predators. In the same way, hybrids are likely to lose competitive battles with the parent populations, more closely adapted to specific niches. This is what Schluter was able to demonstrate with the hybrids he entered into competition with the limnetic sticklebacks.

Reproductive barriers are the feature that makes animal species the most well-defined taxonomic unit. Whereas the rest of the hierarchy is a classificatory convenience for us, many species are distinct and operate as functional units. Reproductive barriers prevent closely related species from blending in to each other. The faster these reproductive barriers are erected the quicker the identity of a species becomes established. The longer the barriers have been in place, the less likely it is that hybrids will form.

Sympatric speciation

With sympatric speciation species are formed by becoming reproductively isolated through genetic change, even though they are living side by side. Gene flow is halted not by a physical barrier but by individuals becoming separated by their genetic differences. This is thought to happen most readily in highly fragmented habitats, where very different conditions are found within short distances, selecting different traits in their resident populations. For example, adjacent valleys can differ in the amount of sunlight they receive or in their geology and soil type. This is a major cause of the rapid speciation and high plant diversity in Mediterranean-type ecosystems around the world (Section 5.3).

In the same way, the adaptations needed to survive in one pond may differ from a second pond perhaps only a few kilometres away. Then an individual moving to the next pond or the next valley has a lower chance of survival in its new habitat, where it is less well adapted. Natural selection then favours the well-adapted in each location and, once again, the intermediate hybrids are not fit for either pond. The same sort of separation can occur temporally, where populations become specialised for flowering or feeding at different times of the day or during the year. As we saw earlier, a population adapted to a different temperature optimum may have a restricted period of activity, reducing its exchange of genes with the rest of the population.

The other main mechanism by which sympatric speciation can occur is through **polyploidy**, most especially in plants. This happens when normal diploid ($2n$) individuals produce gametes with multiple copies of their chromosomes. A gamete that fails to undergo meiosis will stay diploid so that when it unites with a standard haploid gamete a triploid ($3n$) zygote is formed. Tetraploids ($4n$) form when two diploid gametes combine. Whilst polyploidy rarely results in viable or fertile offspring in animals, it is much more common in plants. This may be because plants tend not to have sex chromosomes—save for a few notable exceptions such as cannabis (*Cannabis sativa*), hops (*Humulus lupulus*) and white campion (*Silene latifolia*) (Figure 2.12). In fact, with 95 per cent of all plant species being hermaphrodite, this is wholesale exchange of genetic information may be a useful strategy for an organism which largely immobile.

Polyploids are often bigger than their diploid counterparts and survive despite, and even because of, their unusual genetic make-up. Polyploidy has been an important cause of speciation in crop plants especially cereals (Box 2.5).

The importance of sympatric speciation is still a matter of considerable debate, especially regarding animal species. Many evolutionary ecologists question whether we should regard small changes of habitats as effective barriers, albeit small-scale, making sympatry just another form of allopatric speciation. However, the evidence from Schluter's work on sticklebacks suggests that character displacement can follow when individuals become specialized to feed in one part of a habitat. Other evidence that speciation can indeed follow genetic change alone comes from other freshwater fish, the remarkable cichlids of Africa.

The ribbon of lakes within the African Rift Valley includes both fresh and saline waters. The lakes change dramatically in water level from one year to the next, and over thousands of years have expanded and contracted considerably. Some lakes have been cut off from each other for many thousands of years, during which time their fish communities have speciated into a large number of forms. Even within a single lake, differences in habitat type as well as the range of available habitats created by changing water levels have promoted speciation. Lake Victoria, for example, has over 300 species of cichlid fish that appear to have evolved in the last three-quarters of a million years. In that time they have diverged to feed on the major food sources, and within each category there are specialists that in feed in different ways. For example, there are several species that are adapted to feed primarily on molluscs, using different feeding methods. Yet the similarities in their genetic code and their mitochondrial DNA suggest the cichlids in Lake Victoria have evolved from a single ancestral species.

Outside the Rift Valley there are some lakes which have never been connected to other water courses. Two volcanic crater lakes in Cameroon studied by Ulrich Schliewen and his co-workers have endemic cichlids unique to each lake (11 and 9 species, respectively), despite being very small and with little differentiation of habitats. Again,

FIG. 2.12 White campion (*Silene latifolia*), one of the few plant species to have sex chromosomes.

their molecular biology suggests each group were probably derived from a single colonization event in each lake. Since the lakes are very uniform, with no effective inflow from surrounding rivers the speciation within each was almost certainly sympatric. Indeed, the phylogenetic tree for the cichlids of each lake suggests that their divergence was prompted by niche differentiation driven again by their feeding behaviour.

Some ecologists also recognize another form of speciation termed **parapatric speciation**, which seems to be a feature of small, isolated and rapidly reproducing populations. This is a particular form of sympatric speciation in which gene exchange is confined to individuals occupying a small area. The resultant inbreeding leads to highly adapted local populations, found in distinct and discrete habitats. The immobility of plants with especially restricted gene flow means that parapatric speciation is perhaps most important for them, producing local races or **ecotypes**. A good example of this is the metal-tolerant ecotypes of various grasses (Box 2.4).

In this latter case, natural selection has allowed some individuals to colonize a marginal habitat, which is poisonous to most other members of the parent population. Here they can grow in the absence of competition. Indeed, we often find that highly adapted ecotypes are less competitive forms, so that they invariably lose out in intraspecific battles with the main population. It seems that the costs they have to meet to withstand the high levels of stress (in this case the metals in the soil) put them at a disadvantage when growing in normal soils.

Man-made species

Over the past 10 000 years or so, humans have been the major selective pressure in the lives of thousands of plant and animal species. By means of domestication and cultivation, we have selected those most suited to our purpose (Box 2.5).

There are thought to be around 70 000 cultivated plant hybrids derived from 1100 wild species though some may have an ancestry that includes as many as 35 species. Not surprisingly, all this makes them particularly difficult to classify. In the case of hybrids,

we simply list the most recent parents using an 'X' to indicate the cross.

In the past, cultivated plants were known as varieties, but this term is now used to describe naturally occurring variation within a species. Plants bred by humans and dependent on cultivation for their continued existence are therefore called **cultivars**. According to the International Code of Nomenclature of Cultivated Plants, a plant must be listed according to its genus and species followed by its cultivar name (usually abbreviated to cv.). For example, Red Ace, a cultivar of shrubby cinquefoil is usually written as *Potentilla fruticosa* cv. Red Ace.

Hierarchical classification tends to break down at the level of hybrids and cultivars, primarily because we have so often blended very different genotypes to create new cultivars. Even so, there is considerable value in understanding the evolutionary history of domesticated plants. Their wild-type relatives often hold genes which make them tougher and more resilient, and we use this genetic diversity to produce more resistant cultivars. For example, potatoes are plagued by aphids and the cost of insecticides to protect them is considerable. The wild hairy potato of Central America has been used to produce new aphid resistant cultivars that impale aphid attackers on short, sharp hairs.

Unfortunately, pests evolve too and adapt to our new varieties. Pest-resistant cultivars therefore have only a limited useful life. Most strains of wheat resistant to various fungal ('rust') attacks last little more than 5 years before the fungi itself adapts to the change in its niche. Within a decade the build up of pests and diseases associated with a cultivar may make it uneconomic to grow. Crop breeding programmes represent the continuing evolutionary battle between our domesticated plants and their pests (Section 4.5).

The stakes are even higher now, as genetic engineering has taken some of the guesswork out of crop and stock improvement. By incorporating selected genes from other organisms into the genome of economically important species we can enhance their productivity or confer protection against pests and disease. Gene technology offers the opportunity to move genetic code across species boundaries and even between phyla. This is evolution by design rather than chance.

BOX 2.4   The evolution of the metal tolerance in plants

Large-scale extraction of metal ores has created a series of distinctive habitats in which most plant species cannot flourish. Yet some plants have colonized these tips and they provide evidence of the speed of adaptive change. In Britain, sites which were first exploited 200 years ago have since been colonized by varieties of grasses able to tolerate toxic metals. Several of these grasses appear to have evolved from local populations where no tolerance is evident.

Tolerance seems to evolve fairly readily in certain grass species. Adaptations to lead, copper, zinc, and others have arisen in different species and some varieties are tolerant to combinations of metals. Sowing normal grass seed onto a toxic spoil will usually produce one or two seedlings that are able to survive. This is an indication that the genetic information for tolerance occurs as part of the background variation in the population. Interestingly, while a range of plants are known to have such alleles, albeit at a low frequency, others never show the capacity to develop metal tolerance. Such locally adapted varieties are termed **ecotypes**. Metal-tolerant ecotypes are also found among some animal groups, especially soil-dwelling invertebrates. The grasses, however, have proved particularly useful in restoring spoil heaps, where little else will grow (Section 7.3).

For the grasses, the genetic change needed may not be that large. Some plants produce special proteins that bind these metals and prevent them passing from the roots to more sensitive parts of the plant. It may be that the most important adaptive change is simply an increase in the amount of this protein produced. Others actually accumulate the metal and store it in parts of the cell where it will do the least damage. This includes vacuoles within the cell or storage in the cell wall. Several plants are so effective at accumulating metals in this way that they are being developed as a means of concentrating precious metals from wastes (Section 7.3).

Some clues to the nature of the tolerance mechanism in grasses were gleaned from studying non-tolerant ecotypes of Yorkshire fog (*Holcus lanatus*) (Figure 2.13). These will grow on arsenic-contaminated soil provided they are supplied with additional phosphate. Mark MacNair and Quinton Cumbes found that *Holcus* was unable to distinguish between the poison and the nutrient, simply because the size and shape of their two ions are so similar. By swamping the soil with excess phosphate the uptake of arsenate could be

FIG. 2.13 Yorkshire fog grass (*Holcus lanatus*). Varieties of this species have evolved a tolerance to toxic metals.

reduced to a level low enough for the plant to survive. Tolerant ecotypes of *Holcus* are able make the distinction between the nutrient and the poison. Not only does this tell us how this tolerance mechanism functions but also shows how non-tolerant plants may become poisoned.

Very often, we find that tolerant ecotypes pay a price for their capacity to live in marginal habitats. In normal conditions—say growing in an unpolluted soil alongside normal plants, the ecotypes grow less vigorously than their non-tolerant neighbours. It seems the costs they incur in being tolerant, perhaps the costs of producing metal-binding proteins, mean they are less well adapted to this habitat. This is possibly reason their genotype is only found at a low frequency in unpolluted habitats.

BOX 2.5   The seeds of civilization

Around 12 000 years ago, humans began the move away from hunter-gathering towards a more settled life dominated by agriculture. Wheat, as one of the first plants to be cultivated, was central to this change and remains one of the staple foods of many peoples today. Indeed, the evolution and spread of wheat and other cereals follow those of a number of civilizations (Box 5.1, Table 5.2).

Archaeological evidence from the Near East suggests wheat was being widely grown in the Jordan Valley, Jericho and Damascus around 10 000 years ago, though the first use of wheat may have begun a thousand years before this, at the time of a cooling in the climate known as the Younger Dryas. Perhaps at this time the grain was simply collected from the wild; eventually some seed was saved and sown to ensure the size of the harvest the following year. In so doing, human beings began its cultivation and in the process, its domestication, selecting and sowing those seed with the most useful characteristics.

The cultivation of wheat probably began in the 'Fertile Crescent'—an area stretching across the Near East through to Mesopotamia (the land between the rivers Tigris and Euphrates). The origins of the wild wheat have been revealed by genetic analysis using amplified fragment length polymorphisms (AFLPs), a genome-wide measure of genetic similarity (Box 2.2). The technique identifies einkorn (*Triticum boeoticum*), a grass growing in the western foothills of the Karacadag mountains of southeastern Turkey, as the first cultivated wheat.

By 12 500 years ago, the wild form had been domesticated and was being cultivated by farmers in the Fertile Crescent as a new subspecies of cultivated einkorn (*T. monococcum*, ssp. *monococcum*). Cultivated einkorn quickly spread throughout Europe though by the time of the Bronze Age, it too had been superseded by new varieties.

Unlike the diploid einkorn, more recent wheats are polyploid and are derived from wild emmer wheat (*T. dioccoides*). Emmer enjoys several key advantages from being a tetraploid (4*n*)—it is larger and grows more vigorously; it was also free-threshing, allowing its seed to be readily separated from their ears. Emmer was the product of a hybridization event between another wild wheat (*T. urartu*) and goat grass (*Aegilops tauscii*). By 9000 years ago, cultivated emmer wheat (*T. dicoccum*) had supplanted einkorn to become the single most dominant wheat in cultivation. Within 2000 years it had spread as far as Europe, Ethiopia, and India, and along with barley (*Hordeum distichon*)

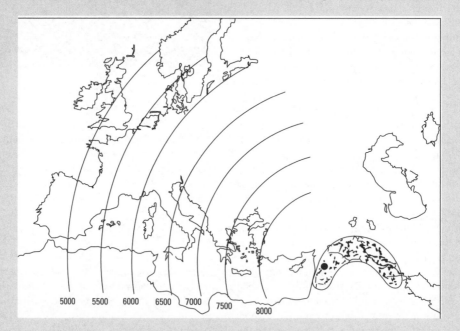

FIG. 2.14 The rate of spread of modern wheat (*Triticum aestivum*) and barley (*Hordeum distichon*) from their origins in the 'Fertile Crescent' (shaded area). Lines show the approximate date of arrival in years before present.

BOX 2.5 **Continued**

had become the staple cereal of the Neolithic Period (Figure 2.14).

Around this time, another new species appeared that would eventually dominate world agriculture. This was bread wheat (*T. aestivum*), a hexaploid with six sets of chromosomes (*6n*). Again, there seems to have been an intergeneric hybridization with goat grass (*A. tauscii*), this time with the tetraploid cultivated emmer (*T. dicoccum*), somewhere in the Southern Caspian Basin.

Although bread wheat has become the most widely grown of all wheats (Figure 2.14), with hundreds of

different cultivars adapted to different soil types and climates, other wheats are still in cultivation. Chief among these is durum wheat (*T. durum*), otherwise known as 'macaroni wheat'. Durum, a descendent of the tetraploid emmers, is extra-hard with an exceptionally high protein content in the form of gluten. It is also rich in the pigment beta-carotene, which helps give pasta its firm texture and golden colour. This has ensured its continued use in traditional agriculture and cookery and, along with the other cereals, has had an important role in the civilizing of humanity.

The first commercial applications of genetic engineering sought to extend the shelf life of tomatoes by manipulating the genes that control the ripening process. Since then, herbicide resistance genes have been incorporated into the genome of soybeans to allow the blanket application of herbicide (in this case *Roundup*) without damage to the crop, and this has been extended to a large range of commercial plants. But it has proved highly controversial. First, it seems contrary to one of the 'green' aims of gene technology, to reduce biocide use. Second, the technology has been seen as a means of tying farmers into buying both seed and pesticide from the same biotechnology company, creating global monopolies. Finally, there is the question of the introduced genes moving into less desirable species, perhaps creating resistant weed species.

The latter is a critical issue, given the ease with which most crop plants will hybridize with their wild relatives (Box 2.5). At least 13 major crop species have this ability and this includes a large number of plant families in which wild-crop hybridization is known to occur (Table 2.2). One possible way round such problems may be to incorporate the engineered genes into chloroplast DNA rather than the nuclear genome. This would ensure that genes do not escape within the pollen, the male gamete: chloroplast DNA, like mitochondrial DNA, is inherited through the maternal line.

Foreign genes can also pose practical and ethical dilemmas to those that consume them. Engineering

high-methionine soybean by incorporating a gene from the Brazil nut resulted in beans that contained protein capable of triggering nut allergies in susceptible

TABLE 2.2 Plant families known to hybridize with wild relatives

Plant family	GM species of potential risk
Asteraceae (*formerly* Compositae)	Lettuce, sunflower
Apiaceae (*formerly* Umbelliferae)	Carrot
Brassicaceae	Broccoli, cabbage, rapeseed (canola)
Chenopodiaceae	Sugar beet
Compositae	Lettuce, sunflower
Ericace	Blueberry
Juglandacea	Walnut
Leguminosae	Alfalfa, soybean, peanut
Malvaceae	Cotton
Poaceae (*formerly* Graminae)	Creeping bentgrass, maize (corn), rice, sorghum, wheat
Rosaceae	Apple, strawberry
Salicaceae	Poplar
Solonaceae	Aubergine (egg plant), potato, tobacco, tomato
Vitaceae	Grape

consumers. The risks to human health were considered so great that the project was eventually abandoned. Clear identification of genetic composition of transgenic material is therefore needed to protect those who have to avoid contact with a species for reasons of health, custom or ethical concerns.

There is also the question of specificity, and whether non-target species can be affected. A gene derived from the bacteria *Bacillus thuringiensis* produces δ-endotoxin—an insecticidal protein that protects plants from insect pests. However, plant material from transgenic maize containing with the endotoxin gene has been implicated in poisoning the larvae of the Monarch butterfly (*Danaus plexippus*). It is not known whether any other non-pest species are at risk, but it seems the insecticidal protein can also leach out of the host plant and persist in the environment for up to 180 days. Traces of the toxin have been found in earthworms and the concern is that it might pass further along the food chain. In trying to escape the problems of conventional biocides—such as toxicity and persistence—we have created a new generation of pest control technologies with both the faults of the old and the fears of the new. Some of these potential risk factors are listed in Table 2.3.

Moving a gene from one species to another is clearly a genetic transformation, but how do we regard copying an existing gene and placing multiple copies in the original owner? Is such an organism transgenic? No foreign code has been incorporated into the genome, but some form of genetic modification has taken place. Joachim Messing and Jinsheng Lai modified the genome of maize to boost the production of the amino acid methionine by simply by altering the DNA on either side of the

TABLE 2.3 Potential sources of foreign genes or gene products of genetically modified organisms

Organism/product	Sources
Animals	Carcasses
	Faeces
	Urine
Plants	Biomass
	Food chain effects
	Pollen
	Root exudates
Microbes	Fermentor malfunction
	Waste media
	Waste microorganisms

gene coding for it, and controlling its production. They claim that this is an improvement on previous efforts, and in many respects might be regarded as no more than a special, accelerated, form of plant breeding.

Gene technology offers the potential to combat disease, hunger and pollution but it seems that in many cases there will be no quick fixes. Barely half a century has passed since DNA was identified as the agent of inheritance and only in the last 30 years have we been able to read genes sequences. By the end of the twentieth century we were starting to understand the language of the genes through ventures such as the Human Genome Project. The challenge is to read, understand and then use the genetic code effectively.

SUMMARY

The binomial system for classifying and naming living organisms not only provides an internationally accepted convention for identifying each species with a unique name, but also its hierarchical structure provides an indication of its phylogenetic relations. Until recently, most classification has been based on morphology, but we now measure genetic differences and arrange phylogenies using molecular biology. The biological species concept is another convention, useful for many higher plants and animals but not readily applicable to all groups, especially those species that hybridize freely.

Variation between individuals within a species can also be considerable, making their classification difficult, though sometimes these differences have adaptive significance.

The ecological niche is the totality of factors, biotic and abiotic, to which a species has adapted, and therefore describes its relations with its living and non-living environment. Species occupying a narrow niche are specialized for a particular part of a resource spectrum whilst generalist species have a broad niche. Two species may not occupy the same niche and competitive exclusion will mean that one species will be lost. Intense competition for a resource can also lead to character displacement, where a species adapts to use a different niche. The fragmentation and divergence of populations leads to the character displacement by which new species may form.

Speciation occurs where gene flow ceases between two populations. This can occur because the sub-populations become physically separated (allopatric speciation) or isolated from each other by a genetic change (sympatric speciation). Speciation follows not only from natural selection but also occurs as a consequence of selective pressures applied by humans in their breeding of plants and animals. Genetic engineering offers new opportunities in the breeding of commercially useful organisms but also presents new challenges in environmental protection.

FURTHER READING

Jeffrey, C. 1977. *Biological Nomenclature*. Edward Arnold, London. (A useful introduction to the rules of naming.)

Jones, S. 1999. *Almost like a Whale*. Anchor, Doubleday, London. (A highly readable review of current evidence and understanding of evolution by natural selection.)

Price, P. W. 1996. *Biological Evolution*. Saunders, Fort Worth. (A systematic and comprehensive introduction to current thinking on evolution.)

Weiner, J. 1994. *The Beak of the Finch*. Knopf, New York. (Excellent account of a detailed study of the Galapagos finches originally studied by Darwin and other detailed research, including Schluter's work on sticklebacks.)

Zohary, D. and Hopf, M. 2000. *Domestication of Plants in the Old World*. Oxford University Press, Oxford. (Useful reference source.)

WEB PAGES

The Tree of Life site provides a wide range of information and links on phylogeny and biodiversity:
http://tolweb.org/tree/phylogeny.html
The Agriculture Research Service of the United States Department of Agriculture operates a useful site on the biology and genetics of wheat:
http://wheat.pw.usda.gov/index. html
The following is web directory to several aspects covered here, including biological nomenclature:
http://www.biologybrowser.org
The following gives details of gene sequencing techniques:
http://www.biology.washington.edu/fingerprint/dnaintro.html

1 Describe the techniques that can be used to detect differences in the genetic code of two individuals.

2 The Superkingdom Eukaryota includes the following Kingdoms:

(a) Animalia, Fungi, Plantae, and Protista only.

(b) Monera and Protista only.

(c) Animalia, Fungi, Monera, Plantae, and Protista.

(d) Fungi and Prokaryota only.

(e) Animalia and Plantae only.

(f) Animalia, Fungi, and Plantae only.

3 Using Figure 2.5 as an example, devise a key to classify the following insects according to the characteristics listed in the table below. Seek to use the minimum number of steps.

Characteristic	Insect Type					
	Wasp	Beetle	Butterfly	Fly	Grasshopper	Ant
Wings present	Y	Y	Y	Y	Y	N
Number of wings	4	4	4	2	4	0
Hard wing case	N	Y	N	N	N	N
Large dusty wings	N	N	Y	N	N	N
Large jumping legs	N	N	N	N	Y	N

4 Insert the appropriate words to complete the following paragraph (use the list of words below; some words may be used twice and others not at all).

Genetic drift occurs within ____ populations. Here, a ____ number of individuals means that rare ____ are easily lost by chance. This poses considerable risks for rare species that might only have a few remaining individuals with little genetic ____. Their numbers may recover in time but considerable ____ can be lost when the population is so reduced, and is sometimes referred to as a ____ . Genetic drift also results in a ____, where relatively rare ____ found in the isolated population become more ____, and this can serve to distinguish them from the main population. Habitats and ecosystems that are highly ____ are more likely to produce distinctive local populations as a result, and this is one reason why they are often very species-rich.

adapt	isolated	small	homogeneous	genetic variation
alleles	large	species	founder effect	rare
communities	less	variation	genetic fixation	
genetic bottleneck	rare	species-rich	species-poor	

5 Select the *most* correct answer.

Species which specialize…

(a) have a narrow niche breadth.

(b) are bad competitors.

(c) have a narrow niche breadth and are good competitors.

(d) have a broad niche breadth.

(e) are good competitors.

(f) have a broad niche breadth and are bad competitors.

Tutorial/Seminar questions

6 *It wasn't so much that we cultivated grasses, rather it was that they civilized us.* Discuss this statement with reference to the impact of early agriculture on the evolution of human society.

See Box 2.5 and from this starting point use Further reading/references and web resources to explore this topic. Also look at Chapter 5 for more information on the development of agriculture in the Mediterranean. The examples discussed in the text mainly deal with wheat and to a lesser extent barley—so do not forget those other important grasses, rice and maize and investigate how they too have shaped human societies in the Old and New World.

7 Assume that you could readily measure the components of the diet in the various types of stickleback studied by Schluter and his colleagues. How would you set about measuring niche breadth and niche overlap? Could you measure the intensity of the competition between the species based on their stomach contents only?

8 Are public concerns about genetically modified food crops well founded, given the centuries of selective plant breeding we have practised and the frequency with which hybridization has produced new varieties of wheat?

3 POPULATIONS

'Everything should be made as simple as possible, but not simpler.'

Albert Einstein

The fisherman is constantly reviewing the size of the catchable stock.

POPULATIONS

Part of the attraction of ecology is its appeal to our sense of order and balance. Many people see in ecology a set of principles that explain a constancy and regularity in living systems: the cycle of life and death, of eating and being eaten, of continuity between generations. The natural world seems to be ordered by simple rules that maintain some sort of 'balance of nature'. We capture these ideas in familiar phrases such as ashes to ashes, big fish eat little fish, and the cycle of life.

Perhaps there is a natural order out there, but few modern ecologists would try to write out the rules of the game. It is difficult to show that natural populations and communities are actually stable or unchanging for long periods. A variety of ecosystems appear to be similarly organized and regulated, yet differences start to emerge when we look closely at each one in detail.

Close up, chance and variation blur the sharp outline of these principles. To use another common phrase, the devil is in the detail. Our ideas might make intuitive sense, but their simple logic fails to reflect the complexity of the living world. Ecology the science often muddies the waters, showing that nature does not always come as clean as we might think it.

Nevertheless, most of us have some conception of an order in our environment, a common-sense view of the economy of nature. We see that nature works by cycling nutrients, that food and the energy it provides fuels living systems and governs the growth of individuals and populations. Such ideas are part of the knowledge and concepts we all need to exploit and survive in our various environments.

A simple example is relevant to this chapter. A fisherman sitting by a pond knows there are a limited number of fish in the water. With no stream bringing new individuals, the fish caught can be replaced only by the reproductive efforts of those remaining in the pond. The frugal fisherman understands that the population must be allowed time to recover.

Consider the elements in this simple system. The pond has a population of a limited size and the fish can reproduce at only a limited rate. The population is defined solely by the boundaries of the pond, with no stream allowing fish to enter or leave. To maintain a constant population the fisherman can remove fish only at the rate at which they are replaced by reproduction.

The intuitive sense of this example underpins the first part of this chapter. Here we use simple models to examine population growth, the limits on its size, and to estimate how many individuals might be caught without damaging a population's reproductive potential. We shall find that the simplicity of these models is clouded by the detail of real fisheries, when data on species' life histories are needed to manage their exploitation properly.

Later, the same models are used to examine why some populations and species face extinction. The loss of a population seems inevitable when its reproduction does not keep pace with its death rate over the long term. To save a species from extinction we need to understand the detail of its ecology, but we shall find that this detail again tempers the use of general rules in conservation.

3.1 ◼ Modelling

Models are used extensively in ecology and in this chapter, so it is worth saying something briefly about their nature and their purpose.

All sciences use models to do one of two things. First, they can simplify a complicated system by removing non-essential information or 'noise'. Think of a road map. This is a model in which most of the background detail of the real world has been removed to highlight the important information—the pattern and direction of roads. We model ecological systems in the same way by isolating the key factors in a study. The model helps to clarify our understanding of a system by defining its important elements and their relationships. These are **analytical models**.

A second type of model simulates the behaviour of a system. Again, much of the spurious information of the real world is pared away, but now sufficient detail remains to produce accurate predictions. A flight simulator is a physical model (or a computer program) for training pilots that provides enough information to give a realistic impression of flying an aircraft. A **simulation model** attempts to make realistic predictions about a system, rather than analyse its workings. Ecological simulations make predictions about populations, communities and ecosystems but demand large amounts of data. Like flight simulators they also allow us to try out different manoeuvres, with no risk of destroying the real thing.

Models abbreviate reality, enabling us to analyse the workings of a system, or to make predictions about its behaviour. Here, we begin by making simple analytical models of population growth, using the minimum of information. Later, we add details that improve our predictions and help us manage both exploited populations and those in danger of extinction.

3.2 ◼ Simple models of population growth

The number of fish hatched in the pond at the end of the breeding season will depend on the number of adults breeding there at the start. In describing the relationship between these two numbers we create a simple population model. This model has just two elements and has only one function: to make predictions about population growth in future years.

This population is defined by the boundaries of the pond (Box 3.1). To simplify matters, we shall assume there is no migration in or out and no limits on the number of fish in the pond. Another unrealistic assumption is the absence of deaths during the breeding season. Our model is for a fish with a single breeding season each year and this is at least true of several species in temperate waters.

BOX 3.1   Individuals, populations and metapopulations

Ecologists counting individuals and treating them as a population need to define their terms very carefully. Criteria that most people would use to define the individual become problematic when we look closely at different species.

Consider organisms that increase their numbers by budding or some other form of asexual reproduction. Many plants and several groups of colonial animals grow by forming modules that will separate and grow independently. For these organisms we have to distinguish between genetically identical, asexual clones termed **ramets**, and genetically distinct individuals (having arisen from different zygotes) or **genets**. In some species (e.g. potatoes) a count of individuals will include both genets (seedlings) and ramets (tubers).

We then have to decide whether we include all stages in the life cycle in the count: for example, do we include tadpoles or only adult frogs; seedlings as well as overwintering tubers? Our decision will depend upon the organism and the purpose of the study.

Even then, defining a **population** is not always straightforward. A good functional definition is a group of individuals of the same species occupying a particular area at a particular time. Notice we define notional boundaries that delimit individuals who can actually breed with each other—for example, the elephants of the Serengeti in 2002. Elephants in the Serengeti would not normally mate with those many miles to the South. Nor, very obviously, could they breed with the elephants in the Serengeti in 1902. Even so, these spatial and temporal boundaries are only truly distinct in the mind of the ecologist, as a necessary convenience to define the terms of a study.

Most species have a collection of populations alive at the same time which are separated spatially. Within a well-defined region, this population of populations is called a **metapopulation** (Box 3.4) where relatively distinct populations occasionally breed with each other and, therefore, swap genetic material. While some populations may interbreed fairly regularly, others may only have indirect genetic exchange with each other. The degree of interchange within the metapopulation is often crucial to the survival of some endangered species.

You will also come across the term **population density**. This is used where we have no need or have made no attempt to count every individual in a population (for most plants and animals this is impossible). Instead, we count their numbers in a specified area or in some circumstances, a known volume.

We make two other simplifying assumptions—that all fish live for just one year, with the adults dying after spawning, and that all fish have an equal chance of reproducing. If there are 100 adults, each of which, on average, give rise to two adults next year, the population in the next generation would be 200 fish. This is expressed as the reproductive rate:

$$\text{Reproductive rate } (R_0) = \frac{N_g}{N_0}$$

The number of adults at the start (N_0) was 100. The number of offspring reaching reproductive age in the next generation (N_g) is 200. The reproductive rate (R_0) equals 2.0.

Note that R_0 is the *average* reproductive rate per individual in the population; each adult does not produce exactly two offspring. If the reproductive rate remained fixed, we could easily predict the population size in each of the following years.

In this simplest of models, the generations do not overlap and the adults do not survive long after reproducing. More realistically, fish populations include individuals that survive for several years and take part in several breeding seasons. Keeping a count is then more difficult, requiring us to note the age of reproduction and death for each individual in the wild. Instead, we can measure the reproductive rate by taking a census at certain times. Counting the number alive before and after a period of time allows us to calculate the net reproductive rate over that interval:

$$\text{Net Reproductive rate } (R_N) = \frac{N_t}{N_0}$$

The number of individuals alive (N_t) at the end of the time interval (t) is divided by those counted at the start (N_0). If the number alive at the end (survivors and offspring) is greater than at the start, the population has increased ($R_N > 1$). The population declines when the number of deaths is larger than any additions ($R_N < 1$).

R_N is called the net reproductive rate because it includes both deaths and births over the sampling interval. Again it is the rate of change per individual averaged over the entire population and if it remained fixed, we could readily predict future population sizes.

Can you see when R_0 and R_N will equal one another? They are the same when the census interval is equal to the generation time. Both then measure the average rate of change per individual over one generation.

We can follow the change in the population size on a graph. If we set R_N to a fixed value, and then plot the size of the population with time, we get a characteristic curve (Figure 3.1). In this example, we start with 100 individuals and set R_N to 1.2. After one breeding season there will be 120 fish in the pond. Note that some may have died in the meantime, but the net increment is 20 individuals. Next time around, there are 24 additional fish, and in the third generation 29. You may like to derive a second set of data to check the shape of the curve produced with a different value of R_N.

This example illustrates an important point about the pattern of population growth. Despite R_N staying constant, the increase in population size is not the same from one interval to the next. Instead, the size of the increment itself increases because population growth is accelerating—there are more individuals breeding at each successive interval, collectively producing more offspring.

This gives the curve its characteristic shape (termed exponential growth). This shape is true of any population with an R_N greater than 1. Even with R_N values only slightly above 1 the curve is the same, though its rise is shallow. Given time it will still grow to a very large population. Similarly, an accelerating decrease follows when R_N falls below one, and the population tips towards extinction.

This model treats the population growth as a series of discrete steps, using a fixed value of R_N. This is the easiest way to calculate the change from one time interval to another.

Populations with continuous change

The picture becomes more complicated when we model populations where births or deaths occur continuously. Many species of fish grow and reproduce for several years. But death can occur at any

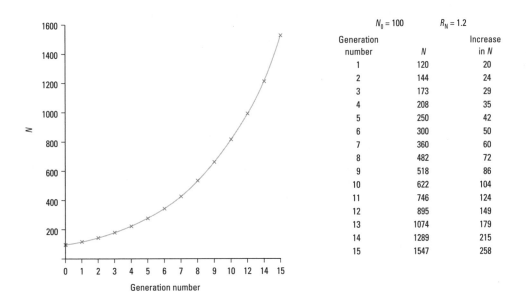

$N_0 = 100$		$R_N = 1.2$
Generation number	N	Increase in N
1	120	20
2	144	24
3	173	29
4	208	35
5	250	42
6	300	50
7	360	60
8	482	72
9	518	86
10	622	104
11	746	124
12	895	149
13	1074	179
14	1289	215
15	1547	258

FIG. 3.1 A model of population growth in an organism with discrete generations, starting with a population of 100 and a constant reproductive rate of 1.2. The results used to plot the graph have been rounded up or down to give whole numbers.

time and reproduction may also be a more or less continual process, not confined to a particular season. Then, both the birth rate (or natality rate) and the death rate (or mortality rate) vary continuously. The difference between them is expressed as r, the rate of change per individual for a particular point in time:

$$r = b - m$$

When the natality rate (b) is greater than the mortality rate (m) the population is increasing at that time. The model of population growth becomes:

$$\frac{dN}{dt} = rN$$

The change in the population size (dN) after the time interval (dt) equals the average rate of change per individual (r) multiplied by the number of individuals (N). It differs from our previous example only in deriving the change over a very small interval of time (dt). These are described as instantaneous rates and r is more properly described as the instantaneous rate of change per individual.

If we fix r at a particular value, the population grows exactly as described before. Again, this model produces exponential growth as long as the instantaneous birth rate is greater than the instantaneous mortality rate. When they balance ($b - m = 0$) the population does not grow ($r = 0$). Note, however, that the rate of population growth (r) can change as conditions change, as b and m fluctuate.

Factors affecting the rate of population growth

The maximum value of r (r_{max}) describes how rapidly a population would grow under ideal conditions when b is maximized and m is minimised. This differs between species according to the details of their life history. Species that take a long time to reach reproductive age, or that produce few offspring on each occasion, have slower rates of population growth.

Elephants and other large-bodied animals typically have low r_{max} values because they produce few offspring over a long time. The gestation period in the African elephant is relatively long (22 months), followed by a juvenile stage lasting 10–12 years,

before the individual becomes sexually mature. They may live for an average of 30–40 years (though females may still be fertile at 60 years old) and produce just one offspring at a time. On average, the African elephant has an r_{max} value of 0.06 per year. At the other extreme, mice live short lives, quickly becoming sexually mature, and have multiple births following a brief (21-day) gestation period. Many small mammals, including mice, have a high r_{max}, ranging from 0.3 to 8.0 per year and their population can increase very rapidly.

Why is r_{max} rarely achieved in the wild? Usually because conditions for growth and reproduction are far from perfect. Natality is reduced with poor nutrition or disease, or when partners simply fail to meet. Mortality rates depend on a wide variety of factors, and any rise in the death rate depletes the numbers reproducing. Nevertheless, the size of r_{max} is of great interest because very different r_{max} values imply very different life history patterns and very different biologies, the product of different adaptive solutions. As we shall see later on, these strategies need not be fixed—some species change their reproductive strategy and biology to suit the prevailing conditions (Box 3.3).

Populations in limited environments

In one famous calculation, Charles Darwin estimated that a single pair of elephants would have 19 million descendents in just 750 years, if all reproduced at their maximum rate. Yet we are not overrun by elephants and common sense tells us that some factor in their habitat must limit their population growth.

Rapid population growth can be sustained only for as long as conditions allow. Resources may become scarce, space may become limited or waste may accumulate in the habitat. These checks on growth, collectively known as '**environmental resistance**', place an upper limit on the population size, and this is termed the '**carrying capacity**' (K).

What will happen in a population close to its carrying capacity? Any shortage of resources will mean competition between individuals with some getting less than they need. A lack of food, for example, might prevent or delay some individuals becoming sexually mature, while others might die. Either reduces the size of the next generation. As a population approaches its carrying capacity, more individuals compete for fewer resources and the intensity of competition grows.

In some cases competition between individuals of the same species (**intraspecific competition**) is the key factor limiting population size (Box 4.2). In these species population growth is said to be density-dependent because the degree of environmental resistance rises with the population density. Population size increases rapidly as long as resources are abundant, but as the number of individuals within a location rise, competition increases and this growth slows down.

We can include these checks from environmental resistance in our model, by setting an upper limit on population size. For example, if our pond can accommodate 500 fish ($K = 500$) and 100 are already present (N), there is 'space' for just 400 more. Put another way, of the total capacity, one-fifth is occupied and four-fifths remain:

$$K - \frac{N}{K} = 500 - \frac{100}{500} = \frac{4}{5} = 0.8$$

So environmental resistance increases as the unused capacity gets smaller—when there are 300 individuals present, the unused capacity is reduced to 0.4. We can build this into our model very easily:

$$\frac{dN}{dt} = \frac{rN\,(K-N)}{K}$$

Figure 3.2a, derived from this model, shows how increasing environmental resistance slows growth by its impact on r. A shortage of food, for example, may limit egg production (b falls) and/or increases mortality (m rises). Growth is gradually checked and becomes zero as the population settles at its carrying capacity.

We can see the effect more easily by plotting net incremental growth (dN/dt) against population size (Figure 3.2b). At first, the number of additional individuals rises as N increases. The greatest increase (23) is at half of the carrying capacity when both N and r are still large. Thereafter, r declines because competition lowers b and increases m. Increment size now falls, until, at the carrying capacity, $b = m$ and no net additions occur.

Consider what happens if the carrying capacity changes. Any expansion in space or resources allows

b to rise or m to fall, and the population grows to a new ceiling. Alternatively, a lowering of K leads to a reduction in the population as mortalities outstrip births. In this simple model, the population is regulated close to its carrying capacity.

This s-shaped pattern of population growth (Figure 3.2a) is described as logistic. The model shows the population rising smoothly and resting perfectly at its carrying capacity. In the real world, time-lags and a variety of other factors cause populations to overshoot their K and oscillate around it.

Not all plants and animals have their populations limited by intraspecific competition. For some, competition with different species (**interspecific competition**) is more important (Section 4.3). Predation or parasitism, where one species is consumed by another, may limit the population growth of both consumer and consumed, sometimes in a density-dependent manner.

What determines the carrying capacity?

An abundance of key resources may mean that competition is rarely significant for some plants and animals. Many insects show little constancy in their population size. Their habitats change too rapidly for their numbers to settle at an equilibrium. Frosts, floods, fires and other, more large-scale catastrophes cause major changes in their habitats. These shifts in the carrying capacity bear no relationship to the size of the resident population, that is, their K is independent of their density. Not surprisingly, species living under these conditions undergo major fluctuations.

Even without dramatic changes in their environment, some population numbers fluctuate wildly from one year to the next. Various pests occasionally undergo explosive population growth, often without any obvious upheaval in their habitat. Several species of locust have periodic population outbreaks, when massive swarms sweep across thousands of miles of Africa. For a long time we were unable to predict such outbreaks, but today we can model swarming behaviour using satellite data on vegetation cover and rainfall in their breeding ranges.

The simple model we have described above is most relevant to species closely dependent on the long-term availability of resources in their

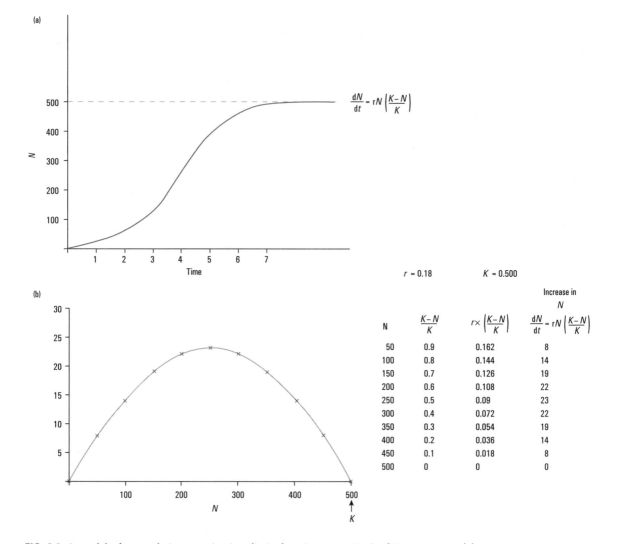

(a)

$$\frac{dN}{dt} = rN\left(\frac{K-N}{K}\right)$$

$r = 0.18 \qquad\qquad K = 0.500$

(b)

N	$\dfrac{K-N}{K}$	$r \times \left(\dfrac{K-N}{K}\right)$	Increase in N $\dfrac{dN}{dt} = rN\left(\dfrac{K-N}{K}\right)$
50	0.9	0.162	8
100	0.8	0.144	14
150	0.7	0.126	19
200	0.6	0.108	22
250	0.5	0.09	23
300	0.4	0.072	22
350	0.3	0.054	19
400	0.2	0.036	14
450	0.1	0.018	8
500	0	0	0

FIG. 3.2 A model of a population growing in a limited environment. (a) In this case we model overlapping generations, with the carrying capacity (K) set to 500. The rate of increase per individual in the population (r) is 0.18. Population growth is reduced by the increasing effect of environmental resistance as the carrying capacity is approached. (b) We can show the effect of environmental resistance by calculating the size of increase at particular population sizes.

For each N, the environmental resistance is calculated as the proportion of capacity still available ($K-N/K$)—for 50 this is nine-tenths ($500 - 50/500 = 0.90$).
Multiplying this by 0.18 (r) gives the actual rate of increase ($0.18 \times 0.90 = 0.162$).
Finally, multiplying the result by N gives the net increase at that population size ($0.162 \times 50 = 8$). Notice that the largest increment is when $N = 250$, half of the carrying capacity. This declines as the habitat fills, so that no increase is possible at K.

environment—particularly the larger plants and animals that grow slowly and live longer. Their numbers do not fluctuate wildly and for these the concept of a carrying capacity has some meaning. What determines the maximum number of individuals that a habitat can support? For larger organisms with relatively stable populations we can make a reasonable prediction based on average body size. A large adult mass needs sufficient resources to grow and sustain itself. Meeting this

demand requires space: a sufficient volume of soil from which to extract water or area over which to forage for food. The amount of space needed to support each individual plant or animal thus determines an ecosystem's carrying capacity for the species. For this reason, the average density of many large species at K shows some correspondence with adult body size. As we see later on, this has important consequences for the large plants and animals we seek to protect from extinction.

3.3 Harvesting a population

Back at the pond, we will assume that growth of our fish population is density-dependent and limited only by intraspecific competition. This pond is a very stable environment and the fish population is close to its carrying capacity. We want to maintain its potential to provide catches from one year to the next. We want to take the maximum number of fish every season, the maximum sustainable yield (MSY), and this requires that the population is able to replenish itself, that is, enough individuals survive each year to reproduce and replace those caught.

What is the maximum number we can harvest? This will be the same as the maximum number that can be replaced each year. We already know this number from the graph of population increase in a limited environment (Figure 3.2b)—the greatest increment was at half the carrying capacity. At this N, the population produces the largest number of offspring, before r begins to be reduced by intraspecific competition. This is said to be its optimal yield—the highest sustainable rate of population increase under a given set of environmental conditions.

Fishery models

Calculating a maximum sustainable yield is feasible if we know the carrying capacity of the environment. That may be easy in a pond, but is highly problematical in the open sea. Besides K, we need to know the size of the **catchable stock**, those fish large enough to be caught in our nets and which are, in effect, the population size (N) of interest to us. Many disputes about

fishing quotas revolve around interpretations of such data and the methods used to estimate the MSY.

Of course, the fish are never actually counted. Commercial fisheries are interested in the yield in weight, and fishery models are based on biomass rather than numbers in the catch. Nevertheless, the principles and basic ideas of the models remain the same—we simply replace individuals with units of weight.

Assuming there is no net migration into the fish population, biomass is restored by the addition of new individuals and also by the growth of new tissues by existing fish in the stock:

$$F + M = G + R$$

Here, the biomass caught by the fishery (F) and the losses due to natural mortality (M) are balanced by tissue growth (G) and by recruitment (R) of new individuals. Fish are only recruited to the stock when they are large enough to be held by our nets.

In most fisheries we have no control over natural mortality. We can only regulate our fishing effort so that fishing mortality (F) does not exceed $G + R - M$, and the catchable stock (N) does not decline—that is, we adjust our effort according to the level of natural mortality. The MSY is achieved when the largest harvest is taken without the catchable stock falling from one season to the next. This is the basis of the **surplus yield model**, so-called because the yield is equivalent to the growth and recruitment, which is surplus to that needed to maintain N. In effect, the population remains close to a constant size by replacing losses through growth and recruitment. By noting the effort (time) needed to catch a fish and size of the catch from one year to the next, the fisherman at the pond is using the same model, though perhaps only intuitively.

This yield is maximized when the population is half of K, when $G + R$ are at their greatest rate of increase (Figure 3.2b). The model estimates whether the yield is optimal by looking at the relationship between yield and fishing effort: any additional effort that consistently produces no increase in catch means the optimum has been passed. The catchable stock is then being depleted and each additional unit of effort brings a diminishing return (Figure 3.3).

For a long time, the size of the catchable stock was estimated from catches landed at the quayside. A decline in the catch per trip (or more accurately the catch per unit fishing effort) over a number of seasons implied that the stock was declining. Fishery managers knew a pattern of falling yields indicated overfishing. Catches then included a proportion of the population whose reproduction and growth were supposed to replace losses to F and M.

In fact, such patterns are notoriously unreliable. Many commercial fisheries are based on highly variable populations, whose size fluctuates with changes in weather, the movements of ocean currents and a range of other factors, not least the behaviour of the fish. Estimating the stock is very difficult under these circumstances. Equally, G and M can be highly variable, as can, most importantly, recruitment (R)—the number of fish which have grown large enough to be caught.

Measuring recruitment rates is crucial in intensively fished populations because these determine the size of catchable stock, more so than the carrying capacity. Invariably such exploited populations are far below their K value, and any depressive effect of competition on r will be small. The critical problem then is to sustain recruitment, to maintain the supply of fish growing up to size.

Because of these uncertainties our simple model is only partially effective. Its analytical simplicity does not include important details governing tissue growth and recruitment rates. In fact, since our prime concern is the weight caught, fish should be harvested only after they had time to grow, not as soon as they have entered the catchable stock. So, for a realistic estimation of maximum sustainable yield, we have to incorporate rates of weight increase as well as recruitment.

A second group of models, **dynamic pool models**, do this. They use growth rates to calculate the biomass harvested from each age group. Many fish grow at a relatively constant rate throughout their life, so size is a direct reflection of age. Average weights can be estimated from the length of time an individual remains in the stock. We can then work out the optimal survival time for an individual to deliver a maximum yield.

Just as a viable population must remain for growth in numbers, so individuals should be given time to grow new tissues. At high fishing intensities, few fish escape capture, and few survive long in the catchable stock. Even if numbers are replenished, biomass is not, because fish have only a short time to grow (Figure 3.4). At the optimum fishing effort, residence times allow individuals to reproduce and to increase their weight. The model estimates the appropriate fishing intensity to give the maximum sustainable yield, based on recruitment and the survival time of the average fish in the catchable stock (the time available to grow new tissues).

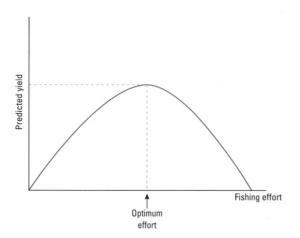

FIG. 3.3 The estimate of the maximum sustainable yield from a fishery, based on the fishing effort. This is derived from Figure 3.2(b). Fishing yields above the optimum will decline with increasing effort because we are now catching the stock needed to maintain itself. Below the optimum, the scope for higher yields are indicated by increases in yields with more effort.

AGE CLASS

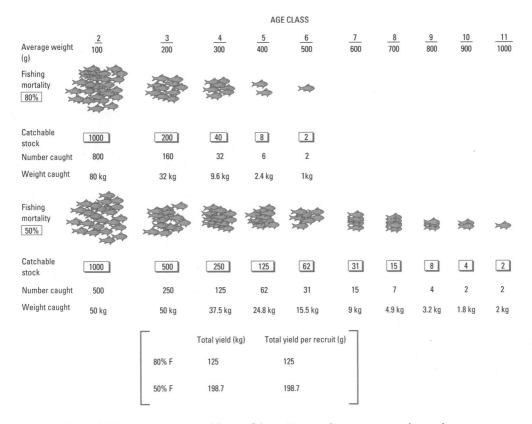

	2	3	4	5	6	7	8	9	10	11
Average weight (g)	100	200	300	400	500	600	700	800	900	1000

Fishing mortality 80%

Catchable stock	1000	200	40	8	2
Number caught	800	160	32	6	2
Weight caught	80 kg	32 kg	9.6 kg	2.4 kg	1kg

Fishing mortality 50%

Catchable stock	1000	500	250	125	62	31	15	8	4	2
Number caught	500	250	125	62	31	15	7	4	2	2
Weight caught	50 kg	50 kg	37.5 kg	24.8 kg	15.5 kg	9 kg	4.9 kg	3.2 kg	1.8 kg	2 kg

	Total yield (kg)	Total yield per recruit (g)
80% F	125	125
50% F	198.7	198.7

FIG. 3.4 The effect of fishing intensity on yields to a fishery. Here we have two examples, each starting with a cohort of 1000 recruits, but with differing fishing intensities. In the first, 80 per cent are removed during each fishing season. The cohort is fished out in 6 years. At the lower intensity the cohort continues for 11 years. The weight of a fish depends upon its age and, at 80 per cent F, most individuals are taken when the fish are youngest. In age class 2 for example, 800 fish are caught weighing on average 100 g, giving a yield of 80 kg. At 50 per cent F the yield is only 50 kg, but more fish are left to grow bigger. The net effect is a larger yield overall at 50 per cent F. A dynamic pool model provides plots of yield per recruit against fishing intensity and allows us to find where the maximum yield is achieved—where the balance between growth and catches is optimum. At low fishing intensities high natural mortality will reduce the catchable stock and we could harvest more—the above (simplified) model assumes no natural mortality and allows all uncaught individuals to grow to full size.

3.4 Growth rates, age and recruitment

Dynamic pool models are a derivation of the surplus yield model, but they divide the population into age groups, and then work out the biomass each contributes to the yield. Such models require considerable detail to make realistic predictions. We need data on the number of individuals being recruited, the rate of mortality in each age group, and the average weight of each age class. Ecologists summarize such data in **life tables**.

A life table estimates the chances of an individual of a particular age dying or surviving for a length of time. You may be familiar with such statistics from comparisons of life expectancies for various age groups or peoples in different countries. Indeed, ecologists have borrowed many of these techniques from insurance actuaries who use life tables to calculate the risk of our dying while covered by their policy.

Life tables come in two basic forms: **cohort** and **static life tables**. In the first, a group of individuals born at about the same time (a cohort) is followed through its life and the numbers surviving in each age class are recorded. From this we calculate the survival probability and life expectancy for an individual of a certain age. A static life table derives the same data but from a census of the whole population, looking at the proportion in each age class, on one occasion.

Life tables tell us much about the dynamics of a population. With the data for several years, we can decide whether the mortality of a particular age group is significant for determining total population size. For example, larval mortality limits recruitment for many fish and it is these numbers which are most important in determining the size of the catchable stock.

Life tables also make plain the **life history strategies** adopted by an organism. All species try to maximize their population growth but they differ in how they allocate resources between the stages of their life cycle. Some produce large numbers of offspring, of which few survive to mate. At the other extreme are species producing just one or two offspring each time, but which are more likely to reach adulthood.

BOX 3.2   Age structures

A population with overlapping generations will have a distinct age structure—that is the proportion of individuals in each age class. We recognize two basic age structures—one associated with a rapidly growing population, and the other in a population close to its carrying capacity.

When there is large scope for population increase and the environment remains relatively constant, the population will increase at a constant rate per individual (r is fixed). It then assumes a *stable age distribution* with birth rates and death rates constant in each age group (Figure 3.5). Such a population will be growing exponentially, with fixed proportions in each age group. However, the highest proportion of individuals are in the youngest age groups, and the distribution is characteristically 'bottom-heavy'. This pattern remains stable as long as there is little net change due to migration and the environment does not change significantly.

When close to its carrying capacity, a population has little capacity for further growth, and the birth rate is equal to the death rate ($r = 0$). Again, the environment must be largely unchanging, but now the proportion of individuals in each age class is more evenly distributed—the birth rate is lower, and the younger age classes do not dominate the population. This is a *stationary age distribution* (Figure 3.5b).

Different human populations approximate to both these types—stationary age distributions are a feature of those countries where the population has remained relatively constant, or even declined slightly. In those nations undergoing very rapid population growth, a stable age distribution is found.

In animal populations, these distributions are not always indicative of the population growth rate. One obvious case is the fish we harvest—these typically have an age structure dominated by younger individuals because our fishing techniques aim to catch only the older, larger fish. Unfortunately, despite their age distribution, these are not populations increasing rapidly.

FIG. 3.5 Age distributions. (a) A stable age distribution has a fixed shape—the proportion of individuals in each age group remains unchanged. This is because additions and losses to each age group are constant. Overall, the whole population has a fixed rate of births and deaths, and is growing geometrically. For this reason it is dominated by the younger age groups. (b) A stationary age distribution is also fixed, but now the birth rate and the death rate balance each other, and the population is close to K, its carrying capacity. This is largely a theoretical distribution (partly because the shape is bound to change as the environment changes), but now, with resources scarce, we see a more even distribution between the age classes.

The structure of a population, the proportion of individuals in each age class, reflects these strategies. They also tell us much about the potential of a population for growth. Rapidly growing populations are typically 'bottom heavy', dominated by the new arrivals in the youngest age classes (Box 3.2, Figure 3.5). Alternatively, a population close to its carrying capacity, when its birth rate is matched by its death rate, has a more even age distribution, with a larger proportion in the older cohorts. Few are being added, since competition favours the old and the large (Figure 3.6).

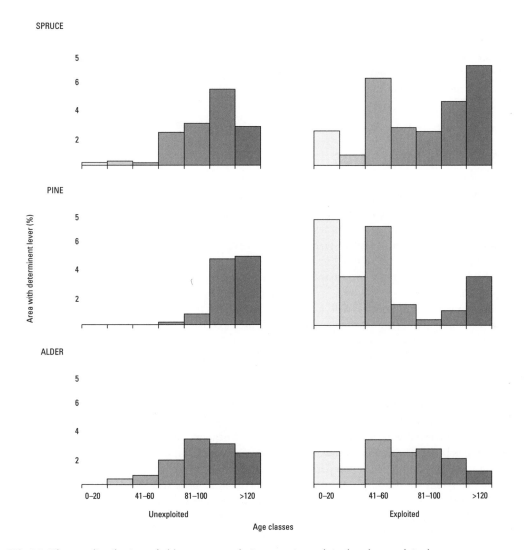

FIG. 3.6 The age distribution of alder, spruce and pine trees in exploited and unexploited areas of the Bialowieza national park in eastern Poland. The unexploited population is derived from a 'primeval' forest that has remained largely undisturbed since the fifteenth century, a community close to the natural forest that once covered most of lowland Europe. The exploited stands have been harvested since 1915. Notice that the age distribution includes a greater proportion of younger trees, established here largely through active re-planting.

You can get a sense of this difference by walking through temperate woodlands of different ages. An undisturbed woodland, several centuries old, is typically dominated by a small number of large, old trees with a few young saplings waiting for a gap in the canopy (Plate 3.1). In contrast, an abandoned field that has begun to revert to woodland has large numbers of saplings, derived from the seeds of a few nearby trees (Plate 3.2). This is a population growing rapidly, when resources, primarily light, are plentiful.

Life table analyses can be crucial for maximizing yields in both fisheries and woodland management. Knowledge of how quickly an individual grows and its life expectancy allow us to estimate the best time to harvest the largest biomass. The dynamic pool model uses these kinds of data with sub-models to calculate rates of growth and chances of survival. Given this emphasis on detail and data from real world populations, it is more a simulation than an analytical model.

The same approach is used in forestry management, though, here, we benefit from a much more complete data set. Managers can measure competition directly, as the effect of tree density on individual growth. They also have more control over the population including its recruitment (by their planting strategy), rates of growth (through the application of fertilizers) and competition (through controlling the density and the species composition of the woodland).

3.5 Life history strategies

The stages in the life cycle of many organisms are very well-defined, marked by different body forms and ways of life. A frog begins as an egg, continues as an aquatic tadpole and metamorphoses into a terrestrial adult. Seeds sprout wiry saplings adapted to life on the forest floor, but they will grow into massive trees if a gap opens in the canopy.

The life history strategy of a species will have evolved to ensure the survival of as many offspring as possible. During their life cycle some organisms radically change their habits and physiology to exploit different niches. For example, many amphibian tadpoles begin as herbivores, grazing the algae attached to stones in their pond, only becoming carnivores as they approach metamorphosis. The adult often feeds primarily on terrestrial invertebrates, some distance from the pond. Similarly, adult dragonflies feeding on insects flying above the pond are not competing with their carnivorous larvae feeding on insect nymphs in the water below.

What are the benefits of partitioning the life cycle in this way? Most likely, making the best use of the resources available. Different stages' feeding in different habitats or on different diets reduces the competition between age classes. Larval stages are often given over entirely to feeding and growth, while the adult, whose role is to disperse and reproduce, may not, in some cases, feed at all: adult mayflies emerge from the pond and fly only to mate and to lay eggs.

Other aspects of an organism's life history are adaptations to maximize survival and have been refined by natural selection. Large numbers of eggs anticipate large losses, but the massive wastage in most years is offset by the odd occasion when conditions allow many to grow into adults. The alternative strategy is for parents to invest time and energy in protecting and supporting the few offspring they have produced.

A female cod can produce four million eggs at a single spawning, but devotes few resources to each egg or effort to its aftercare. On average, perhaps three or four of these will grow to sexual maturity. In contrast, an elephant produces a single calf, which it has fed in the womb, will suckle at the breast, and nurture for several years thereafter. During this time, the young elephant benefits not only from the resources its mother provides, but also from her protection and her knowledge of the environment. We should not think the elephant is more successful than the cod: indeed, there are certainly more cod than elephants in the world. They represent two different strategies—each adapted to their way to life in their particular habitat.

The spectrum from *r*-selected to *K*-selected strategies

A life history strategy consists of several elements: the average length of a life, the proportion of time spent in each stage, the age at first reproduction, the number of reproductive events and the number born on each occasion. Together, these determine the speed of population growth and, consequently, have considerable adaptive significance.

Consistent patterns appear when we compare the strategies adopted by various organisms. In contrast to those mating only once, species that reproduce repeatedly take longer to reach sexual maturity and produce few offspring on each occasion. Their eggs are usually larger and may receive more parental care. Each egg, therefore, represents a much larger investment by the parent or parents.

Life history traits are an adaptation to the permanence of the habitat. Where conditions are transient and opportunities for population growth are short-lived, natural selection will favour species that reproduce rapidly. This implies a rapid completion of the life cycle and a short generation time, which in turn usually means a small body. Being small and producing many offspring are both aids to dispersal, necessary if a species is to colonize transient habitats. As resources become available elsewhere, such opportunist species can quickly move to exploit them.

These are the characteristics of many pest species. Their short life and prolific reproduction mean such organisms have a high r value. These 'r-selected' species are small, short-lived and highly mobile. They also tolerate a wide range of conditions, due in part to their genetic adaptability (itself a result of their fast reproduction rate).

In a more predictable habitat there is little advantage in being adaptable. Nor is there any advantage in being small. Long-standing habitats are dominated by large species with long generation times. Here, the favoured strategy is to stand and fight, to crowd out competitors and invaders. The advantage lies with highly specialized and closely adapted species able to secure the scant resources for themselves, whilst within a population the efficient will be most likely to produce the next generation. These 'K-selected' species are adapted to compete in habitats close to their carrying capacity. Typically, they postpone reproduction until conditions are favourable, but they can reproduce many times. With this strategy, mortality rates are highest among the younger age groups who have difficulty competing with older individuals for the limited resources.

Compare the mouse, the elephant, the oak tree, and the cod. Many small mammals, insects and the annual plants found in disturbed soils show r-selected characteristics. Large mammals and long-lived trees are typically K-selected, growing slowly as individuals and as a population.

This is a very broad schema, with few organisms being readily classified as perfectly r- or K-selected. Many plants and animals show characteristics of each, while some, such as the dandelion, have races following different strategies in different habitats (Section 2.3). The cod, along with many other fish, show r-characteristics in their reproductive strategy because of the uncertainty in recruitment from one season to the next. However, they will grow on as adults for many years and can become very large, which is more of a K-selected character.

Through natural selection, a species will match its reproductive strategy to the prevailing environmental conditions, if these remain predictable over a number of generations. Where conditions are not so reliable, having alternative strategies may be advantageous and some species are able to hedge their bets, with some of their offspring adopting one strategy and the rest another. In this way at least some individuals survive (Box 3.3).

3.6 Genetics and population size

The genetic variation necessary for a species to change its adaptive strategies can only come from mutations. Without mutation, individuals simply represent new combinations of genes. For traits where there is little or no variation there will be few alternative phenotypes by which natural selection could distinguish between individuals. Since mutation rates are low, and many mutations are deleterious, species are limited to the particular environmental conditions for which they have become adapted (Section 1.2).

BOX 3.3 A frog for all seasons

Investing in reproduction or growth are not always simple alternatives for an organism. Natural selection optimizes life history strategies, favouring growth and a delayed reproduction or, alternatively, selecting short generation times, whichever yields the most offspring. In a highly unpredictable world opting to reproduce quickly may be the best option if delay could mean a lost opportunity. Where conditions are more reliable and follow a consistent pattern, an organism can match its activities to the seasons and reproduce when the habitat is most favourable. Using reliable cues that anticipate seasonal change, such as day length, the next phase of a life-cycle may be started to exploit resources as they become available.

In some parts of the world, however, seasonal changes are highly unreliable and adopting a single strategy may be disastrous. Then, natural selection will favour adaptability. Take the case of the West African reed frog *Hyperolius nitidulus*. Like most amphibians, these need freshwater to spawn and to support the tadpoles prior to their metamorphosis into froglets. This requires a minimum period before they can leave the water, and a further interval before they can reproduce. If the rainy season is too short or if they metamorphose too late in the season they will be unable to spawn until the following year.

Kathrin Lampert and Eduard Linsenmair have studied the differences in *Hyperolius* which emerge early in the rainy season and those appearing much later. Late arrivals are, in most cases, unlikely to spawn in the year they emerge, and will have to survive the dry season if they are to reproduce. For this they have to develop a physiology and anatomy to withstand the forthcoming hot and dry conditions. Part of this conditioning is, surprisingly for a frog, sunbathing. By increasingly exposing themselves to the full sun, *Hyperolius* froglets induce changes in their skin to make it more effective in reflecting the sunlight. They then aestivate, ceasing to be active and thus survive without water for an extended period.

As ponds become available at the beginning of the next rainy season these froglets become active again and quickly grow to maturity, and spawn. Their tadpoles can adopt one of two strategies—grow quickly to reproduce before the rainy season ends, or emerge later, grow slowly and condition themselves for the dry season (Figure 3.7). The fast-growing, reproducing frogs make no attempt to adapt for the dry season—their skin has little reflective ability and they hide in the damp vegetation to avoid sunlight. When they mature they tend to be small, especially the males. Fully adult frogs do not

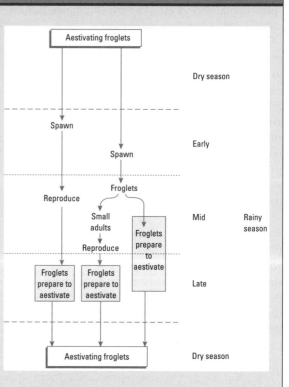

FIG. 3.7 A long and reliable rainy season can allow the froglets of *Hyperolius* to grow quickly and produce a second generation within a year. However, the later they emerge in the rainy season the more risky this strategy becomes. In this case, having some tadpoles grow slowly, preparing to aestivate, is a useful insurance against an early dry season. Froglets which emerge late in the rainy season (not shown) have no option and have to aestivate.

survive the dry season and must lay their eggs before they die. In turn, their offspring must complete their metamorphosis and condition themselves before the dry season arrives. If the rainy season and the ponds remain for sometime this strategy will be successful. However, an early dry season may prevent fast-growing froglets from fully conditioning themselves and then only slow-growing, well-conditioned froglets will survive.

The problem is the unpredictability of the rainy season and the duration of the ponds—for early emerging tadpoles, adopting a rapid reproductive strategy is the best option. Those emerging late in the season have to prepare for aestivation. But those emerging in the middle

BOX 3.3   Continued

of the rainy season produce offspring that follow both strategies, with some tadpoles attempting to reproduce and others preparing for aestivation. These frogs spread their risk, ensuring that some will survive the dry season if it arrives quickly and others will spawn if the rainy season is long enough. In this way, they avoid putting all of their eggs in one basket.

Lampert and Linsenmair have shown how the time of emergence (or more likely, the time of spawning) determines the life history strategy for *Hyperolius*. They describe how one species can adopt different life history strategies according to the most likely environmental conditions that will follow a spawning, or when this remains uncertain, produce offspring to suit either eventuality. It is not known whether shifting the strategy in this way is an example of phenotypic or genotypic adaptation. Whilst phenotypic adaptation is perhaps the most likely, theory suggests that over many generations, a change in the genotype might be favoured when environmental cues are unreliable. Evidence from a series of breeding experiments indicates that this may indeed be true.

Environmental change can be rapid, but genetic change is usually slow and largely determined by the mutation rate for a particular locus.

A single population only has some fraction of the total amount of variation available within a species and a small population can only provide a limited variety of genotypes for the next generation. Endangered species are often in this position, reduced to small numbers with little genetic variation to draw upon and consequently little capacity to adapt to new selective pressures.

This may be offset if there are frequent migrations between populations, introducing new alleles into a population. Usually there will be a number of populations, a metapopulation, composed of relatively distinct populations with different degrees of isolation (Box 3.4). Some of these populations may rarely mate with individuals in other populations and must rely on their own genetic resources to survive. More often there will be migrations between populations, and a gene flow between them, termed **admixture**. So important is this source of genetic novelty that many plants and animals have strategies to ensure that gametes are swapped between populations—male elephants move between groups of females to mate and many flowering plants produce pollen well before their own ovules become receptive. Unless gametes are swapped with genetically different individuals, the full benefits of sex are not enjoyed.

Without this exchange, populations become genetically uniform and small populations lose genetic information very rapidly through **genetic drift** and

fixation (Section 1.3). Remember that a new generation only represents a sample of the genes of their parent's generation because only a proportion of the code will be represented in the offspring. The smaller a population becomes the smaller the total variation to sample from, and the more likely it is that some genes will be lost entirely from the population. In addition, small numbers mean fewer possibilities for mutation. As variation (the number of alleles) reduces at each locus, so the genotype becomes fixed and most individuals show the same phenotype. Fixation happens faster in small populations, and it is the rarer alleles that are lost most readily.

Genes that have major selective significance are less likely to change. Most mutations are disruptive and more often reduce the fitness of the phenotype but even when the change is advantageous, rare alleles have little chance of being sampled if numbers are small. Yet variation, and the capacity to adapt to a changing world, is essential for the long-term prospects of a population and a species. For this reason, much of the effort in conserving endangered species is directed at preserving as much genetic variation as possible.

On occasions, the lack of complete genotypic separation can be used to salvage an endangered species. In 2001 eight female Texas cougars were introduced into Florida to provide mates for a dwindling population (of around 30 individuals) of Florida panthers. The result was a series of successful matings and the birth of kittens that retain panther characteristics. In this case the small genetic difference between the two cats has helped restore the

BOX 3.4 Gene flow on the ice floes

Measuring populations has moved into the space age and, in the case of some large carnivores, demonstrates the extent to which populations within a metapopulation can be defined. Tracking polar bear movements with satellites can also help to explain patterns in the genetic code of the different populations. Together with colleagues from research institutes in Norway, Russia and Alaska, Mette Mauritzen tracked the position of 105 female bears roaming the ice sheet above Norway and Russia, for 3 years (male bears lose their radio

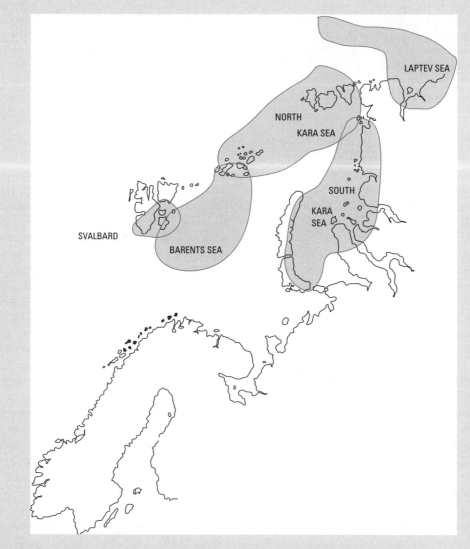

FIG. 3.8 The five polar bear populations identified by satellite tracking in the Norwegian and Russian Arctic. These are for female bears only and say nothing about the mobility of the male bears that may range wider. Four of the populations rely on the extent of sea ice and their ranges expand and contract with the ice. The Svalbard population is near-shore and will retire to the islands when the ice retreats. This is the period of fasting for the bears, when they are unable to hunt from the ice floes.

BOX 3.4   Continued

collars too readily because their neck is larger than their head) (Plate 3.3).

The data was used to map the range of each female and thereby assign her to a population. It also allowed the researchers to record when females wandered into the home ranges of neighbouring populations (Figure 3.8) and, therefore, measure the potential gene flow between populations. Overall, they were able to construct a picture of the dynamics of the whole metapopulation.

The Svalbard population was the largest (containing, on average, 54 females). These bears were highly aggregated and had the smallest home ranges, well segregated from the other groups. This was the only near-shore population, and such bears characteristically return to the same patch of coastline (Box 9.2) each summer. Exchanges between populations were highly seasonal, and not surprisingly, reflected the changes in the extent of the pack ice. Svalbard exchanged most individuals with its neighbour, a population largely confined to the ice floes of the Barents Sea (with 33 females). The Barents population (described as 'pelagic'—literally open sea) migrated over large distances and spent little time on land. Interestingly, the Laptev females,

consisting of just three collared bears showed virtually no exchanges with its neighbours in the Kara Sea.

Some bears were clearly greater travellers than others and individuals were swapped most frequently between the pelagic populations. This is supported by previous genetic studies, which showed that the allele frequencies of populations in this part of the Arctic did not differ between these populations. It seems that these exchanges betweens populations prevent them from becoming genetically distinct.

In contrast, the North American populations are relatively distinct from each other. The greater prevalence of land in this part of the circumpolar region means that most of the populations are near-shore and likely to have smaller home ranges. Either side of the Arctic, it is the distribution of land and ice which defines the degree of aggregation and segregation of the populations.

Ultimately, this reflects the abundance of food resources and the access each population has to its prey. It seems the behaviour of the bears and the size of their home ranges is largely defined by the geography of the landscape through which they move. Perhaps then the genetics of the metapopulation may only be fully understood by looking in an atlas or peering down from space.

fortunes of an endangered ecotype and there is now hope that the Florida panther population can be restored with the help of these surrogate females. What is less clear, however, is the extent to which the distinctiveness of the panther population will be compromised by using cougars as mothers.

More intriguing still are the current attempts to resurrect the thylacine (Section 2.3) from DNA collected from museum specimens. In this case, an entirely different species (but another marsupial) will be used as the mother, both to supply the egg cell and then to raise the embryo.

 ## 3.7   Survival and extinction

The reproductive and genetic characteristics of species we exploit or which face extinction are critical to our attempts to manage their populations. In the case of the North Sea cod (*Gadus morhua*), both apply—commercial stocks of cod have already been lost from one of its most important habitats (the Grand Banks of Newfoundland) and European

Union scientists believe that only a total ban in the North Sea would conserve the populations there. The intensity of our fishing has been a potent selective force on the cod—the population of the North Sea today reaches sexual maturity 2–3 years earlier than it did 50 years ago, almost certainly because of the limited time they now survive in the catchable stock. Under such intense fishing pressure, individuals that mature and reproduce quickly are favoured, and it is their offspring we are catching today.

For both exploitation and conservation we have to maximize a species' reproductive rate and reduce its mortality rate. Adult numbers of *r*-selected species fluctuate greatly as resources wax and wane and they are often able to accommodate change or move to other locations. For *K*-selected species it is juvenile abundance which varies most. This is because births exceed deaths only when there is spare capacity in the ecosystem. It is the young which suffer at other times and their numbers decline if the habitat shrinks or the carrying capacity is reduced. The poor dispersal of some *K*-selected species, closely adapted as they often are to particular conditions, is one reason why they are frequently endangered. Their size is another, implying a low rate of population growth and a low population density.

Thus, much of our conservation effort is directed towards large species whose reproductive rate is low and whose recovery from small numbers is slow. Additionally, *K*-selected species may play a critical role in shaping their community. Browsing by elephant and giraffe is essential to keep the savannah grasslands free of thorn bushes and scrub, allowing other animals to graze (Plates 3.4 and 3.5)—the loss of elephants can lead to changes in the plant community so that many species, including some of the Acacia trees upon which they feed, would disappear with them. Elsewhere, high elephant densities in forests and woodlands can cause major problems and also lead to significant habitat change. Organisms with such critical roles are called **keystone species** (Section 5.2), species whose presence or absence determines the nature of the habitat.

Conserving a species in the wild means conserving its habitat. Of course, life in a zoological or botanical garden may be the only option for a species whose natural habitat has disappeared, but we are then, in one sense, only conserving its genetic code (Box 3.5). For a long-term future a habitat large enough to support a genetically viable population is needed.

Persistence time and extinction

Extinction is a necessary part of the natural process of change. Many people are aware of the abrupt periods in the past when massive numbers of extinctions occurred on Earth (Section 9.2). Fewer realize that we are going through such a period now. Although we are unsure about the causes of past extinction events, today we can be certain that the loss of natural habitats is the prime factor.

As environments are degraded or destroyed the original habitat is broken up into fragments or patches, each with a relatively small carrying capacity. In the process, the distance between patches increases, making migration more difficult (Sections 8.1 and 9.1). Small, isolated habitats are much more vulnerable to change, and less likely to maintain the conditions required by a large or specialized species. A once continuous population is now divided into a fragmented metapopulation whose overall size has been reduced.

Fragmentation, extinction, and colonization

Small and discrete populations can be severely threatened by relatively minor environmental changes. A small patch of habitat provides little protection when conditions deteriorate. For example, a flood will have a minimal impact if resident animals can escape to higher ground within the patch. But these refuges are scarce in small habitats and populations are more easily lost. Even without complete destruction, the carrying capacity of the patch has been reduced. Populations of all sizes may suffer in variable environments, but the implications are far more serious when numbers are low.

Individuals in small populations face other difficulties, not least in finding a suitable mate. Not every adult they encounter will be fertile or free from disease and some will never produce viable offspring. In many species a single male can inseminate many females, but the reproductive capacity depends on the number of females, and in future years, on the number of females born. Together, the size, age and sex structure of a population govern its potential for growth. Consequently the size of the breeding population (referred to as the **effective population size**) is always smaller than the actual population, and often substantially so.

Breeding from a small number of individuals means their genetic differences become less marked

BOX 3.5 Test-tube conservation

Conservation aims to protect rare and endangered species *within* their own habitats, surrounded by the community and conditions under which they arose. However, this is not always possible and in such cases **in-situ** conservation is augmented by **ex-situ** conservation. Here, individuals are removed from their original habitat and held in protective custody until the time is right for their release back into the wild. In the process, many are assisted in their reproductive efforts to maintain or strengthen their genetic fitness and captive breeding programmes have to be carefully monitored to minimize inbreeding. Stud books and pedigree charts keep a record of each mating and their outcomes.

Not so long ago, the private lives of rare species frequently made the headlines, with rare specimens moving between zoos as part of a breeding programme. Today such movements are increasingly rare with sperm, ova and embryos making the journey instead. Using **in-vitro fertilization** techniques avoids many of the problems that can limit reproductive success and even get around some barriers by making use of closely related species. In this way, embryos of the Bengal tiger (*Panthera tigris tigris*) have been raised in Siberian tigers (*Panthera tigris altaica*).

Plants are even more amenable to new reproductive technologies. Plant cells are **totipotent**, that is each cell is potentially capable of regenerating into an entirely new individual as a clone of its parent. This no more than a variation on **vegetative propagation**, used in horticulture and agriculture to produce cuttings and grafts and to obtain many plants from a single specimen.

The ability to regenerate varies from species to species and from tissue to tissue. Some recalcitrant species need to be coaxed into regenerating using techniques of **tissue culture**. Here, a cell or groups of cells are grown on a nutrient gel spiked with plant growth regulators, the hormones that control cell division, enlargement and differentiation. These techniques allow us to propagate rare species of cacti and orchid threatened by collection from the wild. The market for these rarities can then be flooded with millions of legitimately-grown plants, reducing the price and so making the illegal trade in collected specimens uneconomic.

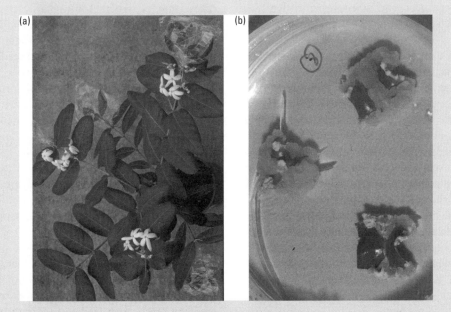

FIG. 3.9 Café Marron (*Ramosmania rodriguesii*) (3.9a). Callus cells (3.9b) are used in research seeking to reverse the genetic defect which prevents successful pollination.

BOX 3.5   Continued

But plants, like animals, can also have breeding difficulties and may sometimes need technical intervention to assist in their survival. An example of this is the Café Marron plant (*Ramosmania rodriguesii*), a relative of cultivated coffee which is endemic to the tropical island of Rodrigues. Like many other island endemics, Café Marron has suffered from the ravages of introduced species. Rabbits and goats almost grazed the plant out of existence whilst the young plants were unable to compete with the spread of another European invader, privet (*Ligustrum vulgare*). By the 1940s there was only one remaining specimen of the plant and after a while even that was declared missing and presumed extinct. Then, in 1980, a boy presented a sprig of the 'extinct' bush for the school nature table. The hunt was on to find and protect Café Marron.

Botanists eventually found a goat-ravaged bush and carefully fenced it off. It was only then that they discovered that the only specimen on earth was unable to reproduce due to pollen incompatibility. In 1986 three cuttings were taken from the plant and a team at the Royal Botanic Gardens, Kew attempted to propagate it by conventional and tissue culture techniques (Figure 3.9). Fifteen years later, in November 2001, they were able to return 11 young plants to Rodrigues, ready for a re-introduction programme.

Unfortunately, these young plants are clones of the original and, therefore, share the same genetic defect that nearly consigned the species to oblivion. However, research continues at Kew to overcome the infertility problem and so give Café Marron a long-term future.

---

over generations. As time passes, the likelihood of mating with a partner sharing much of the same genetic code increases and, with very small numbers, mating may only be possible with close relatives. **Genetic inbreeding** occurs when two individuals with similar genotypes mate—the more ancestors the two parents have in common the more code they will share. Then there is a significant risk of each parent contributing the same recessive gene to their offspring: given this homozygous combination, the coded trait will be expressed in the offspring's phenotype. Since many recessive traits are deleterious, the result can be small, weak, often infertile offspring that ultimately cause a depression of population growth. **Inbreeding depression** is a common problem in domesticated plants and animals bred from a limited stock, when individuals do not grow vigorously and compete poorly with out-bred offspring. We see the same effect in human populations where mating has been confined to small groups, isolated by geography, custom or religion. Even in larger populations, some people assume a status that separates them from the majority and choose to breed with a range of limited partners.

Whilst the size of the metapopulation may be large, the effective breeding population within any single patch can be very small. As habitats disappear, so the metapopulation can become more fragmented and the gene flow between populations declines (Figure 3.12). Alongside inbreeding depression, genetic isolation may also produce locally adapted races: that is, where some characters have been selected within a population to fit them to a specific habitat and which distinguishes them from their neighbours. The result can be **outbreeding depression**, where matings between the two populations are relatively unsuccessful—the offspring are not well adapted to either set of local conditions and cannot compete with locally-bred offspring. This is sometimes a problem with *K*-selected species, particularly various trees that have evolved local races. As we saw in Chapter 2, such local adaptations are the first stages in speciation (Box 2.4). Then, re-united populations may not breed with each other or, as we shall see below, perhaps should not be allowed to.

This combination of difficulties—a susceptibility to environmental variation, the problem of sustaining population growth with a small number of individuals, and the genetic consequences of a reduced breeding stock—can lead to a downward spiral in numbers, where one factor feeds on another, accelerating the species towards extinction. This has been called the extinction vortex (Figure 3.10).

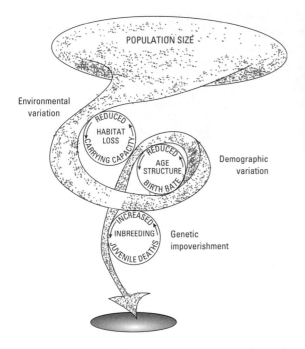

FIG. 3.11 Indian rhino (*Rhinoceros unicornis*) in Chitawan National Park, Nepal.

FIG. 3.10 The extinction vortex. Habitat loss reduces the carrying capacity for the population; demographic factors reduce their birth rate and genetic factors can lead to increased juvenile mortality. Because all of these tend to reduce population size, and smaller populations exacerbate demographic and genetic problems, an endangered population may quickly race to oblivion.

The elephant and the rhinoceros

The creation of the Chitawan National Park in Nepal has saved the greater one-horned rhino (*Rhinoceros unicornis*) from near-certain extinction. This impressive species was found along the margins of lowland rivers in the Himalaya, and until recently it had a population that ran into the tens of thousands. Hunting and habitat destruction during the last 100 years had reduced it to only two groups of about 100 individuals by the early 1960s. The loss of their habitat, as well as the pressure of hunting, both for sport and to prevent damage to tea plantations were the main reasons for the population collapse.

In Chitawan, the rhino was reduced to an effective breeding population of just 21–28 individuals (Figure 3.11). Today, with substantial protection, the total population has now more than 400 animals, while the Kaziranga National Park in India has 1300.

Much of the credit must go to the rhinos themselves: their spectacular recovery after the ending of poaching and hunting has been possible because they have retained a very high genetic variation.

We would normally expect genetic variation to be low in a population reduced to such a small size. Ironically, the one-horned rhinos seem to have held on to much of their variation because their decline was so rapid. Two demographic factors contributed to this—the rhinos have a long generation time (11–12 years) and females produce only one calf every four years. Thus the extent of inbreeding was limited to a short period during their recent decline. Kaziranga has around half of the global population and currently its population is growing at 5 per cent per year. According to the Worldwide Fund for Nature (WWF) the reserve is close to its carrying capacity (Table 3.1), but poaching is still estimated to take 5 per cent of this population each year.

In Asia, the security of the rhino populations and the loss of their habitats are closely interlinked, as forest clearance and agriculture have led to habitat fragmentation. Just 70 Javan rhinos (*Rhinoceros sondaicus*) remain in two populations—in Lam Dong province in Vietnam and in Ujung Kulon reserve in western Java, with the latter (containing 50 individuals) thought to be the only viable population. There have been fierce arguments whether to leave them all in the wild, possibly combining the two populations on a single reserve, or to transfer some to zoos for captive breeding. Combining the

Table 3.1 The global populations of five species of rhinoceros in the world in 1999, based on data from the Worldwide Fund for Nature. The total population size (N) is for wild populations only and the asterisk denotes that a species has a poor record of breeding in captivity. The Sumatran rhino has the most fragmented population and there is some evidence that this suffers from inbreeding depression.

Species	Total N	Number of populations	Number in captivity
African			
Black	2600	75	230
White			
Northern subspecies	25	1	9*
Southern subspecies	8400	248	704
Asian			
Javan	70	2	0
Sumatran	>300	12	17*
Indian	2100	7	126

populations to improve their genetic variation is perhaps even more of an issue for the smallest rhino—the Sumatran (*Dicerorhinus sumatrensis*)—whose population is highly fragmented.

In Africa, much of the rhino habitat remains, and here poaching for horn has devastated the two species (Table 3.1). Despite legislation and a policy in some countries to shoot poachers on sight, the high price of the horn ensures that the slaughter will continue. Because of their masculine associations, the horns are prized as handles for daggers (or *jambiyas*) and, when ground to a horn, as an aphrodisiac. While it probably makes the dagger unusable, the powder has no likely physiological effect on the libido of the human male. As much as half of all traded rhino horn (1.75 tonnes) was going to Yemen prior to 1990 but since 1997, its government has imposed penalties for using horn in daggers and the trade is thought to have dropped dramatically.

Perhaps the conservation of the rhino would be better served by a more radical approach. South Africa argues that the material from de-horned rhino (cut, under anaesthetic, before the poachers slaughter the owner) should be traded. This might lower global prices and make poaching less lucrative. Additionally there would be a requirement to mark and register official stocks and only these would be legally traded. As it is, the high price fetched by the horn (60 000 US dollars per kilogram in the Far East) is a major temptation to local peoples with meagre incomes, who probably see the rhino as a local resource to be harvested, in the same way that other peoples catch fish.

Of the hundreds of thousands of black rhino (*Diceros bicornis*) that once ranged over sub-Saharan Africa, 65 000 were alive in 1970 (Figures 3.12 and 3.13). Twenty years later there were just 3800 distributed among 75 populations but only 10 of these populations had more than 50 individuals. Because of this, individuals are moved between reserves in East Africa, to improve, or at least maintain, the genetic constitution of each population. Based on their horn shape and body size, seven subspecies or local races have been identified. One is now extinct. Present evidence suggests these have become genetically differentiated only recently, and

FIG. 3.12 Black rhinoceros (*Diceros bicornis*).

that breeding between them could be a viable conservation strategy.

The picture is less encouraging for one race of the white rhino. Although the southern race (*Ceratotherium simum simum*) has been through a major reduction in numbers, it has recovered to around 3000 individuals on various reserves in southern Africa. This race was thought to be extinct until a small number were found in a remote area of Natal, South Africa at the turn of the last century. These were protected and have grown to around 8500, with some being translocated to various reserves in the region. The northern race (*C. s.cottoni*), historically the most abundant sub-species, has suffered a much more dramatic fall. By 1960 there were just 2000 left in the wild and this fell to 15 in the late 1980s, confined to a single reserve, the Garamba in the Congo. All African rhinos have since shown some recovery (Table 3.1).

Mitochondrial analysis shows that the two races are genetically distinct probably because their geographical ranges have not overlapped in recent times. Conservation biologists have decided not to interbreed them, fearing the offspring would show outbreeding depression.

Thus, for the different species and the different races of rhinoceros, the details of their decline differ. Conservation efforts must take into account such details of their ecology and genetic history, but we also have to have due regard for their status in the environment, including how local peoples perceive them. A conservation strategy that ignored all of these complicating factors is unlikely to give the rhinos a long future.

As for the horn, so for the tusk. The international trade in ivory has led to the slaughter of large numbers of Asian (*Elephas*) and African (*Loxodonta africana*) elephants. Laos and Cambodia once had 40 000 elephants, but now have less than 4000. Again, much of the decline in the Asian elephant is associated with habitat loss, whereas poaching, has been the prime factor in Africa. Now as agriculture develops in eastern and southern Africa the loss of habitat and the conflict between elephant and humans is becoming a serious environmental issue.

Unlike their Asian counterparts, both male and female African elephants bear tusks and so both are slaughtered. Large tusks attract a larger price, so poachers favour the larger adults, especially the males. An elephant might live between 40 and 60 years, though today most adults are under 30 years old. Unfortunately, males do not begin breeding until they are around 30. Changing the age structure of these populations not only affects their future growth, it also has implications for the social hierarchy of the herds. These are normally led by elderly females, the matriarchs, whose knowledge of waterholes and foraging routes is passed on to the younger members of the group. Today most herds in Africa are led by females in their early twenties and teens.

The loss of the elephant also has important implications for the African landscape. On the savanna, their browsing on shrubs and trees is crucial to maintain open grassland and prevent the invasion of thorn and scrub. The grasses provide food for grazers such as the gazelle, wildebeest and zebra. Less obvious are the many smaller vertebrates and invertebrates, which directly or indirectly, rely on an open bush. Elephant and giraffe are keystone species, maintaining the grassland community of the dry savannah and an open scrub in the wetter parts of Africa.

However, there are complications when the elephant population is close to its carrying capacity. Elephants will push over trees to browse and their trampling damages grassland regeneration, particularly around waterholes (Plate 3.4). Richard Hoare describes how the focus of elephant conservation in eastern and southern Africa has changed, from a concern over poaching intensity in the 1980s to the current conflicts between elephants and humans. Today, elephants are often regarded as an agricultural pest in these areas.

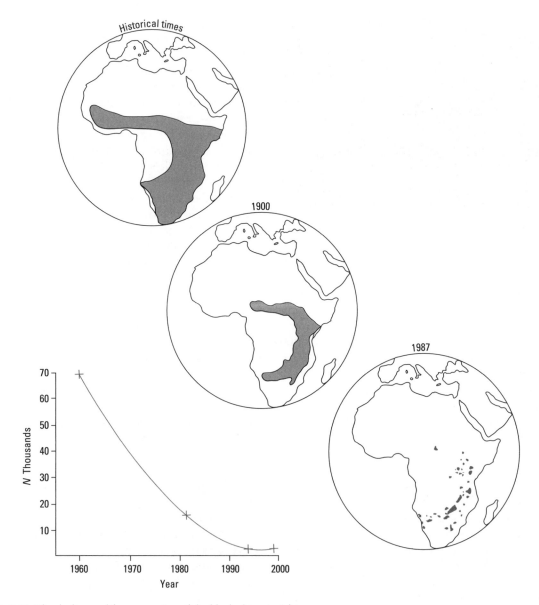

FIG. 3.13 The decline and fragmentation of the black rhino in Africa.

The problem stems from the over-population of elephants in protected reserves spilling over into unprotected ranges where they meet an expanding human population. The density of elephants in some of these reserves is leading to a loss of biodiversity and serious habitat damage. Outside the reserves, a variety of control measures have been used to keep elephants away from crops—from organizing hunts for elephants, to repellent sprays and fencing. Part of the answer may be to designate newer and larger reserves, but coexistence will need the cooperation of local peoples, and perhaps an expansion of the schemes that exploit the elephant, for tourism and for hunting.

In Zimbabwe, the government has designated areas where native agriculture runs side by side with a game management and conservation strategy, in an effort to control poaching. Government marksmen carry out organized culls, taking selected elephants. This helps to maintain the integrity of the elephant herd by managing its age structure while meat from the kill is distributed among villagers and the skin and tusks are sold. The proceeds are being used to fund conservation and to provide compensation to farmers in the form of additional food. Local people are thus being encouraged to tolerate and, indeed, value their wildlife resources.

Richard Hoare describes how organized sports hunting of the elephant has also directly benefited local peoples. In the Addo Elephant National Park in South Africa, each elephant is estimated to generate 300 000 dollars per year in the economic activity associated with tourism. However, culling is not allowed here and Amanda Lombard and her co-workers have measured the impact of an exceptionally high population density (2.2 elephants per square kilometre) has on the rare endemic plants within the reserve. This ecosystem is part of the succulent thicket of the Eastern Cape (Section 5.1), which includes plants found nowhere else in the world. Many of these have declined dramatically in the last 30 years as the elephant population has grown. Now the reserve has a population of around 300 elephants (from a founding population of just 11 when the park was set up in 1931). The elephants now represent 78 per cent of the total herbivore biomass in Addo (presumably based on only large herbivorous mammals), an imbalance that cannot be sustained in the long term. Lombard suggests expanding the reserve and creating botanical reserves, as well as accepting more general conservation priorities for its management strategy. This could include allowing black rhino access to the botanical reserves.

Elephant populations in various African states have been downgraded as conservation priorities and, in 2002, the UN CITES conference in Chile agreed that South Africa, Namibia and Botswana could sell 60 tonnes of their ivory stocks in 2004. The delay is to ensure that adequate monitoring procedures are in place to prevent poached ivory finding its way into the market. One promising scheme is to create a databank on the genetic identity of the different herds to allow the origin of each tusk to be traced back to a legitimate source.

The elephants and the rhino serve to illustrate our ambivalent relationship with such talismanic species. For some they embody the ethos of nature conservation, for others they are symbols of power and masculinity or a resource to be exploited. Different publics make different demands upon them and it may be that the cultural differences between nations will determine these species' future. In the same way, perhaps, that other peoples use their cultural traditions to assert their fishing and whaling rights.

SUMMARY

Simple analytical population models help us to understand the key features of population growth, but lack sufficient detail for us to make useful predictions about real world populations. We can identify the optimal yield as about half the carrying capacity of a population under a particular set of environmental conditions. In a relatively stable environment this also represents the maximum sustainable yield a fishery could expect to take, where the harvest is matched by new recruitment or growth of new tissues. For this, we need information on rates of growth and recruitment and to match our fishing effort with rates of replacement of both tissues and individuals.

For both fisheries and single species conservation, we have to consider the details of age and sex structure within a population and the life history and reproductive strategy of a species. Fragmented and small populations within a metapopulation may suffer genetic inbreeding if there is little migration between populations. Conservation efforts are often concentrated on the large, slow-growing, K-selected species closely adapted to their habitat. These are most likely to suffer when there is significant habitat loss. On the other hand, *r*-selected species are adaptable and more able to move between habitat fragments.

The various species of elephant and rhino in Africa and Asia illustrate some of the conflicting issues in single species conservation. In Asia habitat loss has been the principal cause of decline in both animals, and recovery of the rhino has been particularly effective inside protected reserves. Control of poaching in Africa has also allowed recovery but there are now problems of over-population of elephants on some reserves.

FURTHER READING

Alstad, D. 2001. *Basic Populus Models of Ecology*. Prentice Hall, New Jersey. (A detailed but accessible introduction to the range of basic population models.)

Milner-Gulland E. J. and Mace R. 1998. *Conservation of Biological Resources*. Blackwell, Oxford. (A wide-ranging collection of essays on the conservation of endangered species and habitats, including some of the social and economic aspects of over-fishing.)

Scalet C. G., Flake L. D., and Willis D. W. 1996. *Introduction to Wildlife and Fisheries*. W H Freeman, New York. (A systematic textbook that provides a foundation in both species conservation and resource management.)

WEB PAGES

http://ecology.unm.edu/populus (provides models that complement the Alstad book).
http://www.worldwildlife.org/spec (for details of the current status of
  elephant and rhinoceros).
http://www.irf.org (International Rhino Foundation).
http://www.iucn.org (World conservation Union—home of the red data book).
http://www.cites.org (Convention on International Trade in Endangered Species).

EXERCISES

1 Suggest reasons why an inshore population of polar bears should have a smaller home range than pelagic populations.

2 Use the equation in Figure 3.2 and the following data to derive the maximum rate of increase for a starting population of 50 when

  (i) $K = 1000, r = 0.36$
  (ii) $K = 1000, r = 0.18$
  (iii) $K = 500, r = 0.36$

You will probably find it useful to create a table like that in Figure 3.2(b) for each calculation.

What do you notice about the effect of *r* and *K* across these examples? What is a quick way to calculate the maximum rate of increase in these simple models?

3 When a population has reached its carrying capacity which of the following are true and which are false?

(i) the birth rate is greater than the death rate;

(ii) there are no further births;

(iii) the birth rate and death rate are matched;

(iv) the death rate now exceeds the birth rate.

4 How might the characteristics of a fish population change over many generations if over-fishing leads to too many immature fish being harvested?

5 In a short rainy season, the slow-growing, conditioned *Hyperolius* froglets alone will survive the dry season. However, these froglets will also survive a long rainy season, even though others may have reproduced and produced a second generation. Since growing slowly and preparing for aestivation guarantees survival of the dry season, why do all frogs not adopt the same strategy?

6 Which are more likely to become pests, *r*- or *K*-selected species? Which are more likely to become endangered, *r*- or *K*-selected species? Give your reasons in each case.

Now find examples which are exceptions to each one!

Tutorial seminar questions

7 Should we attempt to conserve single species? Discuss the social, political, economic and ecological implications of setting aside reserves and protecting a species such as the African elephant.

8 Review the various strategies for conserving an open sea fish population that has been intensively harvested. Again, consider the technical, economic and social implications of adopting each policy.

9 Use the various web resources to review the data on the collapse of the New Foundland cod stocks in the late 1980s and the extent to which this fishery has recovered today.

4 INTERACTIONS

They made us many promises, more than I can remember, but they never kept but one; they promised to take our land and they took it.

Chief Red Cloud of the Oglala Teton Sioux

Cheetah after a successful kill.

INTERACTIONS

Red Cloud speaks as a victim of what ecologists call interspecific competition—when individuals of the same species compete for the available resources. We might call it war, because individuals had grouped themselves together and acted cooperatively to acquire or defend a resource. To Jacob Bronowski war was not a human instinct, but rather, a 'highly organized and cooperative form of theft'.

Humanity is not alone in cooperating in this way. Many animals, from ants to chimpanzees, will form themselves into bands to improve their chances of winning a competitive battle. In a world of shortages, individuals and societies need strategies to secure their needs, for this generation and for generations to come.

Of course, it is too simplistic to describe war as a contest for scarce resources. It may be a prime cause, but conflicts also begin when societies seek to promote or preserve their structures and their traditions. Individually and collectively, people make moral and ethical judgements to fight and sacrifice resources to defend an idea. For both humans and animals, defending a society often means defending the system by which resources are partitioned amongst its members.

The problem for the individual is to make a living; that is, to acquire the resources that allow it to survive and to thrive and to ensure that its genes make it into the next generation. This means securing the energy and nutrients necessary to produce gametes and to nourish its offspring. Societies are often organizations that facilitate this process. At one extreme, an ant society is highly integrated with most individuals sacrificing their own reproductive effort to promote that of a single queen. In contrast, mammalian societies are much looser arrangements, but they too increase their efficiency by dividing effort and reducing wastage by partitioning resources amongst their members.

Seen in these stark terms, life appears to be very brutal. Yet the great variety of life on the planet, and the beauty of its plants and animals, has arisen from this struggle for resources. Even so, cooperation can benefit both parties, especially among related individuals that share part of their genetic code, or even between very different species. Cooperative interactions are essential for the survival of whole groups of organisms. Sometimes the benefits are more one-sided, where one species is used by another as a resource. In all cases, these interactions evolve and develop by natural selection and help to tie communities of species together.

This chapter looks at the range of interactions, considering the costs and benefits to each partner. Each individual seeks to maximize the proportion of its genes in the next generation, to maximize its reproductive success, whether this means preying on, or cooperating with another species. In some cases, the trade-offs made by a species may not be immediately obvious and only by unpicking the detail do we find how it benefits from the association.

We begin by exploring some cooperative associations, and then go on to look at the competitive battles between individuals and between species. Some, more direct conflicts, mean that one species loses out because another exploits it—say between a predator and its prey or a parasite and its host. As we shall see later on, these are associations we have used to our own ends in our struggle with pests and disease.

4.1 Acquiring resources

There are some needs which are common to all living organisms. The two most fundamental are food and water: water, because it is the medium in which the chemical reactions of life take place; food, because it supplies both the materials for these reactions and the energy that powers them.

Food and water are often in short supply, either periodically or continually and an organism needs strategies to acquire and to conserve supplies of both. All species require space in which to secure their resources. Plants need soil and air from which to capture nutrients, water, and sunlight. Animals too need space—even sea anemones will jostle each other for room, pushing (albeit very slowly) against each other for a prime location. More obviously, dragonflies dart up and down their stretch of riverbank, defending a territory in which they feed and mate. Even a partner can be regarded as a resource, necessary to produce the offspring that will carry some of their parents' genetic information into the next generation. 'Resources' thus refers to all of those elements of an organism's habitat it needs to survive and to procreate.

Many resources are so abundant that an organism can afford to be profligate, making no special effort to conserve supplies. Oxygen is a prime example. In contrast, energy is limited for most individuals, and they have to allocate its use to best effect (Box 4.1), to partition the supply between reproduction, growth or further energy acquisition. Like the different reproductive strategies we have already examined (Section 3.5), these acquisition strategies help to define the ecological niche of a species.

BOX 4.1   Allocating resources

When a resource in short supply, be it energy or some key nutrient, an organism needs a strategy to make efficient use of what it can acquire. Since its prime objective is to reproduce and ensure the survival of its genetic code, this means allocating its resources to maximize its reproductive success.

The dilemma is often whether to play the short game or the long game. Resources could be used to produce offspring in the short term, with what is at hand. Alternatively, they could be devoted to survival now, perhaps growing larger and seeking further supplies that might support several reproductive events later on.

Natural selection will favour different strategies in different environments, according to how predictable they are (Section 3.5).

Any strategy has to balance the costs of acquisition against the potential return. A predator has to evaluate the energy costs of capturing a prey against the benefit it will get from consuming it. Clearly, if the energy invested in searching for, handling and consuming a prey is less than that assimilated from eating it the predator will show a net loss and have less energy to devote to reproduction or the next chase. An active predator is, therefore, continually reviewing how easily

BOX 4.1 Continued

food is being found and will tend to stay where its prey is abundant. This is an unconscious cost–benefit analysis, with the consumer moving on when search costs start to rise too high.

All that we say of animals we could also say, in slightly different terms, about plants. Deciduous trees shed their leaves in winter because of the cost of respiration. Simply keeping the leaf alive during the short days outweighs the return from any photosynthesis (Section 6.2). Similarly, plants have different strategies for investing in root growth (to acquire nutrients) or timing their shoot growth according to a variety of factors (including other species) in their environment.

In fact, any organism has to meet a series of costs that it cannot avoid, simply to stay alive (Figure 4.1). Only a fraction of the energy in its food will actually be assimilated (A) and then only a small fraction of this is actually fixed in the tissues (Box 6.2). Energy has to be used in metabolism, simply to maintain the tissues, in movement and in growing new tissues. This is the energy of respiration (R) and is eventually lost as heat. The difference between A and R can be used to produce either gametes (Pr) or new tissues (Pg). The balance is thus,

$$Pr + Pg = A - R$$

When food is in short supply and energy reserves are low, this allocation of resources becomes critical. Even if energy itself is not in short supply, a lack of some key nutrient, such as water, may limit its assimilation. Under these conditions an individual may only grow very slowly, often in competition with many others. Long-lived species able to match their growth against resource availability will be favoured in these habitats. Their strategy is to

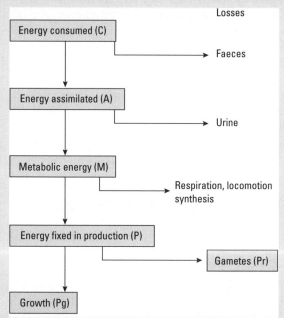

FIG. 4.1 The allocation of energy to growth and reproduction in an animal. Note that the same diagram could be used to describe the allocation of any key nutrients, such as nitrogen or phosphorus.

delay reproduction until its chances of success are most favourable. Alternatively, in an environment that is characteristically unpredictable, brief periods of plenty have to be exploited quickly, favouring short generation times that rapidly produce gametes.

4.2 Cooperation

One strategy is to share some of the costs of securing a resource with the neighbours. A group of female lions combine their strength to overpower a wildebeest. On their own, they have little chance of bringing down a healthy adult; together they are successful and they share the costs (and risks) of killing their prey. This works because the wildebeest represents a meal for the whole pride and far more meat than a single female could eat in a single sitting. A range of animals cooperate in this way, from

whales swimming in coordinated manoeuvres to confine a shoal of fish, to the swarming of ants over a large beetle.

This is cooperation between individuals of the same species, but we need to know how much this benefits each member of the team. One method is to measure an individual's reproductive success. Thus, the fitness of its resource acquisition and its allocation strategy is indicated by the number of offspring an individual leaves to the next generation.

Sometimes, a group consists of close relatives, parents and siblings (primarily sisters and daughters in a pride of lions), which, through their cooperative effort, improve the chances of their shared genetic code passing into the next generation. In other cases, genetic relatedness between the individuals is much lower but there are still benefits—for a wildebeest, living in a large group affords protection and reduces the costs of looking out for potential predators. Further, one individual in a herd of a thousand has less chance of being eaten than when it is the sole object of a predator's attention. For these animals the benefit comes from dividing the workload, being able to feed when others work as look-outs and reducing the chances of being selected by a predator.

Some of the more complex strategies for cooperation are best understood by looking at the genetic relationships between members of a group and the trade-offs that each is making. Working with a brother or sister can improve the prospects for the genes which are shared. Egg production in a colony of ants is delegated to one individual, the queen, and the rest of the females forego any reproductive activity (Figure 4.2).

The queen controls the sex of her offspring by allowing the egg to be fertilized by sperm stored from her nuptial flight. Ironically perhaps, the male develops from an unfertilized egg and therefore carries only the female chromosomes. Males are only produced by the queen when it is time to form a new nest.

The individual worker is a sterile female, but she shares a large proportion of her genetic code with her sisters and her queen. The role of each worker is controlled by what they are fed. Particular chemicals, termed **pheromones**, govern the development of larvae and their subsequent behaviour. The prime source of these pheromones is the queen and these signals move through the nest by contact between workers and with the food. In this way, both the numbers of each caste and their activity is regulated. If the queen is lost, each female larva stands some chance of herself becoming queen, if fed a particular diet.

Although the queen has the power to decide the gender of her offspring, it is wrong to assume that she controls the nest. In fact, the overall coordination derives from the combination of the chemical communication system and a large number of individuals

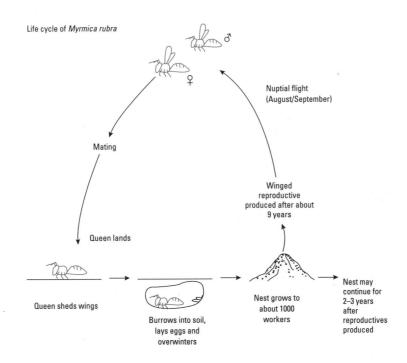

FIG. 4.2 The life cycle of a temperate ant *Myrmica rubra*, a species common in woodland and grassland areas where it can be found in loose soil or rotting logs.

doing a small range of predictable tasks. What may appear as chaotic behaviour from watching a small number of ants can make sense when viewing the colony as a whole. Order appears from the interaction of a large number of repetitive elements.

Together, the collective effort of the female workers ensures that their shared genetic code has a greater chance of survival, better than if they wasted energy and resources competing with each other. Interestingly, all of the sperm produced by a male are identical, so each worker has at least half of her genes in common with her sisters—those she inherited from her father. An individual worker may be dwarfed by her monarch, but this sterile female is not sacrificing herself or her genes for her queen. In fact, the genetic cost–benefit scales are tipped in favour of the code carried by the workers, rather than the queen. The workers all share the code from their father, code not shared with their mother. Their effort is for their brothers and sisters who will eventually leave the nest and carry the shared code into the next generation. This reproductive and resource utilization strategy is found in other insects, including bees, wasps and termites and has produced some of the most elaborate and highly organized of all animal societies.

As you might expect, insect societies are also some of the largest. For example, one ant society from the Jura, in France consists of 300 million individuals divided between 1200 ant hills, covering around 70 ha hectares. The whole arrangement is connected by 100 km of tracks. Information, food and larvae move down these tracks, with the whole colony divided into sectors. There are around 15–20 main nests, and a number of smaller ones used to rear larvae in the summer months. Where food is short, resources in the form of food and labour are moved to give a more equitable distribution.

More recently, another, even larger colony of Argentine ant (*Linepithema humile*) has been discovered along the northern coast of the Mediterranean. Stretching over 6000 km—from Italy to the Atlantic coast of Spain—it comprises millions of nests and is the largest cooperative unit ever recorded. Needless to say, we are still learning how coordination on such an immense scale is achieved. Its success here, outside of its natural range, is due to behavioural changes in *Linepithema*—there appears to be a reduction in inter-nest aggression and this has assisted its invasive spread in the Mediterranean and similar biomes elsewhere.

Large ant colonies are known throughout the world. Some species do not make permanent nests, but instead move *en masse*, continually foraging and setting up bivouacs for the night. Some raid the nests of other species and have castes adapted for fighting the battles. Slave-maker ants rifle the nest, removing larvae and pupae that they then rear in their own nest. The emerging workers become slaves, keeping house for masters whose own workers are mostly warriors. The cost of having specialist fighters is thus offset by not having to produce eggs to make workers. Other species seize chemical control within a nest so that the original queen is replaced when her own offspring are induced to kill her. A parasitic queen is then introduced and protected and fed by these workers, who raise her offspring with their different genetic code. As Richard Dawkins points out, each individual parasitic queen incurs small costs in subverting the chemical control system of the host nest, yet gains immensely from the efforts of the duped workers. It is an interesting question whether we should call this caste war, given that it is two different species fighting over the workers.

Cooperating with different species

Cooperation between individuals of the same species may occur because they both require the same resources. With two different species, however, requirements differ and cooperative associations will only form if there is some benefit each individual can derive from the other.

In **mutualistic** associations individuals of both species benefit. Many insects share truly mutualistic associations with flowers. Flowers are adapted to attract insects to their reproductive organs, so that the insect may serve as a go-between, carrying the male gametes (pollen) to the female parts (style and stigma) of the next flower (Figure 4.3). In most cases, the flower itself is no more than an advertisement that food (nectar) is available, so that many different species of insect will visit. Besides the costs of nectar production (a relatively simple sugar solution), such flowers need to produce abundant pollen to dust the many insects that feed, perhaps only one of which will go on to visit another flower of the same species.

FIG. 4.3 Insect pollination in action.

Abundant pollen is also needed to make up for losses by those insects that consume it. Some plants actually produce dummy pollen to reduce these costs.

Pollination is less haphazard if particular insects can be attracted to the flower. Under selective pressure, a plant which only allows insects of a certain size, or which releases its pollen to those with a certain configuration of mouthparts, will improve the chances of its pollen being passed to another flower of the same species. A range of strategies are used by plants to reduce their pollen costs by concentrating on specific messengers, not only by the configuration of their flower parts but also by the scent used to attract the insects. Sometimes the co-evolution of the two mutualists is so close that each depends completely upon the other.

The yucca (*Yucca filamentosa*) from the deserts of western North America is a dramatic example. The plant can only be pollinated by the female yucca moth (*Tegeticula yuccasella*), which does not feed on the flower, but instead collects its pollen and forms it into a ball. She then flies to another flower and lays her eggs in its seed buds. Thereafter, she climbs up the stigma and applies some pollen, fertilizing the flower, before flying off to repeat the process in other flowers. In doing this the moth ensures her larvae will have an abundant supply of ripening seeds. Much of the seed is not consumed and so the plant benefits from the care taken by the moth. Notice also that the moth spreads her risks by not laying all her eggs in one flower. This mutualism is complete: without each other, both the yucca and its moth would become extinct.

Plants also enter mutualistic partnerships with animals to disperse their seed. Fruit is produced to attract consumers, but only when the seed is ripe and ready to be moved. At that stage, the fruit has produced the sugars to reward the carrier and advertises this by a change of colour. Until that time, the fruit remains inconspicuous and bitter.

Ants make very bad pollinators. Their subterranean lifestyle exposes them to microbial pathogens and they secrete a range of protective substances, including antibiotics, which can affect the viability of pollen. However, an ant colony does make a very effective defence system, not least because of its high degree of cooperative behaviour. The bull's horn acacia (*Acacia cornigera*) of Mexico cultivates an association with the ant *Pseudomyrmex ferruginea*, by producing protein-rich nodules at the tips of its leaves. The ants collect these to supplement the protein content of their diet. The plant also provides carbohydrate from special nectaries located at the base of its leaves. The tree also provides shelter for the ants in the form of large hollow thorns. Here the ants build nests within easy reach of food and from which they readily defend their home and larder. Few herbivores will feed where busy and aggressive ants are swarming. Both ant and acacia depend on each other—they have an obligate association because each needs their partner to survive.

Such close associations can be easily upset and this has happened in the South African fynbos (Section 5.1) where many of the plants within the protea family use ants to disperse their seeds. The proteas recruit the ants by providing seeds with little food parcels (elaiosomes) attached to them. Once home with their 'take-away' meal the ants eat the elaiosome and either leave the seed in the nest or discard it along with the colony's waste. In this way the seeds are, in effect, dispersed and planted by the ants. Within the nests, the seeds are protected from the periodic fires that surge through the low, dry vegetation and are also provided with much-needed nutrients in what is otherwise a nutrient-poor environment.

For one particular species of protea (*Mimetes cucullatus*), this dispersal strategy is now under threat from the invasive Argentine ant (*L. humile*). In a situation similar to that we met earlier in the Mediterranean, the ant has rapidly spread through the South African shrubland following its accidental introduction a century ago. This species of ant particularly favours large-seeded proteas like *Mimetes* and when faced with a protea seed, it simply bites off the elaiosome dropping the seed on the ground, where it can be lost to fire or to seed-eaters. Added to this, the ants nest beneath large rocks, so any seed which might be taken back to their nest has little chance of becoming established.

This behaviour impacts the tight interactions between species of the fynbos where 30 per cent of its flora relies on ant dispersal. The effect of the Argentine ant has been studied by Caroline Christian who has seen a subtle shift within the species composition of the community. Not only does the ant snatch a free meal at the expense of the protea, it has also displaced native ants, especially *Anoplolepis custodiens* and *Pheidole capensis*, both of which specialize in the dispersal of large-seeded proteas. There has been a 94 per cent decrease in large-seeded protea species where *L. humile* dominates (Figure 4.4). Meanwhile, the small-seeded species (which are dispersed by generalist ant species unaffected by

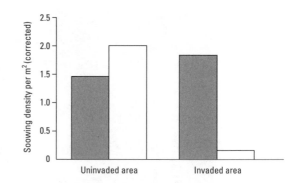

FIG. 4.4 The effect of the Argentinian ant (*L. humile*) on the regeneration of large- and small-seeded proteas in the fynbos of South Africa. The graph shows a decrease in seedling density of large-seeded proteas (white bars) where the invasive ant is present—due to disruption in the *mutualism* between the plants and their native ant dispersers. The small-seeded proteas (shaded bars) remain unaffected as *L. humile* does not compete with the ants that disperse them.

Linepithema) continue to proliferate. The rapid spread of *L. humile* throughout the Mediterranean, the fynbos and the Californian coast, has the potential to disrupt insect communities and their mutualisms on a global scale.

Many mutualistic associations are not obligate, but one or both of the partners may fare less well without the interaction. One example is the association between many plants and soil fungi. Most higher plants readily form these associations, called **mycorrhizae** (Section 7.2), in which the strands (hyphae) of particular soil fungi enter the root system of the plant. The plant benefits from the extension to its root system, improving its drought resistance and its capacity to absorb nutrients (particularly phosphates). The fungi benefits from the readily available sources of energy provided by the plant. Without a mycorrhizal association, the plant grows less well.

The association between photosynthetic plant and fungus is most highly developed in the lichens (Figure 4.5). Around 20 000 types of lichens are known, but theirs is not an obligate association— both fungi and algae will grow in the absence of their partner. Nevertheless, the association is so close that the lichen produces asexual reproductive structures (soredia) that contain both algal and fungal cells. The fungus grows to provide a superstructure (the fungal

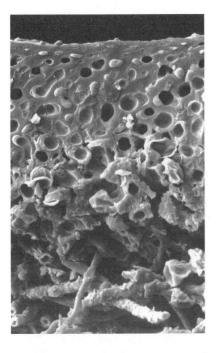

FIG. 4.5 Electron micrograph of a cross-section through a lichen *Lobaria pulmonaria*. The lichen consists of a super structure of fungal mycelium (consisting of thread like hyphal strands) with the symbiotic algal cells embedded within it.

mycelium formed of strands of hyphae) which helps to support the photosynthetic partner, either a green or blue-green algae (cyanobacteria). The fungus also produces complex organic acids which dissuades consumption by herbivores, helps to protect against dessication and may also release minerals from its substrate which the algae can absorb. The fungal partner benefits from the sugars produced by the algae's photosynthesis. In some cases, the fungi may 'cull' older algal cells to absorb their nutrients and some biologists have therefore argued that this mutualism may border on a controlled exploitation of the algae by the fungi.

Not all associations benefit both parties—we saw earlier how the co-evolution of a number of species of orchid with bees and wasps had led to the flower mimicking the female insect (Section 2.3). While female bees show no interest in the flower, the sight, and smell of the flower is irresistible to males who will try to mate with the flower for some time—long enough for pollen to be dumped on his back. Ecologists would regard this as an example of *commensalism* rather than mutualism, since the benefits to the bee are more apparent than real. Very often a commensal is not dependent on a single species, but can derive benefit from several species. The arrangement might be one-sided, yet the beneficiary causes no detriment to the exploited partner. For example, the brown-headed cowbird (*Molothrus ater*) follows behind grazing livestock, feeding on the invertebrates they displace. The birds benefit from the activity of the grazers, but the livestock incurs no costs from the presence of the cowbird. Similarly, scavengers such as vultures make use of the killing power of large carnivores, at no cost to a predator that has fed and since left the remains of the kill.

Commensalism is often a loose association. For the cowbirds, the species of grazer it follows is immaterial. While some dung beetles are relatively specific about whose dung they lay their eggs in, others are not so fussy.

4.3 Competition

When resources are limited, the best strategy may simply be to secure as large a share as possible. Competition is central to the theory of natural selection because the struggle for resources determines which individuals are able to reproduce. The fittest survive and secure the resources to produce offspring.

This sort of competition, between individuals of the same species, is termed **intraspecific competition** (Section 3.2, Box 4.2). **Interspecific competition** is between individuals of two different species. In each case, competition can take different forms. Resource competition is the battle to secure part of a limited supply. Interference competition occurs when an

FIG. 4.6 Bracken (*Pteridium aquilinum*), an invasive fern found worldwide, seen here invading a colony of thrift (*Armeria maritima*). At this time, the *Armeria* still dominates.

TABLE 4.1 Types of competition
Chemical competition
Production of a toxic or deterrent chemical to exclude competitors
Consumptive competition
Competitive use of a renewable resource (e.g. food)
Encounter competition
Physical defence of a resource, possibly by aggression
Overgrowth competition
Successful competitor overwhelms the opponent in size or number
Pre-emptive competition
Rapid colonization of space when it is available
Territorial competition
Defence of territory, breeding, and feeding areas

competing species adopt (Table 4.1). Space is often limited, particularly in the best locations. Plants and sessile animals tend to compete by pre-emptive and overgrowth competition to swamp out potential and existing competitors. One highly invasive weed that uses this form of interference competition is bracken (*Pteridium aquilinum*). Its dense foliage and deep litter layer forces out existing plants and prevents others from becoming established (Figure 4.6). Its success has led to bracken becoming a global pest (Section 4.5).

Intraspecific competition

Chapter 3 explored how limited resources result in density-dependent population growth and the stabilization of a population around a carrying capacity (Section 3.2). We have also seen how differences within a population might, with some degree of isolation, lead to character displacement and the formation of a new species. Specialization on a particular part of a resource spectrum can then result in two new niches being defined which may result in speciation (Section 2.4). Populations of common garden snail (*Helix aspersa*) may not be on evolutionarily separate paths, but they have adapted to local conditions (Box 4.2). The variability in their growth and

individual actively prevents another from exploiting the resource by its activity or behaviour. In scramble competition all individuals have equal access to a shared resource and there is a free-for-all to acquire it.

These competitive battles can be classified according to the way they are fought and the strategies

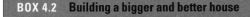

BOX 4.2   Building a bigger and better house

FIG. 4.7 Where the quick and the strong survive—calcium is a valuable resource that determines the outcome of competition between individual common garden snails (*Helix aspera*).

Competition between neighbours—members of the same species—can be subtle but it can also be very obvious. Size matters. Often it is the larger individual that secures the resource—food, space or partners—and in beating its smaller competitors, is able to pass its genes onto the next generation. With so much at stake, competitors often evolve a range of devices and strategies to give them an advantage. For example, individuals of the common garden snail, *Helix aspersa* (Figure 4.7), suppress the feeding behaviour of others with the slime they secrete as they move around.

All snails need large amounts of calcium to build and reinforce their shells. The shell is their principal means of protection from predators and from water loss. Without calcium, the proteinaceous horn on the outside of the shell is little more than camouflage and is thin and weak. If calcium is not abundant in its environment, an individual has to balance the various demands on its small supply—between reinforcing the shell, courtship, producing eggs or the range of metabolic functions that require calcium. If calcium is abundant, young snails will grow large and robust shells quickly.

Interestingly, it seems the rate at which the shell is reinforced is genetically determined and different populations have different priorities for their calcium intake. Populations from sites where calcium is abundant tend to

produce thick, strong shells even when fed a low calcium diet. It seems there has been no selective pressure for these individuals to be more frugal in their calcium usage. Others, typically from calcium-poor sites, are more parsimonious, producing relatively thin shells even when fed a calcium-rich diet.

How do such different strategies work when individuals from two populations compete against each other? In a series of competitive trials, Larry Richmond and Alan Beeby have found one population tends to beat all comers in a race to secure calcium on a limited diet. Individuals from chalk downland in southern England tended to grow larger and denser shells when competing with those from less calcium-rich sites. The chalkland snails may have been just quicker to the food, reducing the food (and calcium) available for their neighbours. Or, perhaps, they made better use of the calcium they did consume.

In most trials between various combinations of other populations there was no predictable outcome and no consistent winner. Invariably each population would have one or two rapidly growing individuals with the majority being quickly left behind. It seems that an early spurt of growth secured the later success of these faster growers.

Perhaps the chalkland snails just feed more readily, especially in the first juvenile stages. Certainly, the rapid growth of their shells cannot be a response to limited calcium.

BOX 4.2   Continued

Possibly the hot conditions which develop on the chalk in the summer requires them to reproduce quickly before soil conditions become too dry for their eggs to hatch.

Alternatively it is their competitors that show the important adaptation. Several of these populations were collected from highly polluted habitats, where individuals may have been selected for their capacity to tolerate toxic metals. Calcium is used by a range of invertebrates to isolate and excrete excess metals, especially zinc and cadmium. It may also be used to accelerate the loss of lead, even though this could deplete the calcium reserves used to reinforce the shell. In such habitats, slow shell growth may be a necessity.

Again, both explanations may apply. When lead was added to their food, the chalkland juveniles accumulated the highest concentration in their soft tissue and had the biggest reduction in shell growth rate. By comparison, a population with a long history of exposure to lead, from North Wales, grew shells slowly but showed relatively little change to this rate when the poison was added to their diet.

However it is the performance of the two populations that makes the race interesting. Over 100 days (covering their fastest shell-building phase) 10 juveniles will collectively increase their shell height by a total of 18 mm. This amount was the same, no matter which two populations were competing, and even if one population dominated. It seems that the total amount of food presented limited the total amount of shell built. Which individuals grew the 18 mm of shell probably depended on who got there first. The chalkland snails were probably just faster than most.

behaviour then affects the outcome when the two populations are brought together in a competitive situation.

With intraspecific competition, direct combat over a resource, be it a mate or food, is both costly and carries considerable risk. Only when the stakes are high—perhaps when a harem of females is to be won—will competitors actually lock horns. Death rates amongst adult males can then be significant. For this reason, many species have mechanisms to diffuse such confrontations, where size, colouration or ornamentation is used to signal the status of the owner or their position in the hierarchy. Even so, the horns of many large mammals are not simply elaborate ornamentation but have evolved for defence and attack (Plate 4.1).

There are also striking examples of intraspecific competition among plants. Competition between seedlings and young plants for resources such as space, light and nutrients is fierce. Whilst many plants produce vast amounts of seed only a small fraction will grow to maturity. The horseweed (*Erigeon canadensis*) shed seeds at a density of 100 000 m^{-2} and these will produce a thousand seedlings over the same area but only a handful of mature plants.

Some plant species show a widely dispersed distribution, with large, regular distances between individuals. In the deserts of North America, the creosote bush (*Larrea* species) is so uniformly spaced that it appears to have been deliberately planted. This spacing is entirely natural. Each plant has a uniform area from which it collects its resources and which it may have established by secreting an inhibitor to suppress the growth of its neighbours. This special form of interference competition is termed **allelopathy** and is not confined to intraspecific competition. For example, allelopathy may also play a major part in the ability of bracken to suppress the growth of other species.

Interspecific competition

In a prolonged competition between two species for limited resources, natural selection would produce one of two alternative outcomes:

- Each species becomes more highly adapted to its part of the resource spectrum and shows the character displacement of a specialist (Section 2.5).

- One species extends the range over which it feeds and makes more efficient use of that resource than its competitor. It may do this by interfering with the growth of its competitor or it may simply produce more offspring per unit of resource. It therefore out-competes its competitor, which must eventually be displaced.

The second alternative is the **competitive exclusion principle** (Section 2.5). Simple mathematical models show that where there is strong competition between two species one must decline to extinction. A number of laboratory experiments using microorganisms, plants and cereal beetles, confirm that one species will be lost where there is significant overlap for a key resource, at least in these simple ecosystems (Figure 4.8). Sometimes, these laboratory fights were resolved by the winner reproducing fastest (resource competition) or where the winner ate the opposition—or at least their young and pupae (interference competition). The result of each battle was not always guaranteed, however, since changing the conditions of the experiment could sometimes reverse the result. Clearly, the outcome of these struggles depended on a range of factors, some external to the species.

We also have evidence of competitive exclusion from the real world, when individuals from two species with similar niches meet for the first time, often as a result of introduction by humans. Sometimes the introduced species have become pests and out-competed the native species, though more often, the introduced species fail to succeed, presumably because they are less well adapted to their new habitat.

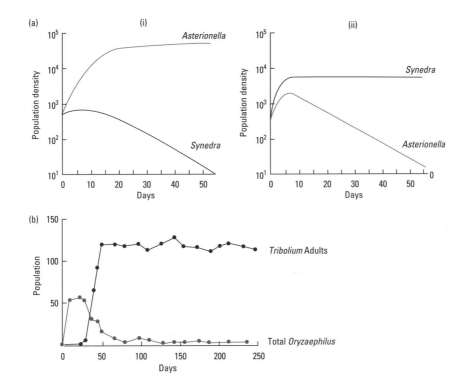

FIG. 4.8 The results of a series of laboratory trials looking at the competitive battles between two species that occupy similar niches. (a) Between two diatom species (unicellular algae with a siliceous case), *Asterionella* and *Synedra*, in cultures where silica is limiting. (i) in most trials between the two species *Asterionella* won, whether *Synedra* had a higher initial population or not; (ii) however, when the temperature was raised to 24°C *Synedra* won. (b) Between two cereal beetles, *Trilobium* and *Oryzaephilus*. These species predate each other's larvae and pupae when placed together in a simple flour ecosystem. *Trilobium* won most battles because it ate more of the opposition. Coexistence was possible if *Oryzaephilus* larvae had a refuge from adult *Trilobium*. In this example, both species start with the same population size; the growth in the *Trilobium* population is matched by the decline in the *Oryzaephilus*.

The common house gecko (*Hemidactylus frenatus*) is native to Asia but was introduced into Hawi'i in the 1940s. Geckos are lizards able to cling to vertical surfaces and are commonly found on the walls of tropical houses, where they feed on insects attracted to the lights. After the arrival of *Hemidactylus*, three other species, previously introduced from various locations around Polynesia, all suffered major population reductions in their suburban habitats. However, the common house gecko has not dominated the whole of Hawai'i. Another Polynesian species, the mourning gecko (*Lepidodactylus lugubris*) survives well in drier habitats and seems to be holding on, even though its numbers have been reduced. The eventual outcome may depend on interference competition again, since *Hemidactylus* is known to eat *Lepidodactylus* juveniles, at least in laboratory experiments.

Hemidactylus seems to be winning most of the battles in other parts of the Pacific too. Its numbers have risen on Suva (Fiji) at the expense of other species. It was unknown on Vanuatu until 1971 but has since become the most abundant species over much of the island. Ted Case and his colleagues recorded the arrival of *Hemidactylus* in Tahiti in 1989, noting that within two years it had spread beyond the docks in a 10 km radius. Case and Kenneth Petren have since shown that urbanization has promoted the success of *Hemidactylus*. *Lepidodactylus* prefers complex habitats in which they can stalk their prey. The relatively simple, open habitats inside modern buildings favour *Hemidactylus'* more pursuit-orientated mode of hunting.

Although the confrontations between *Hemidactylus* and its rivals are far from resolved, its successes and occasional failures, along with the evidence from more controlled experiments, show that competitive exclusion is not the only possible outcome of interspecific competition. In a variable environment, when conditions might change or where there is scope for avoiding competition, coexistence might be possible.

Unfortunately, the success of *Hemidactylus* over much of the Pacific also demonstrates how readily species can be lost when two competitors meet for the first time. Our propensity for global travel has increased the frequency of such meetings between native and introduced species.

Competitive types

Often, highly invasive species, such as the rat, have a particular set of characteristics that fit them to their way of life. A rat approximates to the r-selected type that we described in the last chapter (Section 3.5)—a species capable of rapid reproduction and dispersal, able to quickly exploit a range of opportunities. However, there are plenty of examples of invaders and pest species that are closer to the K-end of the spectrum—species built for more long-term competitive battles. These life history strategies imply very different allocation of resources. The true r-selected species allocates much of its resources to a single reproductive event, whereas K-selected species will reproduce repeatedly and so spread their costs and risks.

Richard Southwood points out that life history strategies are selected not only by the predictability of the environment and the availability of resources, but also according to the presence of stressors—such as the presence of toxins and other factors which might limit rates of growth. There are species that avoid having to compete by occupying marginal habitats, evolving specialist adaptations that enable them to thrive in the most hostile of conditions. Southwood calls this adversity selection (A), one of three dimensions in his 'habitat template' (Figure 4.9). The template describes the selective pressures to which a species has to respond. In addition to adversity-selected species that survive stable but hostile conditions, K-selected species dominate in favourable and stable environments, and r-selected species in unstable environments, provided there is an adequate supply of nutrients.

Philip Grime has developed a similar three-way classification of plant strategies, noting again that some species are adapted to live in poor, highly stressed environments (Figure 4.10). Where resources are limiting, or there are other stresses (such as toxic metals or extreme soil pH), a species may only be able to grow slowly. Those that can survive here are termed **stress-tolerators**. Since few other plants can withstand the adverse conditions, they have few competitors, and slow growth is not a disadvantage. In contrast, **competitors** seek to outgrow or crowd out the opposition where resources are abundant.

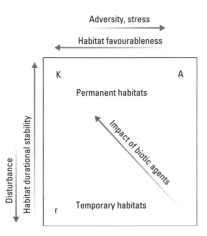

FIG. 4.9 Southwood's habitat template. Life strategies are classified in terms of habitat stability and favourableness. The top left-hand corner represents habitats that are stable and favourable, the top right are stable but unfavourable conditions and the lower left-hand corner unstable but favourable habitats. Habitats can be classified according to these three dimensions and species can also be placed according to the type of habitat they are commonly found in.

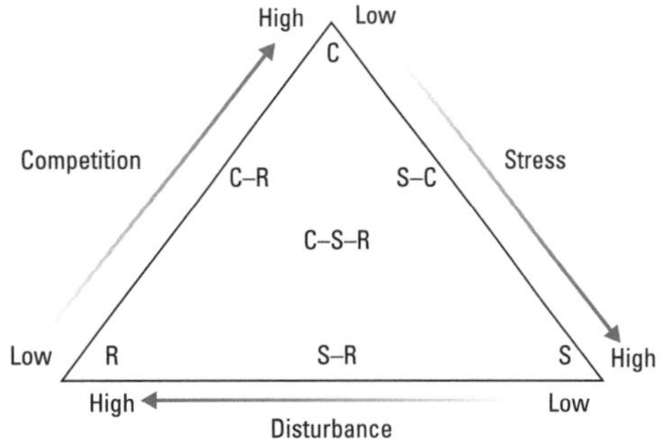

FIG. 4.10 Grime's classification of resource acquisition strategies in plants. Ruderals (R) are roughly equivalent to r-selected weeds (e.g. *S. media*). They are found in habitats that are regularly disturbed and they are typically fast growing and short-lived. Competitors (C) are found in undisturbed habitats where resources are abundant. Their strategy is to try and crowd out other plants by rapid growth (e.g. stinging nettle, *Urtica dioica*). Stress-tolerators (S) are slow-growing species that can tolerate shortage or other forms of stress in the soil; they prevent many other species from establishing themselves (e.g. sheep's fescue, *Festuca ovina*). Between these are various intermediate types.

The final category is equivalent to the weed-type, *r-selected* species we have previously described. **Ruderals** live in disturbed and unpredictable environments but where resources may be occasionally abundant and available. As poor competitors, their strategy is to produce their seed before competitors arrive.

Again, there are intermediate forms between the three main types. This scheme describes a plant's strategy for resource allocation in relation to the available resources, the frequency of disturbance and the adverse features of a habitat. These demands will change during the organism's life cycle, and the conditions under which a seedling flourishes may be very different to those demanded by a mature tree (Section 3.5).

For many adult plants the worst of the competitive fight was fought long before it became established. In a garden lawn for example, species of grasses and broadleaved plants form a tight-knit community which is a far from ideal place for young seedlings. Not only is there fierce competition for water and minerals, there are other constraints on space and access to light. Even when gaps appear, seedlings have

to be tolerant of trampling, soil compaction and the regular grazing of the lawn mower. Gaps with different conditions favour different species. Dry compacted gaps favour plantains (*Plantago* species), moderately damp areas daisies (*Bellis perennis*) and the wetter areas buttercups (*Ranunculus* species). Occasionally, in some lawns at least, the gaps will actually be colonized by grass, but these species vary too according to the nature of the gap. Fine-leaved grasses such as bents (*Agrostis* species) and fescues (*Festuca* species) fare better than annual meadow-grass (*Poa annua*) when water and nutrients are in short supply.

Yet the resources demanded by each plant are relatively similar and many are in direct competition with their neighbours for whatever is available. This presents us with something of a paradox: why should there be such a vast variety of plant species, even within a small area, when they all require more or less the same resources from their environment? Furthermore, how do they manage to coexist without one or two species becoming dominant and excluding the rest?

David Tilman offers one possible explanation. He suggests that the amount of a resource needed by an individual of each species, relative to its availability in the environment, is critical: coexistence is possible when each species is prevented from dominating a resource completely, when the growth of each competitor is checked by a shortage of another resource.

Let us say that the outcome of a competitive battle between individuals of two plant species depends on two key resources—for example water and a nutrient. For each resource, either species would survive if its minimum requirement was met, but it cannot dominate the supply because its growth is checked by a shortage of the second factor. According to Tilman, coexistence is possible if each species is limited by a different factor, but each gets most of that which it finds most limiting. Both can then sustain population growth. Because different resources limit them, each will persist while leaving enough of the shared resource for its competitor. This might help, in part, to explain the remarkable diversity of plants in the tropical forests, on soils that are relatively infertile (Section 9.5). Indeed, there is no reason why this principle should not also extend to animal communities.

Much of the diversity of a plant community derives from the co-evolution between the plants and their associated animals. As we have already seen, the evolution of higher plants has been intimately linked with that of terrestrial animals, as pollinators, seed dispersers and herbivores. Its dependence on animal pollinators or seed dispersers may lead to a plant being only locally abundant or declining at particular times. Equally, the pressure of being grazed or browsed may well be important in preventing it becoming dominant.

Animals too can be checked by their consumers: the attentions of a predator or parasite may prevent competitive battles reaching a definitive conclusion. Overall, the abundance of any species will respond to a range of key factors, including other species, many of which are continually changing. This is one reason why communities are dynamic—variable conditions favour different species at different times. Some species are abundant because of their past successes, others are about to have their moment. We should, therefore, view the forest or the lawn as an on-going competitive battle, with no forseeable end and no winner.

4.4 Consumerism

Herbivores consume plants and carnivores eat animals. Herbivory might be regarded as a form of predation (one organism eating another) but it may not lead to the death of the plant, so here we keep a distinction to avoid confusion. We might also consider the action of parasites or even pathogens as a form of predation, but again the host may not be killed. As ever, we are faced with variations on a simple theme where the demarcations between one category and another are sometimes arbitrarily drawn.

Herbivory

All plants have to maintain some sort of balance between their leaves and their roots: the leaves are net consumers of nutrients and net producers of energy-rich sugars and the reverse is true of the roots. If this balance is not maintained then either respiration exceeds sugar production (when too many leaves are lost) or photosynthesis is inhibited by a lack of key nutrients (when roots are lost—Section 6.1). When they are eaten, plants alter their growth pattern to restore this balance, so leaves or roots are replaced as necessary.

One strategy is simply to grow quickly and produce seeds as rapidly as possible, before being consumed by herbivores. This is the strategy of ruderals, the weeds common in habitats where there is a relatively high frequency of disturbance. An example is chickweed (*Stellaria media*): rarely found in the closed community of the lawn, *Stellaria* is common in the frequently disturbed borders around it.

One alternative is to make a more substantial investment in leaves, roots and shoots, perhaps as a **biennial** (living for 2 years) or a **perennial** (living for several years), possibly with more than one reproductive event or with some provision for asexual reproduction (such as a tuber or a runner). Another is to protect the tissues from herbivores or dissuade their attack. Plant defences are many and include both physical, chemical and biological methods.

Spikes, thorns and stings can be effective against larger herbivores and unsuspecting people, but they are less effective against many invertebrates. Finer hairs, sometimes with sticky secretions can prevent smaller insects from getting close to the leaf surface. A woody trunk protects against slug and fungal attacks at the soil surface. Tough silicate deposits on the leaves of some grasses limit the range of animals prepared to tackle them. Others simply toughen the leaves and deposit distasteful chemicals, such as tannins and phenolics within them.

Plants contain some of the deadliest chemicals known (Table 4.2). The origins of these defences are still debated. Many of these chemicals, known as secondary plant metabolites, probably started out as being waste products stored in leaves, to be lost when the leaf was shed. If herbivores found such compounds unpalatable then there would be some selective advantage in concentrating them in living leaves. Deciduous oaks accumulate tannins and phenolics in their leaves during the year, gradually making them less palatable and less nutritious.

However, plants appear to allocate their chemical defences according to their growth pattern, and lower their production when no longer browsed, presumably to avoid the cost of production whenever possible. Thus, even if these chemicals started out as metabolic waste products, they must have since been refined by natural selection so that they can now be produced on demand to protect the plant. Indeed, the presence of secondary plant metabolites can indicate if a plant is currently or was recently under attack.

This would also explain the vast range of chemicals synthesized. A large variety of alkaloids, terpenes, phenolics, amongst others are responsible for a myriad of effects on herbivores. Not only does their taste seem to dissuade some herbivores they also act as antifeedents, nerve poisons, carcinogens and, in human beings at least, hallucinogens. Some of the clover and rose family release deadly hydrogen cyanide when chewed. Phenolic compounds and tannins bind to digestive enzymes, reducing protein assimilation. Animals grazing such vegetation lose condition and, once weakened, begin to lose their own competitive battles. However, while this deters many herbivores, others have hijacked these defence mechanisms for their own purpose (Box 4.3).

The co-evolution of defences between plants and their herbivores is as impressive as that between plants and their pollinators. For example, ecdysone, the hormone that initiates moulting in insects, is produced by bracken, upsetting an insect's moult cycle and maturation. Several plants produce precursors of hormones able to disrupt the reproductive cycles of the animals that feed on them. We have used this for our own purpose: the first contraceptive pill was based on naturally produced progesterone from Mexican yams. This is just one example of the many thousands of pharmacologically active compounds derived from plants. Traditional medicines based on plant extracts are used throughout the world and are still the primary form of treatment for many peoples.

Plants that make these compounds tend to be long-lived and need to reduce the cost of replacing leaves. Many come from relatively undisturbed habitats, typically competitive or stress-tolerant species, such as the plants of the Mediterranean maquis (Section 5.2). These are stress-tolerant species, adapted for surviving a long summer drought. Typically low bushes, most retain their leaves throughout the year which become very tough and leathery with age. They are also responsible for the characteristic perfume of the Mediterranean. Thyme, rosemary and lavender produce a range of secondary metabolites which dissuade most insect herbivores but which we find attractive as flavourings or scents.

Some plants are far from passive in their interactions with animals. Carnivorous plants, more properly the insectivorous plants, are typical of environments where nitrogen is in short supply—often very damp places. Yet nitrogen is abundant in the insects that visit their flowers to feed and, using various mechanisms, some plants have found ways of exploiting this nutrient source (Section 7.1).

TABLE 4.2 Secondary plant products

Class	Number of of compounds	Occurrence	Physiological effect
Nitrogen compounds			
Alkaloids	10 000	Found in many flowering plants in roots, leaves, and fruits	Toxic and unpleasant tasting
Amines	100	Found in many flowering plants principally in flowers	Unpleasant smelling, some hallucinogenic
Amino-acids (non-protein)	400	Found in seeds of many flowering plants, particularly legumes	Toxic
Cyanogenic glycosides	30	Occasional (e.g. clover, Rosaceae rose family)	Toxic (hydrogen cyanide)
Glucosinolates	75	Cruciferae (cabbage family) and 10 other plant families	Bitter tasting, acrid smell.
Terpenoids			
Monoterpenes	1000	Found in many flowering plants, principally as essential oils	Strong smelling (not unpleasant)
Sesquiterpene lactones	5000	Mainly in Compositae (daisy family) and other groups of plants	Toxic, bitter, allergenic
Diterpenoids	2000	Found in many flowering plants mainly in latex and resins	Toxic, sticky
Saponins	600	Widely found (in 70 plant families)	Haemolytic—damage blood cells
Limnoids	100	Mainly in the Rutaceae (citrus), Meliaceae (mahogany), Simaroubaceae and Quassia families	Bitter-tasting
Cucurbitacins	50	Cucurbitaceae (cucumber) family	Bitter-tasting and toxic
Cardenolides	150	Mainly in Aponcynaceae (Periwinkle), Asclepiadaceae (milkweed), and Scrophulariaceae (figwort) families	Bitter and toxic
Carotenoids	500	Widespread in fruits and flowers	Coloured pigments
Phenolics			
Simple phenols	200	Widespread in plant tissues	Antimicrobial
Flavenoids	4000	Widespread in higher and lower plants	Coloured pigments
Quinones	500	Widespread (particularly in Rhamnaceae—buckthorn family)	Coloured pigments
Other			
Polyacetylenes	650	Mainly in Compositae (daisy) and Umbelliferae (umbellifer) families	Some toxic

Predation

The most direct effect of a predator on its prey is to reduce its numbers. But if this prey species is the only food source for the predator then prey numbers will in turn govern the abundance of the predator. Who is in control, the predator or the prey?

A series of mathematical models, based on the logistic equation (Section 3.2) have been used to examine predator–prey relationships (Figure 4.11). In their simplest form, these models assume that the predator population can only grow if there is enough prey to support its reproductive effort. When prey numbers are high, the predator has abundant resources from which to produce offspring and then any increase in the predator population must mean that more prey are consumed. If the reproduction of the prey species cannot match the losses to the predator, the prey population must decline.

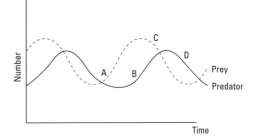

FIG. 4.11 Predator–prey relationships. Changes in the population of a predator and its prey based on mathematical models where the predator exploits one prey species only. At point A the prey can increase because of low predator numbers. Predator numbers begin to rise at (B) as their prey become more abundant. Eventually predation checks the population growth of the prey (C). The decline in the prey is followed by a decline in the predators (D) and the interaction begins to cycle again.

BOX 4.3   Objectionable behaviour

Caterpillars are soft-bodied, often fat, always slow-moving eating machines which seem to be an easy target for predators. Not surprisingly, this vulnerability has led to them evolving various ways to avoid being eaten. Some produce a mass of irritant hairs that defy all but the most persistent predator. Others swell their bodies or produce sticky strands to make them awkward to swallow, or flash false eyes to appear part of something larger and much more able to defend itself. Meanwhile, some use cryptic coloration to camouflage themselves as leaves or twigs and others use warning coloration to advertise the threat that they pose to potential predators (Table 4.3).

Bright and stark coloration signals that the owner is distasteful, can sting or will poison. Once experienced, a predator will readily associate those markings with an animal to be avoided. Both vertebrates and invertebrates use yellow and black stripes as a warning. An example are the stripes of the caterpillar of the cinnabar moth (*Tyria jacobaeae*), signalling the poisonous alkaloids it has accumulated from its diet of the ragwort (*Senecio jacobaea*). The animal gets its name from the striking red and black colouration of the adult. Similarly, the yellow–black–white stripes of the monarch caterpillar (*Danaus plexippus*) warn that it contains an alkaloid concentrated from its diet of milkweeds. The adult, with its stark orange and black wings, will be readily disgorged by any bird that tries to eat it. Its shares this wing pattern with the viceroy butterfly (*Limenitis archippus*) that is also distasteful to birds. Together the monarch and viceroy represent an example of **Mullerian mimicry**, where the common pattern means that predators learn more quickly to avoid any butterfly with these markings.

In contrast some species bear the warning colouration but avoid the costs of producing or concentrating any poison in its tissues. This is termed **Batesian mimicry**. Hoverflies that sport black and yellow livery are often mistaken for wasps even though they have no sting. Edible species copying distasteful species may save resources in the short term, but they also risk lessening the impact of the signal on a predator. If their numbers are large, and a predator has few or no encounters with the original distasteful species, an aversion to the warning colouration may not develop. Given long enough, natural selection would then favour predators not fooled by Batesian mimics.

When stripes and false eyes fail, the defensive behaviour of caterpillars can be equally colourful. Some will vomit on their attacker. As if being pelted with the half-digested content of its gut was not bad enough, what was on the menu can make matters even worse. For

BOX 4.3   Continued

TABLE 4.3 Animal defence strategies

Aggression
  Threat behaviour, intimidation displays, and overt aggression are used to frighten off potential predators or competitors.

Aposematic coloration
  Warning colours signal danger. These tend to be stark colours and patterns which are used for predators to associate with stings, poisons, or being distasteful.

Chemical defences
  This can include passive defences such as the secretion or accumulation of poisonous and distasteful compounds within or on the body, or active in the case of venomous or unpleasant bites, stings, and sprays.

Crypsis (camouflage)
  Merging into the background to avoid attention of predators. *Catalepsis* (frozen posture) adds to the effectiveness of the defence.

Masting
  A strategy which relies on strength in numbers, by producing numerous progeny which either overwhelm, confuse, or swamp out predators.

Mimicry
  A strategy in which animals mimic living and non-living things. *Batesian mimicry* involves innocuous prey impersonating either a dangerous or unpalatable organism. *Mullerian mimicry* reinforces aposematic coloration. Dangerous and distasteful organisms independently evolve similar forms of patterns and coloration.

Polymorphism
  Groups of populations within a prey species avoid being eaten by looking sufficiently different from most of their species to go unrecognized by predators.

example, one species feeding on the coca plant (*Erythroxylum coca*) will shower their assailants with a cocaine-laced vomit.

Stephen Peterson studied similar behaviour in Eastern tent caterpillars (*Malacosoma americanum*). These seek out young leaves of black cherry (*Prunus serotina*) and then carefully mark out pheromone trails leading to the leaves with the richest supply of a compound called prunasin that generates cyanide. When attacked by ants, caterpillars simply regurgitate the part-digested leaves, which release hydrogen cyanide, dissuading the ants from continuing their attack.

This strategy can have a much wider impact. In the spring of 2001, a spate of sudden and unexplained miscarriages in horses on central Kentucky stud farms were eventually attributed to a population explosion of *Malacosoma*. Terrence Fitzgerald analysed the potential concentrations of cyanide consumed, regurgitated and excreted by the caterpillars. He found they actively chose young leaves containing between two and three times more hydrogen cyanide than was present in the other older leaves. Between 63 and 85 parts per million (ppm) of cyanide could be found within their dried faeces. Whilst this may not be toxic on its own, Fitzgerald suggests that combined in quantity with leaves and young seedlings of *Prunus serotina* it could well prove harmful to vulnerable grazers and browsers.

In turn, this will mean that food eventually becomes limiting for the predator. Then, its population growth is checked, and as the prey population declines further, predator numbers too start to fall. Once again, some point will be reached when predation is no longer limiting the prey and the prey population can start to grow. Their increased numbers will eventually support growth in the predator population and so the cycle repeats itself as the two populations stay locked in their deadly waltz.

In effect we are running two population models together, where the capacity for growth in one is determined by the population size of the other, and vice versa. In these simple terms, we can see that the answer to our question is that prey and predator control each other. Because these populations begin out of step with each other, they continue to oscillate around each other, never settling down to constant population levels. If we do impose other checks (such as harvesting) on the population growth of either species, then the oscillations begin to disappear.

In fact, it is hard to show such cycles exist in nature and even more difficult to create them in the laboratory. Simple bench-scale ecosystems invariably lead to the predator eating all the prey and then itself starving to death. Adding refuges for the prey, where some can escape predation, allows for oscillations to become established but they rarely last more than a few generations before one or other of the populations goes extinct.

So, what is it about the real ecosystems that allow predator and prey to coexist? Complexity is part of the answer—adding detail to the models helps to defuse the simple relationship between predator and prey numbers.

- First, we should not assume that each predator or each prey is equivalent. Some predators are better killers than others and some prey are more likely to get eaten. We know that predators often take the weak and the infirm, including older prey that may have already ceased reproducing.

- Second, many predators do not confine themselves to one prey species. Predators are often opportunists and will take what is easily available to reduce their costs in finding and acquiring their food. Then a predator population can maintain itself when one prey species is in short supply. By the same token, a predator will compete with several other species to utilize a prey resource. Together, this complex of interactions can mean the abundance of either the predator or the prey is not closely coupled to each other but reflect changes in the larger community.

- Third, a predator does not only respond numerically to an increase in prey numbers. Each individual may consume more (termed the **functional response**) and, only later, will it produce more offspring (the **numerical response**). If the predator only shows the functional response, then prey numbers will decline without any change in predator abundance. Most often, the predator will respond both functionally and numerically, but each type of response will have delays associated with it. It may take time for the predator to increase its feeding rate (and perhaps switch its attention from an alternative prey).

- How quickly a glut of food translates into new offspring depends on the generation time of the predator, or whether they are young alive whose survival depends on the food supply. The significance of such time-lags is often seen with major insect outbreaks in temperate forests, where considerable damage may be done before birds begin to feed exclusively on the pest. The prey abundance may not translate into an increase in the adult population of the birds for several months or even a year.

- Finally, real ecosystems are not uniform in space but have patches where prey may accumulate or where they may escape predation. To reduce its search costs, a predator will stay where prey are abundant or of higher nutritional quality (Box 4.4), so that some prey go undiscovered or undevoured. Escapees are crucial if both the prey and the predator populations are to continue from one generation to another.

Taken together, these different factors mean predator and prey numbers may not be closely linked so these simple oscillations of the models are relatively rare in nature. Nevertheless, there are plenty of real-world examples where predator abundance is the best explanation of variations in prey numbers and conversely, some where predator abundance follows that of one of its prey (Box 4.4, Figure 4.12).

Although predator and prey numbers are linked we should also recognize that prey rarely sit around waiting to be picked off by a predator. An aphid feeding may seem to be oblivious to the ladybird consuming its sister a centimetre away, but neither it nor its genes have given up the battle. Like the ants, their individual fate is almost immaterial because their genetic code is likely to be passed on by one of their sisters, mothers, or daughters that avoid being consumed.

BOX 4.4   Seals wax and wane with the bears

The picture of a predator population chasing prey numbers up and down their peaks of abundance suggests a simple relationship between consumer and consumed. The reality, as you might expect, is far more complex, even when the assumption made in the simplest models—of a single predator feeding on a single prey species—is largely met. Take, for example, polar bears (*Ursus maritimus*) feeding on ringed seals (*Phoca hispida*).

Estimates of both populations in the Beaufort Sea of the Canadian Arctic in the mid-1970s and 1980s provide the close correlation we might expect from the traditional predator–prey models. Indeed, the numbers of one seem to be a reliable predictor of the other. Ian Stirling and Nils Are Øritsland used aerial surveys of seals resting on the ice and counts of bears to estimate their population sizes and suggested that the close match indicates some sort of equilibrium between the two. However, their work also showed these simple counts hide important details.

When the ice breaks up in the summer the bears cannot feed. Consequently, predation rates vary according to season: a bear will kill a seal on average every 3 days in April and May, but this falls to once every 5 days in July and probably none at all at the height of the summer in August and September. Overall, the average bear is estimated to kill 43 seals each year in the Beaufort Sea region.

But not all seals are the same. Newborn pups have little fat and represent much less energy (perhaps 40 000 kJ) than a fully grown adult (627 000 kJ). Adults may represent more calories but they are also more wary and more difficult to catch. In Barrow Strait, Hammill and Smith found that polar bears are responsible for virtually all predation attempts on ringed seals, and that 90–100% of these are directed at the ice lairs used by pups. Indeed, on occasion, the bears ignored holes used by adult male seals—presumably the costs of the chase were too high, especially when easier pickings were available elsewhere.

When the lairs become exposed or easily found, successful predation attempts are much higher—for one season on Baffin Island they were as high as 33% when a warm spring meant little snow cover. More usually the figure is 11%. Because of this, the thickness of the snow and the extent to which the pup's lair is hidden is one factor governing loss rates to the bears.

Occasionally, the bears will feed opportunistically on other species, but most of their predation attempts are directed at older and fatter pups, those that are relatively large though still naïve and easily caught. In fact, the bear's biology follows this closely—their annual cycle of fat accumulation and depletion matches the calendar of pup births and growth before weaning.

Stirling and Øritsland estimated how many seals were required to maintain a population of polar bears, with a roughly equal sex ratio of bears, given that adult male bears require around 1.5 times the energy intake of adult females. If the bears fed solely on the youngest seal pups, many more individuals would have to be killed to supply the population's energy needs—around 15 times more than if they were feeding entirely on adults. Given their preference for fattened pups, a more realistic estimate is that 4.6 times as many seals are taken compared to a diet consisting entirely of adults. In fact estimates suggest that predation rates vary according to the region of the Canadian Arctic—in some areas, almost half the unweaned pups are taken each year, but more normally the range is 14–27%.

Because it is mostly the young that are taken, these high levels of predation are sustainable since the reproductive potential of the population is not reduced by significant adult losses (Section 3.3). In some populations as much as 65% of pups could be taken each year with no overall change in adult ringed seal numbers. Perhaps the high degree of correlation between the two populations is, as Stirling and Øritsland suggest, because the polar bear populations are close to their carrying capacity, matching the maximum sustainable yield of the seal population. In this way, each population fluctuates with the fortunes of the other.

In self-defence

The stakes are not the same for the predator and the prey: for while a predator may give up on a chase and only be short of breath, the prey has much more to lose. Richard Dawkins and John Krebs call this the 'life–dinner principle': for one it is a matter of life or death; for the other it is just a matter of dinner or no dinner.

Although a predator cannot keep failing indefinitely, it need not win every chase. The prey however cannot afford to lose a single chase. Prey that escape may get to pass on their genes, as will those predators that succeed most often. This selective pressure is so direct, so intense, that a wide variety of dramatic defence and attack strategies have evolved (Box 4.3, Table 4.3).

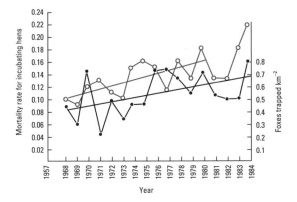

FIG. 4.12 Foxes and hens: changes with time in the number of foxes trapped in five farms in Sussex and its correspondence with the number of grey partridge hens lost to predation by foxes. The level of predation is closely linked to the number of foxes, but this study showed that the size of the prey population was primarily regulated by two predators—foxes and the crows that take their eggs. However, when the partridge population is growing, abundance also depends on their food supply, primarily insects, and its effect on chick mortality. Blue circles = mortality rate for incubating hens, primarily due to fox predation; black circles = number of foxes trapped within the study area.

The pressure for improvement is likely to be greatest for predators and prey whose populations are closely coupled, where change in one species will have a direct effect on the reproductive success of another. In this way predators and prey help to shape one another. The most dramatic cases of this are parasites that have to be highly adapted to the environment represented by their hosts.

Parasites and pathogens

Parasites have special strategies that avoid some of the risks of wiping out the species on which they depend. Most only consume a small part of their host. They can do this because they are often much smaller than their host. Classically, parasitism has been viewed as the weak attacking the strong.

The range of parasite strategies and the intricacies of their life cycles are remarkable. They range from highly specific associations—where a parasite is locked into a single host—to generalist species that exploit a range of hosts (Box 4.5). Some of the most specialized

parasites are the **parasitoids**—insects from the Hymenoptera (bees, ants, and wasps) and the Diptera (the true flies) that practice a form of parasitism bordering on predation. These lay a single egg inside the host insect, often a pupa or larva such as a caterpillar. As the larva grows, so does its parasite and all the effort of the caterpillar will ultimately benefit its passenger, in a form of controlled predation. The parasitoid then pupates inside the husk of its host, using this as protection until it can emerge as an adult.

The effects of most other parasites on their host are far less dramatic, as long as their numbers do not rise too high. In a particular form of intraspecific competition, an existing parasite may prevent other individuals becoming established in their host. The host is then protected from too severe an infection and the parasite thereby safeguards the resource it has colonized.

Unlike predators that devour their prey and quickly move on to the next victim, parasites tend to establish long-term associations whether they live on the outside (ectoparasites) or inside of their host (endoparasites). During this time they will reproduce frequently and produce eggs or larvae that attempt to migrate to another host (Box 4.5). A parasite will typically exploit the host's associations with other species to pass between hosts (Box 4.5).

While parasitoids blur the distinction between predator and parasites, some protozoan parasites blur the line between parasite and pathogen. A **pathogen** is any organism that causes a disease and, at high levels of infection, this would include many multicellular parasites. There is, of course, a massive range of bacterial and fungal pathogens that infect us, alongside the viruses that make the most direct attack on cellular life by hijacking its gene replication machinery. Pathogenic disease may lead to the death of the host, and a parasite can only afford to adopt this strategy if it can transfer readily from one host to another. This is possible in many unicellular parasites because of their short generation times and rapid population growth. Malaria is caused by a series of sporozoan parasites that attack a wide range of animals and which represents the biggest killers of humanity throughout our history (Figure 4.13, Box 4.5).

At the other extreme, some parasites have sought to reduce the cost of passing their genes onto the next generation by usurping the reproductive efforts of

BOX 4.5   The human ecosystem

Any quick review of the range of organisms that live within each of us makes it plain that we represent a valuable resource to other species. Over 100 animals are parasitic on human beings. In addition, a wide range of fungi and bacteria also make their home in or on us, with our bodies offering a variety of niches to be exploited.

Some of these associations are benign with little overall effect on our well being. Others are positively beneficial—many of the microorganisms inhabiting our gut and other tracts are important for maintaining their internal environment. Indeed, they provide some protection against invasive pathogenic species. Such associations are mutualistic, benefiting both the resident flora and ourselves. Other organisms are true parasites, exacting a cost to the host. Occasionally, an organism may change from one to the other, from a benign form to a life-threatening infection. Amoebic dysentery, for example, occurs when *Entamoeba histolytica*, a normally well-behaved resident, switches to a form that attacks the wall of the large intestine, in a transformation we do not fully understand.

The variety of animals that live within us is truly impressive, even if we confine the list to parasites alone (Table 4.4). Different species attack different parts of our bodies in different parts of the world. Which and where depends not only on the local climate, but also on the possible routes for infection. Local habits and hygiene determine our parasite burden, as do our associations with domesticated animals, soil and water.

Improvements in our understanding of these sources have enabled us to prevent infection, interrupting the parasite's life cycle, in some cases leading to their eradication. A number of viral diseases have been controlled in this way, most notably smallpox that now only exists in laboratory cultures. Poliomyelitis is facing a similar fate. Other diseases have not succumbed so easily. Malaria continues to be the greatest killer of humanity, just as it has been throughout our history, killing 2.7 million people each year and affecting a further 500 million. The disease persists despite the chemical war of insecticides waged against the mosquito vectors that carry the protozoan parasite (*Plasmodium* spp.) (Figure 4.13) or the drugs used to fight the parasite itself. Both mosquito and parasite have continued to outwit humanity by evolving and re-evolving resistance to the chemicals used against them. We are making progress, however, as the gene sequence of *Plasmodium falciparum*—the principal malarial parasite—has now been described along with a draft sequence for its mosquito vector *Anopheles gambiae*. This breakthrough comes after a six-year Anglo-American research project and brings with it the promise of a new generation of drugs and vaccines to combat the disease.

With any parasite the ability to break the cycle of reproduction and transmission is the first step in control. All parasites eventually face the problem of moving from one host to another. Most animal parasites are from ancient animal groups with relatively simple body plans. This gives them a large capacity for asexual reproduction and regeneration and means they can produce large numbers of infective stages at various phases in their life cycle. In this way they improve their chances of moving between hosts.

As an ecosystem we are relatively short-lived and have a limited carrying capacity. Besides our behavioural or technological defences, our biology also has a highly adaptive immune system the parasite has to overcome to establish itself. Many parasites have sophisticated mechanisms for 'fooling' or suppressing the immune system.

One other strategy is to infect a number of different species and reproduce in a variety of hosts sharing the same physiology. For example, trypanosomes (Sleeping sickness, Chaga's disease) and *Leishmania* (Leishmaniasis) will attack most mammals and the final or **definitive host** (where the adult develops) depends on the species of blood-sucking insect (the **vector**) that carries the parasite from one host to another. Usually the vector, or intermediate host, transfers a larval form that will only mature in the definitive host.

Many parasites have resistant stages that can survive in the outside world. Human beings are the sole definitive host for the tapeworms *Taenia solium* and *Taeniarhynchus saginatus* though the first arrives in pork and the second with beef. The beef tapeworm may produce 600 million eggs during its life, necessary because this is the resistant external stage which has to be consumed by cattle if the life cycle is to be completed.

Most internal parasites use the stages of their life cycle to adapt to the different environments they encounter as they move between hosts. Life cycles so dependent on such transfers could only evolve where close associations between hosts are maintained. Indeed, many of our parasites use domesticated animals as vectors to move between us (Table 4.4). This implies a rapid evolution on the part of the parasite, within the 2 million years of *Homo* or the 10 000 years of *Homo sapiens*, the pastoralist.

More generally, parasites may play a fundamental role in the evolution of many species. The prevalence of sexual reproduction (Section 1.3), rather than using less costly, less risky asexual reproduction, may be a response to the threat posed by parasites. Within a population, the argument goes, individuals with the most common genotype are the ones most likely to be attacked, to which most parasites will be adapted. Variation can mean a parasite is not adapted to the environment, the phenotype produced by the new genotype. Novelty has a selective advantage and sex is the prime means of generating it.

BOX 4.5 Continued

TABLE 4.4 Some animal parasites of humans[a]

Species/group	Other hosts	Vectors	Principal sites in host	Distribution
Protozoa				
Trypanosoma (sleeping sickness)	Mammals	Flies, bugs	Blood	Tropical Africa, S. America
Leishmania (leishmaniasis)	Mammals	Sandflies	Blood	Tropics
Entamoeba (amoebic dysentery)	—	Housefly helps to transmit cysts	Large intestine, various organs	Tropical and warm temperate areas
Toxoplasma	Mammals	Domestic animals	All organs especially brain	Global
Plasmodium (malaria)	Mammals	Mosquitoes	Blood/liver	Tropics, warm temperate
Flatworms				
Fasciolopsis (Large intestinal fluke)	—	Resistant external stages. Aquatic snails	Intestine	S.E. Asia
Clonorchis	Dogs, cats	Resistant external stages. Aquatic snails fish	Liver	S.E. Asia
Paragonimus (lung fluke)	Mammals	Aquatic snails Crabs, crayfish	Lung	Asia Americas
Schistosoma (bilharzia)	—	Resistant external stages Aquatic snails	Blood vessels of liver	Africa
Dyphyllobothrium	Dogs, cats	Resistant external stages. Aquatic snails, fish	Small intestine	Most of northern hemisphere
Hymenolepis	Dogs, rodents	Resistant external stages. Fleas	Small intestine	Global, warm areas
Taenia	Pig	Cow, pig	Small intestine	Global
Roundworms				
Trichnella (trichinosis)	Mammals	Other mammals used as food	Small intestine and migratory	Gobal
Ancylostoma, Necator (hookworms)	—	Resistant external stages	Migrates through different organs during development. Adult in small intestine	Tropics
Ascaris (Intestinal roundworm)	—	Resistant external stages	Small intestine; migrates through organs during development	Global
Wucheria, Loa (elephantiasis)	—	Various biting flies	Connective tissue, blood; migrates during life cycles	Tropics
Onchocerca	—	Black flies (gnats)	Connective tissue beneath skin	Africa
Dracunculus (Guinea worm)	—	Aquatic crustacea	Deep connective tissue	Topical Africa, Middle East

[a] This table gives only those animals whose association with humans is semi-permanent and for which humans are a definitive host. It excludes those who may feed on humans, but for whom the association may be intermittent. It thus omits external parasites such as the human headlouse, bedbugs, ticks, and mites, as well as fleas, leeches, and vampire bats that show less and less dependence upon humans as their main host. You may notice that this latter list gradually strays into the grey area between parasite and predator.

BOX 4.5 Continued

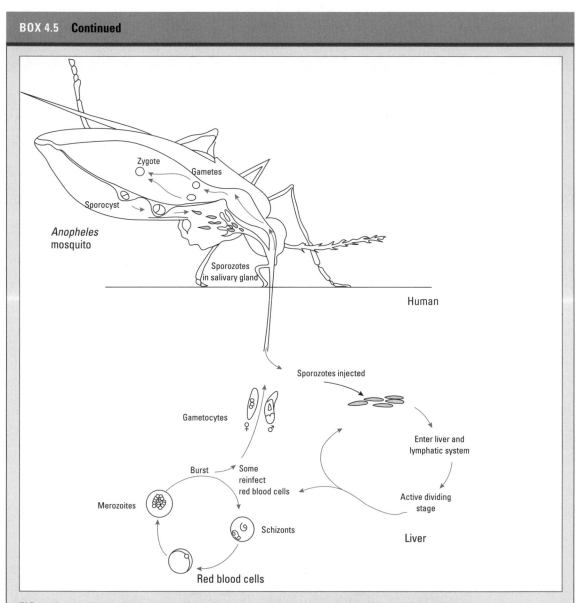

FIG. 4.13 The life cycle of *Plasmodium*, the genus of protozoan parasite that causes malaria in humans and other mammals. The timing of its life cycle depends upon the species, but the stages are the same. **Sporozoites**, injected when the mosquito takes a blood meal, enter the liver and undergo rapid division. After they burst from the host's liver cells, some re-enter liver cells but others enter red blood cells. Here they become **schizonts** and undergo further asexual multiplication. When the **merozoites** are released from these cells, some reinfect other red blood cells, and some form sexual cells. If these **gametocytes** are then taken up by a mosquito they form gametes within its gut and fertilization produces a **zygote**. This encysts in the insect gut wall, eventually releasing the sporozoites that invade the insect's salivary glands, to be injected at the next feed. The insect itself is acting as an ectoparasite by taking a blood meal (or at least the female is, since only she feeds on blood). *Plasmodium* is parasitic on both the mammal and the insect: all its sexual stages take place in the mosquito and it is entirely asexual in the mammal, so we might be regarded as vectors, with the mosquito as the definitive host.

other species. Cuckoos (*Cuculus canorum*) are **brood-parasites**, manipulating the behaviour patterns of other bird species to improve their reproductive potential. Not only are its eggs laid in another's nest and reared by foster parents, the young cuckoo also automatically practices interspecific competition by easing their chicks out of the nest. Its parent, meanwhile, freed from the demands of rearing her young, can go on to lay further eggs in other nests.

Several species of duck also have a form of intraspecific brood parasitism, where eggs are laid in the nests of neighbours. All the birds benefit from the protection of numbers, but set against this is the risk of raising the offspring of a neighbour. Since many within the group will be relatively closely related, the genes shared with close neighbours offset this cost. Some insects also practice brood-parasitism—several species of wasp have parasitic strains of queens who will lay their eggs in the nest of others. Taking this further, ants have also learnt the value of regicide. The queen of *Lasius reginae* will kill the queen of another species (*Lasius alienus*) so that its workers care for her larvae. Eventually the *L. alienus* workers die, but by that time the *L. reginae* workers have taken control of the nest.

Plants can be parasitic too and these are divided into two main types. **Holoparasites** such as broomrapes (*Orobanche*) and their relatives have abandoned photosynthesis and must, therefore, acquire all their resources from their host plant (Plate 4.2). **Hemiparasites**, on the other hand, hedge their bets and will photosynthesize, both to supplement their energy needs or keep them going in the absence of a suitable host. Some, such as yellow rattle (*Rhinanthus minor*), are able to exploit a range of species (mainly grasses), slowing down their host's growth (Figure 4.14). This helps to prevent vigorous grasses from out-competing other plants in a grassland community and for this reason yellow rattle is sometimes included in 'wildflower' seed mixes used to produce low-maintenance grassland.

The most famous parasitic plant is undoubtedly mistletoe (*Viscum album*). A hemiparasite, it taps into the living vascular tissue of trees, but photosynthesizes itself. Bad infestations can reduce the growth of a tree by almost 20 per cent. Despite this, *V. album* is not a major economic pest, though there are more

FIG. 4.14 Yellow rattle (*Rhinanthus minor*), a hemiparasite.

troublesome members of the mistletoe family (Loranthaceae). *Scurrula cordifolia* has spread from the Himalayas into northern India, where it attacks commercial orchard and forest trees. Control is difficult, because herbicides could harm the host trees as much as the mistletoe itself. However, *Scurrula* has an enemy of its own: another mistletoe, *Viscum loranthi*. In the wild, *V. loranthi* is unable to live directly on trees, but hyper-parasitizes *Scurrula*, weakening and sometimes killing it. Now introduced into plantations and orchards, *V. loranthi* has proved to be a safe and efficient means of control because it has the most desirable property of the ideal biological control agent: it only attacks the pest.

Controlling pests

Understanding the close association between species and the details of their interactions enables us to manipulate these to our own advantage. Sometimes, this means correcting our past mistakes, where our actions have led to a species escaping the checks imposed by their natural enemies and competitors. Released from these constraints, species we have introduced have frequently become pests. However, native species can also be pests—many weeds are r-selected opportunists, which grow rapidly when there are opportunities and resources available.

A species is usually designated a pest because of it effect on human health, wealth or well-being. The term pest has no biological meaning—it is merely a label that we apply to a species which causes us to incur costs. Increasingly, this includes expenditure to safeguard threatened species. As we saw earlier, some introduced aliens are so effective as competitors or predators that they have devastated the native species. Sometimes, in our attempt to reverse our mistakes, we make matters worse. The Indian mongoose, released onto several Caribbean islands to control the brown rat, has had a disastrous effect on their native bird, reptile and amphibian fauna.

There are numerous examples of introduced plant species flourishing in their new homes. Sycamore (*Acer pseudoplatanus*) and *Rhododendron ponticum* are non-native species which have slowly invaded large areas of native vegetation in Britain and their competitive staying power has made them difficult to control. Similarly, species from the Old World have established themselves in New World habitats: cheatgrass (*Bromus tectorum*) arrived with European settlers and was first reported in Pennsylvania in 1790. Within a century it had reached the west coast of North America, covering perhaps 200 000 km^2, and excluding many native grasses in the process.

However, not all introduced species become pests. Many fail to establish themselves in the new habitat, either because they are poorly adapted to the conditions or fail to out-compete the better adapted flora or fauna. Mark Williamson and Alistair Fitter

suggest that, on average, 10 per cent of introduced species will become established in their new habitat and of the successful colonists, 10 per cent will become a pest. This has become known as the 'tens' rule and although it cannot be used to predict the outcome of individual introduced species, it does indicate a regularity that seems to underlie invasions: Charles Boudouresque found that out of the 85 naturalized plant species within the Mediterranean Sea, nine had since become pests—one of which is the invasive algae *Caulerpa taxifolia* (see Box 4.6).

Pest control requires difficult choices and hard work. Often the fastest (and cheapest) way to control a pest is to use chemical methods. Insecticides, herbicides, molluscicides, and so on can rapidly bring a pest outbreak under control, but many of the chemicals used in the past were general-purpose poisons that killed non-target, as well as the target, pest species. Many have also been highly persistent—killing or impairing species long after they were applied (Box 6.4). This is one reason why several pests have developed resistance to some of these compounds. DDT resistance in malarial mosquitoes is one famous example and an impressive demonstration of natural selection producing new ecotypes within tens of years.

Nowadays, a more detailed study of the biology of pest species is undertaken to find out how a poison works and the best way of using it to control a pest. Increasingly, we are able to produce highly specific pesticides that have low persistence. While these may be part of future solutions to pest outbreaks, they will not be the final answer. Modern pesticides are expensive to develop and add considerably to the farmer's costs. There is invariably some ecological cost: even if the evolution of resistance can be avoided (and the experience from many case histories is that it cannot), there are a host of other impacts from chemical control methods—ranging from the risks to workers, to subtle shifts in species interactions.

Biological control offers one alternative approach, where another species is used to control the pest. It can amount to simply re-uniting a pest with its

BOX 4.6   Algal attack

Some visitors overstay their welcome in the Mediterranean. The exotic alga *Caulerpa taxifolia* (Figure 4.15) is such a visitor, thought to have been accidentally introduced into the sea from an exhibit in the Oceanographic Institute of Monaco in the early 1980s. In 1984 a small patch was found growing just off the coast by the Institute. It has now spread throughout the Mediterranean (Figure 4.16), covering 13 000 ha of the seabed, from southern Spain to Croatia. As it spreads over the seabed it swamps other seaweeds and because it contains a potent toxin (caulerpanyne) fish will not eat it. Alexandre Meinesz and his team are trying to find potential biological control agents for the weed.

In its native environment, *C. taxifolia* seldom exceeds 25 cm in height and will die when the sea temperature drops below 20°C. However, the plants in the Mediterranean grow two to three times larger and survive with temperatures as low as 10°C. Meinesz suggests that the Mediterranean population might be an ecotype that arose either in or outside the aquarium environment. It shows no evidence of sexual reproduction and it seems to be an exclusively male form, so Meinesz suggests that the Mediterranean infestation might be due to a single clonal population. Indeed, he speculates that it might be a new species (*Caulerpa xenogigantea*, literally 'large stranger').

Initially, control consisted of volunteers physically removing the plants. More recently there has been a search for a suitable biological control agent. In 1992, Kerry Clark of the Florida Institute of Technology reported that a number of marine molluscs feed exclusively on *Caulerpa*. Meinesz's laboratory screened a number of molluscs from the Caribbean, *C. taxifolia*'s home range, and identified *Oxynoe azurepunctata* and *Elysia subornata* as potential control agents. Both species are voracious feeders on *Caulerpa*, though *Elysia* will graze any species of the genus *Caulerpa* and therefore poses a risk to the native species of the genus. Additionally, *Elysia* is not likely maintain a viable population in the Mediterranean where conditions for most of the year are unfavourable for its feeding, growth and reproduction.

A more practical solution might involve enhancing populations of existing Mediterranean molluscs that are known to feed on *Caulerpa*—in particular *Oxynoe*

FIG. 4.15 *Caulerpa taxifolia*—alive and spreading throughout the Mediterranean Sea.

olivacea and *Lobiger serradifalci*. These are currently found at low densities but feed on all species of *Caulerpa*. It may be that *Lobiger* will be more of a hindrance than a help, as it shreds the algae as it eats it, leaving fragments that can regenerate into new plants. *O. olivacea*, however, is more thorough and consequently more promising as a control agent. Of course, there is still the risk that boosting the population of a native mollusc might, in some way, alter the food web within its ecosystem. However, using these native species is less risky than using an exotic to catch an exotic.

BOX 4.6   Continued

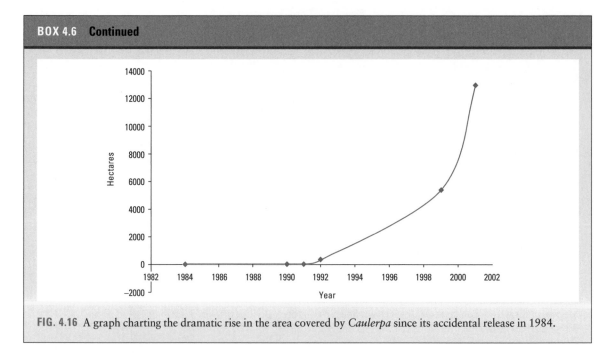

FIG. 4.16 A graph charting the dramatic rise in the area covered by *Caulerpa* since its accidental release in 1984.

natural enemy (Box 4.6), though considerable research is needed to find such a species and to ensure that it will indeed control the pest. This requires us to find the home range of the pest and identify a species which might regulate its numbers there.

Research for a new biological control agent begins with a detailed study of the ecology of the potential natural enemy and its chances of establishing itself in the new habitat. In particular, we need to know whether it will compete with the native fauna and if it can maintain a population without itself becoming a pest. Ideally, it will only consume the pest species, and that usually implies a specialist predator or parasitoid. To avoid the need for re-introduction, the natural enemy should not kill all the pests: rather, it should allow a low (though economically unimportant) population to survive. In that way both species remain at low numbers, with the natural enemy present as some safeguard against future outbreaks (Box 4.6). Thereafter, our prime concern is that the natural enemy should not affect any other species in the community nor become a nuisance or create

TABLE 4.5 Ten desirable attributes of a predator or parasitoid as an effective biological control agent

1. Rapid functional and numerical response to a rise in pest abundance
2. Able to find pest quickly (high searching efficiency)
3. Able to maintain a low pest population and sustain itself over the long term
4. Able to survive competition with native predators and parasitoids
5. High prey specificity with minimum impact on non-target species
6. Its activity and its life cycle show a close match to that of the pest
7. Easy to culture in the laboratory
8. Easy to release in the field
9. Cheap to use with rapid results to inspire confidence
10. No social nuisance

other problems. Ideally, it will be cheap and easy to rear and to release (Table 4.5).

We are not restricted only to manipulating predator–prey interactions in a biological control programme—competitive interactions can also be used. An ingenious use of intraspecific competition is used in the control of a highly destructive parasite, the screwworm (*Cochliomyia hominivorax*). The fly, native to central America, lays its eggs inside the open wounds of mammals, including humans. The larvae hatch and feed on the flesh, enlarging the wound and preventing it from healing. The stench from the wound attracts other adults, who add their eggs to the seething mass. In cases where there is a very high level of infestation the host may die.

Here the biological control technique attacks a weak point in the life cycle of the screwworm—that the female mates only once. Flies are bred and reared at a factory in Mexico where the males are separated and irradiated as their testes are forming. Large numbers of these sterile males are then shipped to outbreak areas and dispersed as pupae ready to emerge as properly formed adults. The female cannot distinguish sterile males from wild males, so swamping an area with millions of factory-bred individuals means that most copulate to no purpose. The reproductive rate plummets and the population crashes. This sterile insect technique has eradicated screwworm from the southern United States and most of Central America with the most recent success being in Nicaragua in 1999.

The same strategy is now being used in Africa to control the tetse fly (*Glossina* spp.), the vector of *Trypanosoma*—the protozoan parasite that causes sleeping sickness (trypanosomiasis) (Box 4.5, Table 4.4). Trypanosomiasis occurs in 36 countries in sub-Saharan Africa where it affects over 500 000 people and 50 million cattle. There have been many attempts to eradicate it, mostly with limited success, but one successful project in Zanzibar finally achieved eradication in 1997, 3 years after the release of around 70 000 sterile male flies each week.

Sterile insect techniques might also prove useful in the control of West Nile Fever, a form of encephalitis caused by an RNA virus. An Israeli strain of the disease recently crossed the Atlantic emerging in New York in 1999 and has since spread to 23 states in the eastern United States. The disease is spread by birds which in turn infect mosquitoes (principally *Culex pipiens pipiens*) and these can transmit it to mammalian secondary hosts, including humans.

With agricultural pests we can also check the growth of a pest by changing cultivation techniques, such as the timing of our crops. Some cotton farmers grow a 'trap crop' to catch the over-wintering boll-weevils (*Anthonomus grandis*), beetles which otherwise would decimate the main crop in the new season. These early plants are destroyed so that the main crop has a reduced chance of infestation. Another method is to grow strips of alfalfa to attract the weevils and other pests such as aphids away from the cotton.

The sequence of crops grown on an area of land can also help to avoid pest outbreaks, especially for those pests closely tied to one food plant. Removing this plant for one or more years, or indeed, part of a year, then makes it difficult for the pest to maintain a resident population.

Hedgerows and patches of wild vegetation are important for their role in harbouring generalist predators that can check pest growth before it takes off in the crop. In Britain, trials of specially planted grass banks in large fields have proved to be a useful control measure. In effect, increasing the patchiness of the habitat reduces the distance between predator and pest at important stages in their life cycle (Section 8.1). Similar methods are used to attract helpful specialist predators or parasitoids. In the vineyards of California, an introduced pest, the grape leafhopper *Erythroneura elegantula*, is controlled by using a native parasitoid *Anagrus epos*. This lays its eggs inside the leafhopper egg, but overwinters as an adult, feeding on a native leafhopper that lives on bramble. By leaving bramble around the edges of the vineyards, *Anagrus* can sustain itself over winter and be waiting to attack the pest in the new growing season. The surrounding scrub vegetation increases the parasitoid population by as much as 5 per cent and this can be further enhanced by introducing a network of corridors with and between vineyards. These not only improve the spread of the parasitoid, they also attract the pest leafhoppers away from the vines and onto the more diverse vegetation beside it (Section 8.1).

The use of biological, cultural, and chemical methods together in a control programme is termed **integrated pest management** (IPM). This uses the principle advantages of each method to reduce costs and to give effective long-term pest control.

Biological and cultural methods require a full understanding of the ecology of the pest and the natural enemy and will, ideally, give a sustained, low-cost control. Inevitably there are delays before an introduced natural enemy has an effect on pest numbers and we may then need to use chemical methods for a short period to prevent an outbreak. Chemicals may also be needed to allow a natural enemy to survive, perhaps by reducing predator or competitor numbers, or even to attract control agents to a pest outbreak. Some of these techniques have been improved by developing pesticides that are highly specific to a target species. As a result, IPM is being used as an economic and viable alternative to the simple application of general-purpose poisons.

As our knowledge of predator–prey interactions increases so does the possibility of developing a new generation of agrochemicals designed to control rather than kill. Pheromones are already used in agriculture. For example, traps baited with the scent of the female codling moth are routinely used in orchards to lure males to their death. The recent discovery and synthesis of the female screwworm pheromone makes this a possibility for its control. We might also use pheromones to confuse: Jeremy Thomas and his colleagues recently isolated a panic-inducing pheromone which ichnuemon wasps produce to confuse ants when they raid their nests. If it can be synthesized such a compound might prove a safe and effective ant-deterrent.

These interactions between individuals and between species maintain the structure of natural communities. Each individual is attempting to ensure its own reproductive success and natural selection has explored a range of strategies by which its interactions with other individuals or species promote this. The subtlety of some of these strategies should not surprise us—competition and predation are powerful selective forces and are, therefore, likely to favour the novel. Understanding their detail enables us to control pests, and to reconstruct and protect communities.

SUMMARY

Organisms interact in various ways to secure resources essential for their survival and reproductive success. These take various forms. In mutualism, individuals of two species interact to the benefit of both partners. In commensalism, an association benefits one species but with no advantage or detriment to the other.

Intraspecific competition is the struggle for resources between individuals of the same species. Interspecific competition describes the interactions between individuals of different species competing for a limited resource or which inhibit each other. In simple laboratory experiments, competition for the same resource often leads to one species excluding the other, but interactions in the real world are far more complex, and competitive battles do not lead to a clear resolution. Predation, herbivory and parasitism are interactions where one species exploits another as a resource, and in which the performance of a host is impaired or the prey is killed. In all cases, interacting species have evolved elaborate strategies for reducing the costs or for maximizing their advantage from these interactions. In simple models, the cycles of abundance of predators and their prey indicate that neither species controls the other, but again, the detail of real predator– prey relations shows how a range of factors serve to control numbers.

Some introduced species have caused the loss of native animals but there are also many cases where the native fauna have been able to exclude the invader. Species often become pests because they have escaped the checks imposed by the interactions within their natural community. Biological control uses a range of techniques to re-establish these interactions and to limit pest damage.

FURTHER READING

Krebs C. J. 2001. *Ecology: Experimental Analysis of Distribution and Abundance*, 5th edn. Benjamin Cummings, London. (A comprehensive text which provides a useful insight into species interactions with each other and the environment.)

Beeby A. N. 1993. *Applying Ecology*, Chapman & Hall, London. (This book reviews ecological aspects of pest control, including integrated pest management.)

Horn D. J. 1988. *Ecological Approach to Pest Management*. Guilford Press, New York. (A comprehensive account of biological control methods.)

WEB PAGES

The World Health Organization has a central web site that contains up to date information on human pathogens and also has links with other health related web resources:
www.who.int/en

Cornell University College of Agriculture and Life Sciences operate a useful site on pests and biological control:
www.nysaes.cornell.edu/ent/biocontrol/

The Commonwealth Agricultural Bureau International (CABI) operate a site as part of their Global Invasive Species Programme which contains useful links concerning invasive species worldwide:
www.cabi-bioscience.ch/wwwgisp/gtc2b1.htm

EXERCISES

1 Classify each of the following animal defence strategies as either: aggression, aposematic coloration, chemical defence, crypsis, masting, mimicry or polymorphism, according to the descriptions given below.

  (a) A strategy which relies on strength in numbers, by producing numerous progeny which either overwhelm, confuse, or swamp out predators.

  (b) Strategy in which animals look like other living and non-living things.

  (c) Threat behaviour, intimidation displays, and overt aggression are used to frighten off potential predators or competitors.

  (d) Groups of populations within a prey species avoid being eaten by looking sufficiently different from most of their species to go unrecognized by predators.

  (e) The secretion or accumulation of poisonous and distasteful compounds within or on the body, or active in the case of venomous or unpleasant bites, stings, and sprays.

  (f) Warning colours that signal danger, such as the black and yellow stripes of wasps

  (g) Merging into the background to avoid attention of predators.

2 Fill in the missing words using the list below. Note that some words may be used more than once and some not used at all.

  A ____ predator, able to switch to a variety of prey species, would be useful where the aim is to ____ a pest species, such as an insect, that undergoes periodic outbreaks and shows large ____ in numbers. A pest, like____, with highly seasonal population fluctuations may then be checked, at least to some extent, by a ____ that has over-wintered by feeding on other insects.

A specialist, closely tied to ____ prey species, (such as many parasitoids) may be more appropriate where a pest is ever-present, at least in____ numbers. Long-term control will mean that some pests will escape, but that should allow the ____ of both prey and predator. Alternatively, we might use a specialist in a programme where the aim is the ____of the pest in a very short space of time and when its re-introduction is not likely. In that case, we could allow the ____ to eventually go extinct (locally) as well.

aphids	fluctuations	maintain	pigeons	eradication
crop	generalist	natural enemy	predator	
coexistence	eradicate	one	prey	
control	low	outbreak	specialist	

3   Match the codes (a)–(f) shown on the Southwood habitat template below with the following series of habitats

   (i) Desert
   (ii) Grassland
   (iii) Heathland/moorland
   (iv) Temperate woodland
   (v) Tundra
   (vi) Weedy field margin

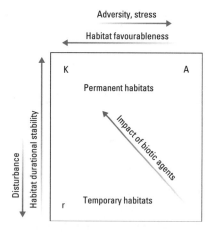

4   Look at the table of characteristics (Table 4.5) of the ideal pest control agent. For each of the ten attributes, identify, as far as possible, how far *Anagrus epos* (referred to in Section 4.5 of the text) meets each of these criteria.

Seminar/tutorial topic

5   Suggest a series of experiments you might complete to test whether a natural enemy might be safely released to control a pest in a new habitat.

6   Consider the ways in which an extended ant colony, consisting of a large number of nests spread over a large area, confers an advantage over a colony that is not so expansive.

7   What are the dangers of comparing the organization of our societies with those of the social insects? In what ways do they differ, both in terms of organization and in the role of the individual? In what ways are they similar?

5 COMMUNITIES

'You are left . . . with something rather like the skeleton of a body wasted with disease; the rich soft soil has all run away leaving the land skin and bone.'

Plato: *The Critias*

Remnant holm oak (*Quercus ilex*) amidst the garrigue that now dominates the limestone of the Corbières, in the Languedoc of southern France.

COMMUNITIES

Emerging from an aircraft after flying any distance, we always compare the temperature of our destination with that we left behind. For those of us from the colder and wetter latitudes, this is nearly always a pleasant change.

On leaving the airport, we soon realize that the weather is not the only difference. Beyond the cultural, social and architectural contrasts, those with an eye for the plant-life around them quickly spot species unlike those back home. Some growing in the wild here may only be found in greenhouses there. Others we may have never seen before.

Of course, this is no surprise to us. We expect this wonderful climate to support a different plant life. If we give it rather more thought, we would also expect the plants to show adaptations to the local heat and dryness. Indeed, we expect to see certain plants in certain climates—the almond blossom of the Mediterranean or the heather of the Scottish moors. Some plants are indicative, even evocative of their habitat, signalling the presence of whole communities of plants and animals.

This expectation underpins a fundamental question in ecology—to what extent are these collections of plants and animals predictable? Are such species assemblages inevitable for a particular climate or a particular soil? And what links them together—is it the climate and soil, or their interactions with each other—the insects that feed on the plants and the birds that feed on the insects? An ecological community may be no more than a loose collection of species, found together because they are adapted to the same environment. But perhaps it might be something more highly organized, a complex network of interactions that only allows certain combinations of species to coexist together.

We have seen evidence that different taxa can converge on similar solutions to selective pressures—we compared the marsupial thylacine with the placental dogs in Section 2.3—but can whole communities also converge to the same endpoint? If they did we should expect the same sort of community to form under identical conditions or to re-form following some major upheaval. Such a self-organizing community, repeated in time and space, would be strong evidence that only certain configurations of species, or at least, the niches they represent, can persist. This idea of a natural community, ready to reassert itself, is, perhaps, what many people assume to be part of the 'balance of nature'.

In this chapter, we look at an experiment the Earth has been running over the last million years which allows us to test these ideas. Across the globe, where

FIG. 5.1 The location of the five Mediterranean biomes around the globe, between 30° and 40° north and south of the Equator. The arrows indicate the cold oceanic currents which help to generate the drought conditions during the summer, a key feature of their climate.

the tropics give way to temperate regions (Figure 5.1), five discrete areas have the climatic conditions typical of the Mediterranean basin. Following the period of the ice ages, each region had a different pool of species from which to draw colonists. Today, their plant communities share a characteristic appearance and a similarity of form that prompts us to ask if they also share the same community structure. Here, we examine the extent to which the same patterns have been recreated in each region and whether this experiment gives any clues on how communities are constructed.

For many people, the Mediterranean climate is the closest to the ideal. The Mediterranean basin was the cradle of Western civilization and the birthplace of the European package holiday. Under its benevolent influence some of the most influential cultures have grown up, from the Nile to Jericho, Athens and Rome. Going further back, early remains of modern humans are found in the Near and Middle East predating the present climate. Human hands shaped the ecology of these lands and the fate of several major cultures is written in its soils—from the annual rise and fall of the Nile to the doomed irrigation schemes around Babylon. Plato observed the wasting of the hills of Greece 2500 years ago, though by this time, their agriculture was already failing to feed the people and the Greeks had taken to the sea, leaving Arcadia to trade and to conquer.

Today, all Mediterranean-type communities throughout the world, and especially those in the Mediterranean basin itself, are under threat, whether from tourism, agriculture or industrialization. Having changed with human cultures for hundreds and thousands of years, the current shifts in land-use may yet be the most drastic, just as the climate of paradise itself is forecast to become less benign.

5.1 Mediterranean communities

The five regions that we designate as 'Mediterranean-type', defined by their climate, began to develop around 3 million years ago. However, these conditions have not been continuous during this time: colder intervals associated with each glacial advance have meant these communities have reconstructed themselves more than once, especially in the northern hemisphere. The communities of today's Mediterranean basin are largely man-made and at most 7 500 years old.

Mediterranean climates represent transition zones between the moist temperate areas and the semi-arid regions of the sub-tropics. They are the most restricted of all climatic zones, confined to narrow bands around 30°−40° either side of the Equator (Figure 5.1), on the western edges of continents. Here, cold currents induce coastal fogs and mists so that moist air never progresses far in land—the coastal fringe may be moist and green, but beyond this there is often a desert. The prime climatic feature is an extended drought during a hot summer. Plants have to conserve water, so many cease to grow and some even shed their leaves in the summer. Winters are cool and wet, though rainfall is highly variable from one year to the next. Plant growth is largely confined to the spring and autumn when it is warm and water is available.

Not only is the climate very similar in the five regions, three of them also share similar topographies. California, Chile and the Mediterranean basin have rugged landscapes with steep-sided valleys close to the coast. This creates gradients of moisture and exposure, and pockets of different soils with contrasting plant communities forming a finely-divided mosaic of habitat patches. The Mediterranean basin is dominated by limestone, but this is much less common in the other regions and is absent from California and Chile. In South Africa and south-western Australia, the landscape is more ancient and rounded by a long history of erosion. Here, there are fewer contrasts and although all regions have relatively infertile soils, these have soils with particularly low levels of phosphate.

The vegetation of the five areas share a common physiognomy, that is, the dominant plants have similar appearances, and a similar physiology. Many are **xerophytic**, meaning they are adapted to minimize water loss through transpiration (Plate 5.1) necessary for the summer drought. Their leaves are characteristically small, tough and leathery (termed **sclerophyllous**) and are retained throughout the year. Typically, the plant community forms an open woodland dominated by short, evergreen trees; local names include maquis, chaparral and matorral (Table 5.1).

Sclerophylly could also be an adaptation to the low nutrient level of the soils. Evergreen shrubs tend to dominate in poor soils because it would be costly to re-grow leaves each year and to secure the nutrients to

TABLE 5.1 The main forms of Mediterranean-type vegetation communities

Location	Name
General type. Typically, low woodland (trees 2–5 m), evergreens with sclerophyllous leaves, beneath which is an under-storey of annual and herbaceous perennials	
Mediterranean	Maquis (France) Macchia (Italy)
California	Chaparral
Chile	Mattoral
South Africa	Renosterveld
Australia	Mallee
More arid or disturbed types. Low and open communities (trees 0.5–2.0 m or low tussock bushes), often with drought-deciduous species, thorn bushes and aromatic species	
Mediterranean	Garrigue (France) Phrygana (Greece) Batha (Israel)
California	Coastal sagebrush
Chile	Jaral
Low-nutrient soils supporting a heathland-type community. Low and open communities (between 0.2 and 1.5 m high), frequently showing a high degree of diversity and species endemism. Dominated by species of *Protea* in Africa and *Banksia* in Australia	
South Africa	Fynbos
Western Australia	Mallee heathland

build them; better instead to maintain existing leaves, even if these will use more energy than they produce at times of drought. Indeed, evergreen trees also dominate in poor montane soils and in nutrient-poor temperate areas, so leaf retention may well be an adaptation to nutrient shortages (Section 8.2). However, deciduous trees become dominant in Mediterranean ecosystems when water is available throughout the year, such as in high coastal mountains; so, water must be a determining factor. Another possibility is that the tough sclerophyllous leaves are a means of dissuading insect attack, though it is clearly worth protecting leaves from herbivores if they have to last a long time (up to 7 years in the case of kermes oak).

Where conditions are drier still, the sclerophylly becomes even more extreme. Then, a shorter plant community is found, often consisting of tussocks or low shrubs, or even tight cushions of thorns. These are called **chamaephytes** (literally 'dwarf plants'— Plate 5.2). Their thorns not only prevent grazing, but also help to create a microclimate of still air within the cushion, helping to reduce water loss. Some of these plants shed their leaves in the summer. Other adaptations, particularly seen in the Californian coastal sagebrush and the Chilean Jaral, include the use of succulent leaves and tubers to store water against the summer drought.

These shared forms are not due to the same plant species being found in each region. Each has its own particular collection of plants (and animals) which have colonized since the climate first developed and which re-established itself after each ice age. South Africa and Australia were colonized from tropical communities to the north. Most of the plant species of the Mediterranean basin came from the temperate north, though several are derived from tropical Africa and the near east. Interestingly, several important species were derived from the Cape region of South Africa, including the most emblematic plant of all, the olive (*Olea europaea*).

Given the different pools from which they have drawn their species, we should not expect the same species filling the same niches in each region. But, we might ask whether equivalent niches are found, and if so, whether these are defined by the prevailing abiotic conditions, or by the interactions between the species that assemble in each location. Certainly these communities have had long enough to organize themselves.

Following their colonization, each region enjoyed a rapid speciation as these new communities developed. Of the five, the highest diversity is found in the Cape Province of South Africa followed by south-western Australia (Figure 5.2). The fynbos is one of the most diverse plant communities in the world, with a vast range of endemic species. Both these regions have open plant communities, sometimes described as 'heaths', where soil nutrients are low and exposure is high (Table 5.1, Plate 5.3). Besides their infertile soils, these two regions have a high frequency of fire and a more predictable annual rainfall. This association of low soil nutrients, high environmental predictability and high plant diversity

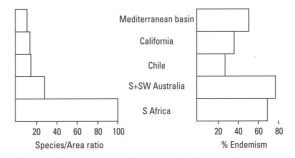

FIG. 5.2 Plant species richness in the five Mediterranean regions. All areas show a high species richness, but this is most marked for the Cape Province of South Africa and south-western Australia.

is a pattern we shall encounter again in Chapter 9, and which is indicative of considerable niche differentiation (Section 3.5).

In the other regions, a large altitudinal and latitudinal range allows for several community types within a short distance of each other, from the valley floor to the top of a hill. Their relative isolation along steep abiotic gradients has prompted local speciation and a high level of endemism. Today, all Mediterranean regions are recognized as 'biodiversity hotspots'— ecosystems with a disproportionately high number of species; for example, the Mediterranean basin has 10 per cent of known vascular plants in less than 2 per cent of the land area of the planet.

The role of fire

This high diversity follows when no single species or group of plants becomes dominant, when any competitive advantage is kept in check. Along with the summer drought, fire is a key feature of many of these habitats, creating gaps, which colonizing species can occupy.

Short-lived summer fires are commonplace in these communities, though rarely devastating. Because they shed their leaves infrequently, small amounts of litter accumulate beneath sclerophyllous shrubs and the fires are not sustained by a burning litter layer. Fire passes quickly through the canopy and is soon extinguished. Most plants, such as cork oak, laurels and olives are unharmed by the fast burn whilst others sprout rapidly from crowns protected at soil level.

Many also produce seeds that need scorching to induce germination, a useful trigger that indicates when space and nutrients are available (Figure 5.10). The Chilean matorral has species well adapted to frequent fires, with seeds able to germinate within days of the fire passing, and a self-replacing community develops. A high fire frequency in California invariably leads to a community dominated by chamise (*Adenostoma*). As the frequency of fires increases in the Mediterranean basin so its species composition also changes; in the wetter north-western corner oaks are replaced by more resistant pine species, such as the aleppo pine (*Pinus halepensis*), whilst at the highest frequencies a low aromatic scrub community or *garrigue* dominates (Table 5.1, Figure 5.3).

Fires have always occurred naturally in these regions, but ecologists recognize that human disturbance, including the deliberate setting of fires, is key to many of these plant communities. Much of the uplands bordering the Mediterranean were wooded until

hominids began to use fire. Humans discovered its value in clearing the scrub to improve both their hunting and gathering (Table 5.2) and the earliest indications of hominids using this strategy are found from northern Greece dating back 1 million years. By 10 000 years ago, it was used routinely in the eastern Mediterranean. The gaps encouraged rapid re-growth by a range of edible plants that attracted potential game for the hunter and eventually led to the early domestication of grasses for the farmer (Box 5.1).

Where fires are very frequent (Figure 5.3) the plants become very uniform and the canopy largely continuous. At lower frequencies fires help to maintain the high diversity of these areas by creating gaps for different species to invade. Without fire, competitive exclusion occurs (Section 4.3), and those species forming the densest canopy come to dominate. Grazing or some form of human disturbance can also serve to create opportunities for invasive species (Box 5.3).

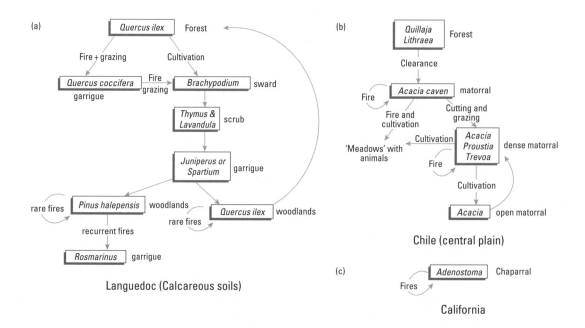

FIG. 5.3 Changes in the composition of three Mediterranean-type communities under different disturbances. In the Languedoc of Southern France and the central plain of Chile forest gives way to a series of more open communities depending on the nature of the disturbance and its frequency. The matorral is now largely confined to steep slopes where there is little grazing pressure. The abandonment of grazing in the Languedoc has allowed the native holm oak (*Q. ilex*) forest to return, especially in upland areas. The chaparral of California is actually dominated by two species—either Chamise (*Adenostoma fasciculatum*) as shown here or scrub oak (*Q. dumosa*) and each will tend to sustain itself under the influence of fire alone.

BOX 5.1   Mankind and the Mediterranean

Much of western culture has its origins in the Mediterranean basin—Jewish, Christian and Islamic traditions can trace their origins back to the lands of the middle Earth—and the ecology and geography of the region helps to explain much of its history. Indeed, Francesco Di Castri suggests that there is a Mediterranean culture that has evolved in step with the ecology of the Basin, something which may be true of other regions sharing the same climate: co-evolution features are present in a number of ecological and cultural characteristics of these regions. Certainly the plant and animal communities of the Mediterranean basin have been selected, directly and indirectly, by human activities (Table 5.2).

The current climatic conditions first appeared after the first glaciation of the Pleistocene, around 1.64 million years ago, as *Homo erectus* arrived in the area. A series of glaciations followed as did, much later, a new species of *Homo*—the Neanderthals, the first people to colonize Europe and Asia. Around 40 000 years ago these were joined in the Mediterranean by a new sub-species, modern humans. With the end of the last Ice Age, the coexistence of Neanderthals and modern humans comes to an end, and it is the older residents which lose out.

Thereafter, from about 12 000 years ago, the lands surrounding the Mediterranean Sea begin to change. This follows the arrival of widespread agriculture and the development of various tool technologies. We know that human beings had some measure of control over what was growing in different regions long before they actively cultivated the land. Wild barley may have been collected from the Nile Valley perhaps 18 000 years ago and humans were using a sequence of burning and clearing more generally to encourage useful grasses. Evidence that sheep, goats and gazelle were being husbanded exists from the Near East at the beginning of the Neolithic, around 10 000 years ago (Figure 5.4).

Cultivation proper began in the 'Fertile Crescent' of the Near East at this time and developed independently in China and Mexico at later dates. Primitive cereals were exploited for the first time in Greece and the Levant, around 8000 years ago and this encouraged a more settled way of life (Box 2.5). The division of labour this required promoted a culture and tradition which operated with the seasons of the agricultural year. It also

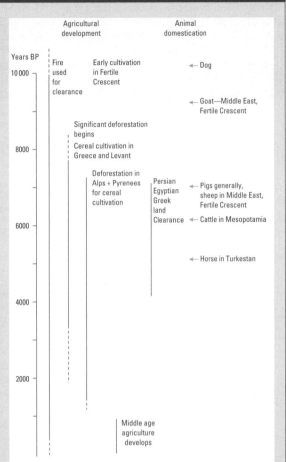

FIG. 5.4 The main agricultural developments through the history of the Mediterranean basin. Deforestation started soon after the end of the last ice age as humans used fire to clear areas, probably for hunting and gathering at this early stage. Much of the vegetational pattern seen today started with the expansion of agriculture in the Middle Ages, and this extended much of the deforestation to the western half of the basin.

led to establishment of markets and towns, where produce could be traded.

The general trend in the Mediterranean basin was for technologies to originate in the East and slowly spread west. These developed and spread

BOX 5.1   Continued

TABLE 5.2 The impact of three species of *Homo* on the ecology of the Mediterranean basin. The development of agriculture, as hunter-gathering gave way to a more settled way of life had a profound impact on Mediterranean ecosystems. Humans used fire to clear scrub and simple tools to expand food production. Use of selected grains promoted conditions that favoured the evolution of the earliest cereals. At the landscape level, human disturbance led to widespread soil erosion. It also led to established settlements and trade

Timescale (thousand years BP)	Homonid development	Technological advances	Ecological impact on the Mediterranean basin
0.500	*Homo erectus*	Lower Palaeolithic Hunter/gatherers using hand-axes	Relatively minor—creation of gaps— later widespread using fire. Gaps important for the spread of some grasses, especially *Avena sterilis* and *Hordeum spontaneum* the precursors of modern cereals. Gaps also attract game. Fire may be used to hunt
100	*Homo sapiens neanderthalensis*	Middle Palaeolithic Hunter/gatherers; flaked tools, first torch for carrying fire found in S. France	
40	*Homo sapiens sapiens*	Upper Palaeolithic Bone/antler tools; leaf blades, fire can be kindled	Fire used extensively; creating gaps allow food plants to flourish
20		Animal husbandry sheep and goats	Forest clearance on a large scale for grazing and cultivation
12	Population growth starts	Neolithic Revolution Agriculture, pottery weaving	Primitive cereals to increase rapidly found in Greece and Near East
9		Agricultural cultures in Egypt, Greece and Persia	
7		Grapes and olives cultivated	
6		Shepherding in S. France	
5		Sequence of cultures flourish in E. Mediterranean and Mesopotamia; later W. Mediterranean	Increased aridity as plant as cover is lost. Widespread soil erosion begins
1.4	Population growth slows		Forest clearance stops
0.1	Population growth confined to undeveloped nations	Industry and, recently, tourism	Desertification in nations dependant on agriculture. Woodlands used for fuel; cultivation

BOX 5.1 Continued

through trade and travel, with the sea as the highway between different centres. The human population grew considerably but the consequent expansion of agriculture took its toll. Evidence from Greece suggests that major soil erosion began about 1000 years after the onset of significant land use, a process that continued until about 600 AD. These losses were already substantial when Plato described them nearly 2500 years ago.

The history of the Mediterranean is of cultures meeting, often leading to confrontation and attempts to grab resources. These conflicts continue to this day. Because of the different traditions and languages packed into this small area, the Mediterranean is both blessed and cursed by its cultural heritage. Yet, the fruits of this treasury have been bequeathed to the rest of the world, from the grape and the olive, to mathematics, philosophy and the pizza.

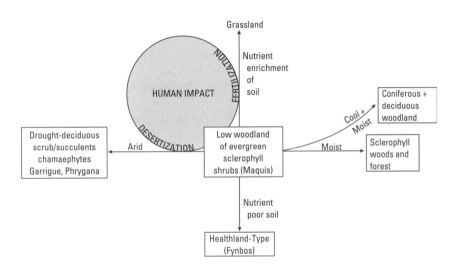

FIG. 5.5 The main factors governing the plant composition of Mediterranean-type plant communities. Human impact, primarily the loss of cover (by fire and the grazing of animals) increases aridity and the fertilization of the soil, conditions which favour grasslands, or under drier conditions, low-growing chamaephytes.

The scale of disturbance

These regions differ by their scale of human disturbance. Widespread human impact can be detected in the Mediterranean basin from around 7500 years ago, but our influence can be traced to each of the three species of *Homo* (Table 5.2). The typical garrigue-type community which dominates upland coastal regions today follows from changes in land use in the thirteenth century. Francesco Di Castri shows how these factors produce plant communities which are characteristic of each region (Figure 5.5).

About half of the land in the Mediterranean basin can be described as undisturbed, confined primarily to steep or remote uplands. The cool and moister areas support cork oak (*Quercus suber*) and sweet chestnut forests (*Castanea sativa*). On the lower hills, much of the land had been used for grazing, resulting in soil erosion and more arid conditions. As this was abandoned, maquis scrub developed, dominated by evergreen species, including kermes (*Quercus coccifera*) and holm oak (*Q. ilex*). At lower levels, other trees and shrubs become important including broom (*Cytisus*), *Genista*, olive (*Olea europaea*), strawberry tree (*Arbutus andrachne* and *A. unedo*), usually with an understorey of xerophytic grasses.

With significant disturbance, grazing and trampling by sheep and goats, or the setting of fires, grasslands may develop though more often a garrigue forms. This is a low open scrub of evergreen oaks, juniper (*Juniperus oxycedrus, J. phoenicea*) and aromatic herbs, including thyme (*Thymus mastichina*), lavender (*Lavandula pedunculata*), and rosemary (*Rosmarinus officinalis*) (Plate 5.4). The large patches of bare soil and rock separating the plants are evidence of its long history of erosion.

An equivalent community, but one with different species, has developed under similar pressures in Chile (matorral) and been subject to human disturbance. This is not so in California where one or two species form large even-aged stands. Here, the open evergreen oak community has its counterpart as chaparral with scrub oaks (*Q. dumosa*) and aromatic shrubs (principally *Salvia mellifera* and *Artemisia californica*) (Plate 5.5). However, the massive urban development in southern California, along with changes in agriculture and the prevention of the natural cycle of fires have led to a major reduction in the areas of the natural sagebrush communities (Box 8.1). In South Africa and south-western Australia human impact has not been so extensive and each region retains patches that can be broadly regarded as 'natural'.

Early agriculture had a significant impact on the vegetation of the Mediterranean basin, removing the plant cover that otherwise protected the fragile soil from the searing summer heat and drying winds. The increased grazing and trampling from their stock led to further soil loss, though upland areas were later terraced to conserve soil and to create groves of olives, figs, almonds and pomegranates. The grape, its cultivation and its fermentation, has since been exported to each of the other Mediterranean climatic regions and for each, viticulture has become increasingly important. Today, the area under the vine has expanded, the livestock reduced, and the abandoned uplands left to regenerate a cover of maquis or garrigue (Box 5.2).

In effect, humans exacerbated the summer aridity of the Mediterranean basin, thereby favouring plants that could survive long droughts, simplifying

the plant community in each location (Figure 5.5). In such communities, it is the abiotic conditions which are the prime determinants of the species present.

Animal communities

Other characteristics of Mediterranean communities result from the interactions between its species. For example, the composition of the chaparral of California is, in part, determined by seedling consumption by small mammals. Generally the seedlings of California lilac (*Ceanothus*) are preferred to those of chamise (*Adenostoma*) so with any large scale herbivory the plant community quickly becomes dominated by chamise. Where seedling consumption is reduced or indeed prevented, *Ceanothus* dominates. In the Mediterranean basin, intense grazing activity by voles can allow plants other than grasses to dominate in years of large vole populations.

Obviously, the animal community is closely tied to the collection of plants upon which it relies, directly or indirectly, but it too has to adapt to the abiotic conditions. Animals resident in the soil particularly have to survive the long summer drought, and much animal activity is timed to match the availability of food and water. Indeed, this is one reason why these regions are visited by a large number of migratory species. Much of the insect and bird life of the Mediterranean basin is shared with the rest of Europe as a result of these seasonal movements.

Their powers of dispersal and also their ancient lineage probably explain why the insects show the greatest similarities between the regions. Another ancient group, but far less mobile—the earthworms—also show remarkable similarities across the globe. Indeed, the larger community of invertebrate decomposers within the soil is very similar between the five regions (Table 5.3). Australia has the most distinct soil fauna, probably a consequence of its early isolation (Section 8.2), but all regions are notable for their lack of beetles and their high diversity of woodlice, neither of which is readily explained by the detail of their biology. Similarly, mites are more prevalent than springtails (Collembola). Not only do the soil invertebrates live to a great depth in each area, they also share the same patterns of seasonal movement up and down the soil profile.

BOX 5.2 Drying the basin

The Mediterranean basin is, over most of its area, a far from natural ecosystem, principally because of past agriculture and clearance of the hills. Most rural areas have assumed their present form since the Middle Ages, as a series of practices developed which allowed some form of sustainable production. Today, the pressure of tourism, the demands for more agricultural land or just new agricultural methods (Box 7.1) are conspiring to accelerate further clearance of the scrub, woodland and forests. Together with the threat of increasingly hot and drier summers, these changes may mean that the Basin will become less verdant and more desert-like.

But it did not start with us. As Jacques Blondel and James Aronson note, the forests have waxed and waned and, as major civilizations have waned so forests have managed to wax in some regions. Fragments and patches disappear and then re-appear at different times over the last 10 000 years, largely following agricultural developments (Figure 5.4).

The forests of the Palestine, including the extensive cedar groves for which Lebanon was famous, were exploited by the pharaohs, by the Judean Kings and even by the Phoenicians themselves. On their ships the Phoenicians built a whole trading empire, but at the cost of much of their woodlands. Much of North Africa and the Middle East were dense forest as late as 200 BC and Julius Caesar used Tunisian forests to re-build the Roman navy in 26 BC. The Iberian Peninsula lost large areas of forest as late as the fifteenth to seventeenth centuries, as Spain and Portugal in turn built their navies.

Different invaders had different impacts on the landscape, and in some areas, forests began to recover and regenerate themselves. The forests of the western Mediterranean—southern France and northern Spain, began to recover with the fall of the Roman Empire and the arrival of the Visigoths. Land clearance here did not begin again on a large scale until the Middle Ages, especially in upland areas.

Today, forests cover just 10% of the land area of the basin. A general recovery of Mediterranean forests has been under way since the 1939–45 war, at least on the Northern shores, but natural forest loss has continued at an accelerating rate in North Africa. There is now little or no natural forest in most regions and regrettably, much of the re-planting has been of pine and introduced eucalyptus (Plate 5.11).

As a general trend across the basin, much of the deciduous forest in the wetter regions has been replaced by sclerophyllous scrub, dominated by the evergreens adapted to the drier conditions. This is because deforestation brings in its wake increasing aridity. Removing trees allows rainfall to strike the soil directly, running off and failing to percolate to any depth. Higher surface velocities loosen and wash away particles, and erosion of the upper soil layers increases, especially on the steeper slopes. This gulley erosion (Plate 5.6) is a feature of much of the Eastern Mediterranean, of overgrazed upland pasture or long abandoned agricultural lands. Soil erosion rates are around a hundred times higher in areas of the basin where the trees have been lost.

This reflects the depths to which some plants send roots in search of water. These soils contain many plant storage organs, which, along with their invertebrate community, provide an extensive larder for burrowing mammals.

Rodents or their equivalents are important in all regions where soils are easy to burrow. Rabbits are thought to have originated in the western Mediterranean and were probably instrumental, along with goats and sheep, in preventing trees from dominating its drier areas. A number of equivalent grazers, filling similar functional roles, are found in the chaparral of California (brush rabbit and mule deer) and in the matorral (llama, alpaca).

Similarities

So, to what extent are the patterns of community structure repeated in these five locations? Some of the obvious comparisons are summarized in Table 5.3. For the most part, the similarities in appearance are not due to shared plant species but because their different species share adaptive features.

First, we should recognize that some regions have more in common than others. Two groupings stand out as distinct: Chile/California and South Africa/Australia (Figure 5.6). In each pairing, the similarities stem not only from a shared geological history but also shared latitudinal and altitudinal ranges (Figure 5.1). In the New World, Chile and California

TABLE 5.3 Comparison of characteristics of community structures

Ecological characteristics of Mediterranean-type ecosystems

Similarities across all the regions

A high level of plant endemism which has arisen since the Pleistocene
    Each region seems to have enjoyed rapid speciation of plants and some animals during this time

Sclerophyllous shrubland dominating large areas

The physiognomy of sclerophyllous plants seems to favour certain insect types

Counts of native bird species are similar for each region as are population densities amongst different feeding categories
    This indicates that resources and resource utilization is comparable between the regions, at least for birds

The soil invertebrate community has a relatively high endemism
    This is attributed to rapid speciation during the Pleistocene, though there are greater similarities here than in the plant community

The soil invertebrate community is formed into discrete layers in all regions and extends to a great depth with a large vertical movement

Differences across all the regions

Fire has a different impact on the species composition of each community
    The Mediterranean basin changes radically as fire frequency increases, California and Chile less so

Most groups of plants and animals are phylogenetically distinct in each region
    This is because each region drew on a different pool of species in each colonization

Plant species richness differs markedly from one region to another and some communities change relatively little following fire or disturbance

Some ecologists suggest that sclerophylly may be a response to different factors in different regions
    The diversification of sclerophyllous plants in South Africa and Australia may have been a response to poor soils rather than climate

The feeding ecology of South African birds is much less specialized than those of the other regions

Comparative ecological characteristics of the regions

Chile and California

Similarities

The plant community changes in a similar way with latitude and altitude in each region
    Both have a similar seasonal growth pattern, concentrated in the spring

The plant communities have identical carbon-gaining strategies (patterns of photosynthetic activity)
    Similar leaf forms are found in comparable habitats in each region

Both regions have been severely affected by weeds and other pests invading from the Mediterranean basin

Birds, lizards, and mammals fill similar niches in each region, but with different species derived from different stock
    Birds foraging in the vegetation for insects appear to fall into equally well-defined niches (which are also comparable to the Mediterranean basin)

Differences

Californian chaparral has higher productivity, greater litter production and is more prone to fires

Large areas of chaparral are dominated by a single shrub species, all of a similar age, and which does not develop into forest
    The low shrubs seem to need fire to promote germination

Chile is more intensively disturbed by grazing
    The matorral also has a well-developed herbaceous layer below it which is missing in California

The matorral has a more fertile soil, a legacy of its recent deforestation
    Its coniferous and deciduous forests are better developed and has greater diversity of plants

TABLE 5.3 Continued

Australia and South Africa

These woodlands are taller with a grassland under-storey
  South African woodlands have been invaded by several species from Australia

Both plant communities have a summer-dominated growth season that probably echoes the tropical origins of their floras

Both form distinct heathland communities on less fertile soils, with high levels of plant endemism

Each region has high termite activity
  They also share a large proportion of invertebrate genera. Australia has the highest soil invertebrate diversity of all regions

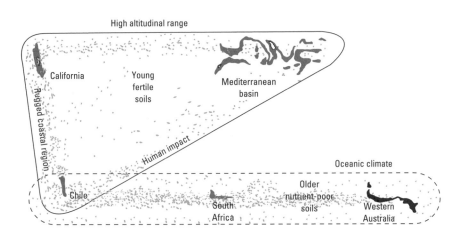

FIG. 5.6 A division of the five Mediterranean regions according to their abiotic factors. The major division is between those with oceanic climates and those with a high altitudinal range (of which Chile fits into both categories). The greatest similarities are between the communities of Chile and California, and between South Africa and SW Australia. The Mediterranean basin is somewhat different from the rest because of its large-scale human impact, its east–west extent and the prevalence of limestone rocks in this region.

both have high mountain ranges close to the coast, whereas South Africa and Australia have a small altitudinal range. The Mediterranean basin stands out as different because of its long history of human disturbance, its large longitudinal extent, and the dominance of its limestone geology.

The main plant species of both Chile and California have evolved equivalent carbon-gaining strategies: growth is confined to the spring, but the plants continue to photosynthesize at a low rate throughout the year. California chaparral has little human disturbance and will restore its dominant species readily

(Figure 5.3); the matorral is now heavily grazed and exists in its least-disturbed form only on the steeper slopes. Elsewhere in Chile, animals both trample the native plants and fertilize the soil, encouraging different species to grow. Within each region, the plant communities change with soil types: in South Africa and Australia the main distinction is between the more fertile soils, now largely used for grazing, and the poor soils that support native fynbos or heathland mallee.

Studies on the animal communities suggest that the larger groups, especially birds and lizards, have several similarities between regions. A key factor for the birds is the vertical development of the shrubs and the density of their cover (Figure 5.7). Similar numbers of bird and lizard species are found in each region and comparable population densities occur in equivalent positions in the vegetation. Feeding on seeds and insects also requires similar strategies in each region. Clearly, these parallels represent equivalent species interactions between the dominant plants and the birds, across the different regions. Martin Cody has described the close match between the insectivorous birds of the Californian chaparral and Chilean matorral. Although they contain different species, each niche seems to be occupied by a bird of equivalent size and shape, according where they feed in the canopy. The birds are said to comprise a **guild**—a group of species with similar functional roles exploiting the same resource; there is some experimental evidence that the number of species within a guild may be limited (Section 9.3).

Outside the woodlands, differences start to emerge. The fynbos and its dominant plant group, the Proteas, provide opportunities for nectar-feeding birds. Likewise, lizard diversity and abundance is higher in Chile because of the well-developed herbaceous layer absent in California. Overall, the complexity of the vegetation in all Mediterranean-type regions seems to be a good predictor of bird diversity and also the degree of niche separation of lizards (Figure 5.7).

Some have argued that this may be a misleading picture for the birds, at least for the Mediterranean basin. James Aronson suggests any mature northern temperate woodland in Europe, whether from around the Mediterranean or further north, will tend to have the same number of birds species and more or less the same species list. The basin is also a crossroads of migrating birds: not only birds from northern Europe stop there to feed on their way to tropical Africa, so do a large number of Siberian and Asiatic species, preferring not to cross the Himalayas. Such movements, back and forth, occur through much of the year, so there is a massive turnover of bird species, more so than the other regions.

A rapid speciation has been a characteristic of each region as the modern climate became established. Possibly the higher endemism and higher biodiversity in South Africa and Western Australia results from their higher frequency of fires (Figure 5.6)

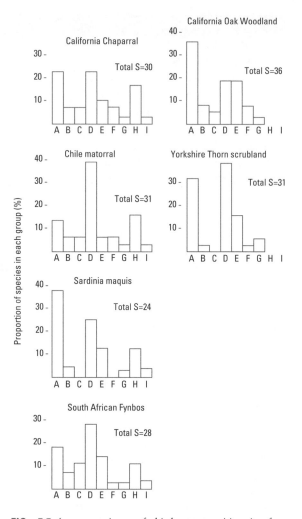

FIG. 5.7 A comparison of bird communities in four Mediterranean regions and in two very different woodland communities. In each case, the birds are grouped according to their feeding habit (A = foliage insectivores, B = sallying flycatchers, C = nectarivores, D = ground foragers, E = seed/fruit eaters, F = trunk/bark foragers, G = aerial feeders, H = raptors and scavengers, I = crespuscular insectivores). Notice that the proportion of bird species in each category shows the same sort of profile in all communities. However, the Mediterranean-type communities do have groups (H & I) which are not found in the other habitats, and there is a greater similarity in the profiles between these regions. This is some indication that the bird community has developed to a similar result in each region, with similar proportions of species in roughly equivalent niches. The most different Mediterranean community is Sardinia, which has the smallest number of species and an absence of birds in two categories, most probably because it is the one island among this group.

BOX 5.3   Building a community

Different species have different roles and demand different things from their environment. So we expect to find different niches occupied by different species. In fact, being different is a good way of finding space for yourself in a community: being adapted to an empty niche may enable a species to establish itself. However, every invader has to cope with the abiotic environment and the local regimes of heat and light, of moisture and nutrients.

These conflicting pressures to conform or to be different are resolved by natural selection, according to the dominant abiotic factors and the associations formed with other species. Based on his studies of the bird communities on islands in South-East Asia Jared Diamond suggested there were 'rules of assembly' that only allowed certain species to coexist. Only certain combinations of birds persisted together because, Diamond suggested, resource utilization and competition (niche overlap) precluded other combinations. We know that some species are excluded from the plant assemblage of the Chilean matorral if insect herbivory is intense; in California, it is the feeding preferences of small mammals which are critical. In the hills above San Diego, James Mills has shown that consumption of *Ceanothus* seedlings allows chamise (*Adenostoma*) to become dominant in regenerating chaparral.

Diamond's basic principle has since been extended. Based on his studies of desert rodents, Barry Fox proposed a 'guild assembly rule' which says that the next species to enter a community is most likely to be from a different functional group. So each category is equally filled—if there are two herbivores and two omnivores present and only one insectivore, the next successful invader is likely to be an insectivore. This 'rule' has been hotly debated by ecologists, some of whom think it too simplistic and probably only applicable to desert rats.

Another view is that collections of species will self-organize, according to circumstance, chance, and prevailing conditions. Clearly, species poorly adapted to the abiotic circumstances will not persist, but thereafter the competitive and cooperative associations of the rest, as well as their adaptability, decides community structure. Some outcomes are more likely than others but there is no fixed result. This view acknowledges that communities can be pushed into various configurations, according to the accidents of history. In this case, convergence on similar outcomes will be largely a product of the dominant abiotic factors and the selective pressures they exert.

Any ecological community is the current solution to the various interactions between its species and there are often alternative solutions, so that some communities cycle between different states.

and the higher species turnover this promotes (Section 5.3). Each region started with a different pool of potential colonizing species, but now the question is whether these have produced an equivalent set of niches—of equivalent functional roles for each regional community. If they did, we might suspect there were constraints on how these communities are organized, **rules of assembly**, rules that determine how a community can be put together (Box 5.3).

5.2   Convergence and integration

Our task is to distinguish between a tightly integrated community and a looser collection of species living together because they can survive the abiotic conditions. For plant communities, especially, Bastow Wilson notes the difficulties in distinguishing common adaptive features from well-defined rules governing how (or which) plants coexist. We do see repeating patterns and these suggest there is a limited range of community structures, but ecologists have yet to agree on how to distinguish a rule of assembly from a shared adaptation or the 'noise' of chance variation (Box 5.3).

It is certainly not chance that makes California appear similar to Chile, and the fynbos comparable to the mallee. Given the climatic regime in each locality there are only a small number of viable plant strategies to cope with the seasonal conditions and especially the summer drought. And animals that exploit these plants will also need to survive the drought and overcome plant defences that protect the long-lived leaves.

However, shared characters in the plants or animals are some way away from a tightly knit community in which niches are matched across the regions. The differences in detail are likely to create differences in niche space so the community jigsaw does not map perfectly from one site to another. A fair test of whether these communities converge on the same configuration would require us to account for all the variables of geology, frequency of fire, history of human disturbance and so on. Certainly, the closest similarities do occur between regions that share many of the abiotic features in common (Chile and California; South Africa and Australia—Table 5.3), but it seems that even minor variations can become magnified into major structural differences, producing different trajectories of community development.

Perhaps the simple truth is that no two habitats or regions are sufficiently alike in their abiotic factors. The similarity in Mediterranean communities is largely attributable to their plants reaching similar adaptive solutions to long hot summers on poor soils. Thereafter, their different ages or their different degrees of human disturbance mean they are separated today as much by their histories as their ecologies. This natural experiment comes to no conclusion, and, as ever, leaves many questions unanswered. But it does provide us with some important insights into community organization. The differences in detail matter, so that the superficial resemblance between the Mediterranean plant communities is no indication of repeating patterns of community organization. If there are shared assembly rules governing their construction we need more incisive methods to identify them.

The ideal experiment would allow communities to assemble themselves, starting with the same abiotic conditions and to draw from the same pool of species. Then, given long enough for species turnover to reach a minimum, perhaps the same collection of species, organized in the same way would result. Or, perhaps, chance events would push in them different directions.

What are communities?

We might regard communities as highly integrated if their species assemblages came in bundles—collections which cannot be unpicked. So to have species A the community must already be occupied by species B and C, or perhaps D must be absent. A tightly knit community, built according to strict rules of assembly, would be highly predictable for a climate, showing readily defined characteristics. In fact such dependences between species are known for most ecological communities, and the presence of some plants, animals or microorganisms are critical for the character of the community. Indeed, these associations can be used to classify community types (Box 5.7). Yet we also observe that communities grade into one another—grassland gives way to woodland, woodland to high forest—and communities are not entirely quantal. Thus, ecologists now ask which configurations of functional roles (niches, rather than particular species) are critical to define the nature of a community (Box 5.4).

Central to the debate is the role of species within the community: to what extent does the occurrence of one species determine the presence of others? Clearly, some species are crucial for the structure of the community and ecosystem function. **Keystone species** are those whose presence determines the nature of the community and without which the community would change to a different configuration. An example is the African elephant, whose browsing activity on the savanna maintains the open grassland and suppresses encroachment by scrub (Section 3.7). We have also seen how the loss of a keystone species, the ant *Anoplolepis custodiens*, resulted in a dramatic change in the species composition of the South African fynbos (Section 4.2). Yet, within any community there will be a number of species that can be lost with no appreciable effect on the community as a whole (Section 9.5). In such cases, the key roles are undertaken by guilds of species in which no single species is important (Box 5.3).

BOX 5.4 The right type

In studying communities we frequently find situations in which no one species is key. Instead, functional groups of several species undertake key roles. Because of this, techniques have been developed to classify species according to functional types and guilds. So, instead of being a mere species list, a community is described on the basis of the types of organisms that characterize it.

Plant functional types are central to defining communities. Functional types are based on the life-form and physiognomy of plants first described by the Danish botanist Christen Raunkiaer (Figure 5.8). Here, plants are categorized into five categories according to the position of their perennating buds—the buds that give rise to the following season's growth. Phanerophytes (literally 'visible' plants) are trees, shrubs and other woody plants with buds 25 cm or more above ground. Low growing woody plants with buds below 25 cm are classified as chamaephytes, whilst those with buds at ground level are hemicryptophytes, and cryptophytes have their buds buried beneath the soil. Finally, there are therophytes—annual species that sit out the unfavourable season as dormant embryos within seeds and do not have persistent buds.

This classification becomes useful when we begin to compare different communities. Often the profile of Raunkiaer's life-forms are characteristic for a community type. Figure 5.9 shows the life-form spectra of six Mediterranean shrub communities across three geographical regions. Notice that coastal communities have fewer phanerophytes than their inland counterparts, almost certainly a response to the stresses of living by the sea—such as exposure to salt-laden winds.

Jon Keeley and William Bond investigated the degree to which fire-adapted germination occurred in the Californian chaparral and South African fynbos by categorizing the species according to their germination strategies and life-form types (Figure 5.10). Many phanerophytes use fire as a cue to germination. There are also proportionally less cryptophytes and hemicryptophytes but when they do occur, their seeds tend not to be fire adapted.

Equivalent patterns suggest that the plants are responding to similar conditions with similar adaptive strategies. In this way, we can compare communities not on the basis of shared species but on shared adaptations, indicative of similar responses to the abiotic conditions though perhaps also to species interactions within the plant community.

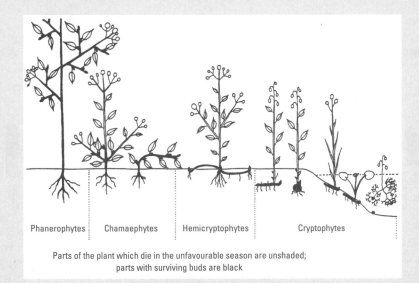

Phanerophytes | Chamaephytes | Hemicryptophytes | Cryptophytes

Parts of the plant which die in the unfavourable season are unshaded;
parts with surviving buds are black

FIG. 5.8 Raunkiaer's life-form classification where plants are grouped according to the position of the buds that survive from one year to the next. This classification is a useful way of comparing plant communities not by their species composition but by their shared adaptive forms.

BOX 5.4 **Continued**

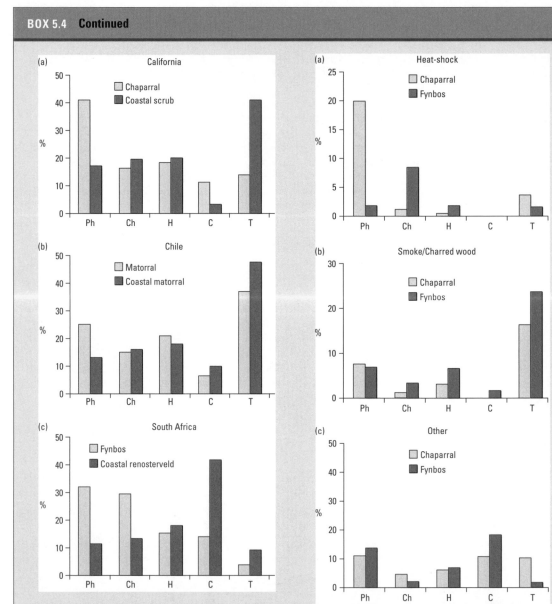

FIG. 5.9 Patterns of plant life-forms in three regions with a Mediterranean climate. Raunkiaer's classification of plant forms is shown for coastal and inland zones for each region, with the percentage of each type in the two zones. Drier inland areas tend to be dominated by phanerophytes— tall, perennial trees and shrubs whereas annuals (therophytes) tend to dominate the moister coastal zones. In South Africa cryptophytes—perennials that protect their buds by keeping them underground survive the very harsh conditions in the coastal renosterveld. (Ph—phanerophyte; Ch—chamaephyte; H—hemicryptophyte; C—cryptophyte; Th—therophyte).

FIG. 5.10 The pattern of plant life-forms in chaparral and fynbos under different germination regimes. Phanerophytes germinate especially after the heat shock of a fire (a). Therophytes instead require to be primed by the effects of smoke or an abundance of charred wood in the seed bed (b). Most other types require other cues (c). (Ph— phanerophyte; Ch—chamaephyte; H— hemicryptophyte; C—cryptophyte; Th—therophyte).

5.3   Change in communities

Human disturbance is key to the structure of many Mediterranean-type communities and we continue to play an important role in maintaining these assemblages. Yet, even in our absence, species are lost and others invade. The competitive battles and other interactions that develop within a community left to its own devices can quickly lead to a procession of species, a sequence of invasions and replacements known as a **succession**. Eventually, the turnover in species will become small and a succession ends, usually, when colonization and loss is minimal. This stable configuration is termed a **climax community** and the plant community assumes a persistent structural form.

Studying succession can tell us which species interactions are important to a community and also explain how they can change so readily (Box 5.5). There are two basic types: a **primary succession** develops on a site where there has been no previous occupation and where colonizing species must initiate ecological processes. These are generally newly exposed sites such as volcanic larva, the stony remains left behind by retreating glaciers, or the shifting sands of developing sand dunes. **Pioneer species** are those able to colonize the recently created space (Plate 5.7), even though nutrients and organic matter are lacking. By their growth and activity these resources can begin to accumulate and thereby support later colonists.

Secondary succession is a process of re-colonization following disturbance. For example, a site that has been previously occupied may retain some of its original species as a seed bank in the soil. Most importantly, the soil also contains organic matter and nutrients that can be used by invading species. Colonization is thus faster than in primary succession. Which species colonize depends on what remains of the original ecosystem, including the species already present, and also the distance invading species must travel to reach the site (Section 9.1).

What makes a good colonist? Most obviously, the capacity to disperse and to reach an unoccupied site is critical. Both plant and animal pioneers are typically r-selected opportunists, species which can produce vast amounts of rapidly dispersed seed or offspring. Many weedy annual plants and insects are typical pioneers, able to reproduce prolifically within a short generation (Section 4.3). Colonizers also need to survive a wide variety of conditions so that their reproductive success does not depend on a particular resource or on other species. For this reason, early-successional communities typically consist of loose associations of short-lived species, each able to survive a range of habitats and exploit a range of resources.

In most cases, plants have to first establish themselves if animal colonists are to survive. Here is one simple rule of assembly for most communities: plants ultimately provide the resources needed by both herbivores and carnivores (Section 6.1). Similarly, the development of the microbial and fungal communities, key components of a living soil, follows as organic matter is added by the pioneer plant community. This is known as *facilitation*—one species paving the way for others. Facilitation includes any modification of the environment and the creation of conditions amenable for later arrivals, such as providing shelter or hastening the release of nutrients from rocks. For example, in many coastal dune communities, marram grass (*Ammophila arenaria*) is an important pioneer, stabilizing the loose sand blowing landward off the beach. By its growth it adds organic material to the sand, producing a soil that other species can invade. Marram can thrive despite frequent burial and different species have various adaptations to the shifting sand (Table 5.4, Plate 5.8).

Similarly, estuarine saltmarshes are inundated at high tide, and here sand and silt are also being deposited and move at a rapid rate. Plants also have to survive periods of inundation by salt water, followed by exposure to air at low tide. An important pioneer in Mediterranean saltmarshes is perennial glasswort (*Arthrocnemum macrostachyum*) which readily colonizes the open salt pans of the marsh flats. Alfredo Rubio-Casal and his co-workers have found that once established *Arthrocnemum* provides the necessary shade and shelter for a series of successors, including annual glasswort (*Salicornia ramosissima*) (Figure 5.14). This pioneer also reduces the salinity and increases the nutrient content of the silt that enables later plants to establish themselves.

BOX 5.5 Studying succession

Ecologists investigating succession are dogged by one problem: time. Gradual, long-term change makes it difficult for one ecologist to follow the successional sequence (known as a *sere*) within a single site. One possible solution is to match successional patterns across several comparable sites but at different stages of development, and so give a composite picture of the successional sequence. In some cases, an individual site may consist of areas at different stages. An example is the hydrosere, the transition which occurs from open water to dry land, where each stage of community development can be seen over a short distance (Figure 5.11).

Shallow lakes have a relatively short ecological life. The colonizing activity of reeds, rushes and sedges and the sediment that collects around their roots means that lakes gradually fill in and shrink. Close to the water's edge, trees such as willows (*Salix* species) can tolerate the damp conditions. These are excellent facilitators since their large roots and fast transpiration rate pumps water out of the soil. Further back on drier deposits, willows give way to late-successional species, in the form of canopy trees such as oak (*Quercus* species) and ground flora that can only survive relatively dry conditions.

Sand dunes also show seral change as a linear sequence. Ecologists can thus substitute change in space for change in time to follow the succession. Going inland from the shore represents progressively older dune communities (Plate 5.8, Figure 5.12). In the Mediterranean, the stabilizing effect of marram (*Ammophila arenaria*) and the nitrogen-fixing activity of sea medick (*Medicago marina*) are the first to appear on the young dunes. These facilitate the arrival of more demanding species and a scrubby garrigue will develop on dry ridges. On the more sheltered landward side of the dune ridges, assemblages of xerophytes such as species of *Cistus* and lavender (*Lavandula stoechas*) form. On nutrient-poor sites the result is a scrub dominated by juniper (*Juniperus phoenicea*) and mastic tree (*Pistacia lentiscus*). With richer soils and less disturbance, trees colonize the oldest dunes and woodland of oak (*Quercus* species) and pine (*Pinus* species) develops (Plate 5.10). An equivalent pattern of development occurs in other regions with a Mediterranean climate. In California, a healthy sagebrush (*Artemesia californica*) or pine forest may form, whilst in Australia the soil conditions determine whether a mallee heath or *Eucalyptus* forest will develop.

Very few sites have been studied for long enough to give a complete picture of community change. One frequently used method is to collect fixed-point photographs taken from the same position over many years, to

FIG. 5.11 The hydrosere: an area of open water gradually being encroached upon by vegetation.

BOX 5.5   Continued

record the advances and retreats of the major species. Ecologists have also used archaeological techniques to uncover the history of sites and to reconstruct the development of its community over the longer term. Some sites reveal their past in vertical cores taken from their soil, especially in aquatic or well-layered sediments that can be readily sequenced. The pollen found in layers of a known age can be particularly instructive (Figure 5.13).

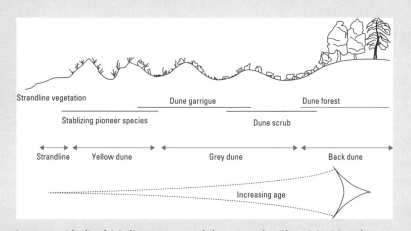

FIG. 5.12 Succession on an idealized Mediterranean sand dune (see also Plate 5.8). New dunes are colonized at the seaward end, whilst woodland and forest occupy the oldest part of the site, consisting of late successional species that have replaced the earlier colonizers. Note that the changes occur as transitions rather than abrupt zonations, with overlap between the various species assemblages within the succession. The overall trend is one of increasing abundance of woody species over time and this is reflected in the transition from pioneer communities, through garrigue and scrub into woodland.

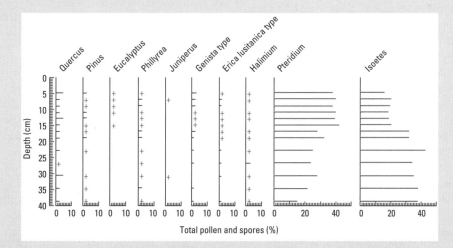
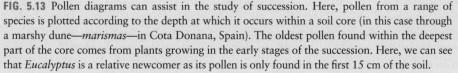

FIG. 5.13 Pollen diagrams can assist in the study of succession. Here, pollen from a range of species is plotted according to the depth at which it occurs within a soil core (in this case through a marshy dune—*marismas*—in Cota Donana, Spain). The oldest pollen found within the deepest part of the core comes from plants growing in the early stages of the succession. Here, we can see that *Eucalyptus* is a relative newcomer as its pollen is only found in the first 15 cm of the soil.

TABLE 5.4 Burial responses in two species of *Ammophila*

Species	Response to burial by sand
Ammophila arenaria	Increased internode length
	Increased tiller production
	Increased adventitious rooting
Ammophila breviligulata	Increased vertical growth of tillers
	Increased vertical growth of rhizomes
	Increased net photosynthesis
	Increased leaf thickness
	Increased above- and below-ground biomass
	Increased leaf area
	Increased tiller production
	Increased total chlorophyll concentration
	Increased shoot emergence time
	Decreased shoot density

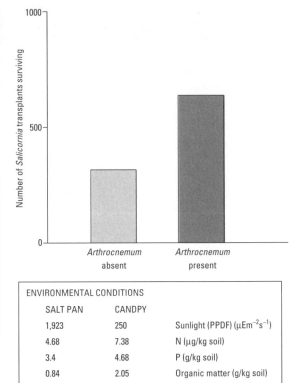

ENVIRONMENTAL CONDITIONS		
SALT PAN	CANDPY	
1,923	250	Sunlight (PPDF) ($\mu Em^{-2}s^{-1}$)
4.68	7.38	N (μg/kg soil)
3.4	4.68	P (g/kg soil)
0.84	2.05	Organic matter (g/kg soil)

FIG. 5.14 The role of the perennial glasswort (*A. macrostachyum*) as a facilitator within Mediterranean salt pans. The graph shows survival of the annual glasswort (*S. ramosissima*) with and without the presence of *A. macrostachyum*. Environmental data for conditions in the open salt pan and under the canopy of *A. macrostachyum* indicate that life under the canopy is less stressful than life on the open marsh.

Facilitation is most obvious in the early stages of a succession, but not exclusively so. Later in the sequence, well established, perhaps keystone species, create conditions or provide resources which much of the rest of the community depend upon. The long-lived oaks of the maquis or chaparral play this role in several Mediterranean communities throughout the world, not least in the bird life they support (Figure 5.7).

Early-successional facilitators often sow the seeds of their own destruction and are rarely a part of late-successional communities. Species which arrive later are typically more competitive and these squeeze out the pioneers. Plants such as *Arthrocnemum* give way to species that become increasingly dominant as the community develops. These late arrivals invest for the long term, and have strategies to out-compete their neighbours for space and resources, often inhibiting their growth or establishment. This **inhibition** slows down colonization by other species or excludes them altogether. An example found on Mediterranean coastal dunes is actually due to a recent invader, *Carpobrotus edulis* (Plate 5.9). Having been introduced from South Africa, this plant has, within a few centuries, become dominant on more sheltered early-successional dunes. Here, it forms a thick mat that inhibits native colonizing species. *Carpobrotus* will also stabilize the sands of young dunes but it cannot delay the colonization of late successional plants indefinitely. Eventually, its hold is loosened by more competitive invaders.

Again, we could see this as a general rule of assembly—late-successional communities, without significant disturbance, become dominated by slow-growing competitive species. A succession is thus a process by which the species composition of a community becomes increasingly fixed. Towards the end of a succession species turnover—arrivals and departures—slows and the community composition is largely governed by the interactions between its species. Where community development is primarily driven by these internal processes it is termed an **autogenic** succession. If, however, some external, abiotic factor determines the outcome, it is described as an **allogenic** succession. Ordinarily, the sequence of plants on a sand dune is principally an autogenic succession (Box 5.5), but if there is periodic water-logging of a dune slack (the hollows formed behind the main ridge) a very different plant community becomes established. The plant and animal species found here are adapted to the wetter conditions and the community would represent the result of an allogenic succession.

Changes in the species assemblage are matched by changes in ecological processes. As nutrients and organic matter begin to accumulate, so a reserve of resources and a decomposer community can develop (Box 6.2). As before, this facilitates the arrival of other species and nutrients cycle more tightly within the community. Towards the end of a succession, nutrients released by the decomposers are quickly utilized by competing plants so that little is lost from the ecosystem (Section 6.2). We thus expect ecological efficiencies to improve from early- to late-successional communities.

Competition for these resources is often so great in late-successional communities that plants have to be able to survive shortages of nutrients, light, or even water. Plants that survive these conditions are said to show **tolerance** and are typically slow-growing and infrequent reproducers (Section 4.3). Theirs is a waiting game, maintaining themselves in times of shortage, ready to exploit times of plenty. As others fall away, tolerators come to dominate. This again may be a rule of assembly, favouring such strategies in habitats where nutrient supply is highly restricted.

Disturbance and succession

Any disruption that causes the loss of most or all of the resident species will re-set the community clock and these processes will start ticking again. The rate of change can be dramatic: volcanic ash and lava can wipe out a community in a matter of hours, while the advance of an ice sheet is slower but equally pervasive. In both cases, some species may escape when most others are lost.

The likelihood of any species being lost will depend on the availability of refuges. A gradual environmental change can be tracked by the colonizing species, so that pioneers are continually advancing into new territory followed by late-successional species as a community becomes established (Box 5.5).

Communities are often adapted to a particular frequency of disturbance and are said to **incorporate** this disturbance (Section 8.1). This is particularly true of Mediterranean-type communities that are dominated by plants able to withstand short fires. Indeed, a certain frequency of fire is necessary to maintain their community structure (Section 5.1). Most of these communities never reach a stable configuration because fire initiates a cycle of change. This postponement of a climax community is termed an **arrested succession** or a **plagioclimax**. Thus, a garrigue represents a plagioclimax of maquis, with the woodland scrub arrested by a high fire frequency (Figure 5.4).

The structure and diversity of the whole community may depend on the frequency of such disturbances. In his **intermediate disturbance hypothesis**, Joseph Connell suggests that a community supports its greatest number of species when the disturbance is relatively frequent—not too frequent to cause major extinctions but frequent enough to prevent the competitive dominants squeezing out other species. Disturbance, in whatever form it takes, creates gaps which invasive species may occupy. Competitive species cannot then dominate and some plants, known as **fugitive species**, are able to maintain a population just by colonizing one gap after another. The perennial glasswort of the saltmarshes is an example, surviving by colonizing in the open pans where new silt is still being deposited or where the existing vegetation has been scoured away by a change in the tidal flow.

A community that is largely the result of an autogenic succession is likely to be very elastic, always returning to a similar species configuration after a disturbance. A stable configuration is the product of its species interactions and these will tend to lead towards a particular endpoint. We get some sense of this predictability by looking along the line of a beach and seeing, at each point, the same vegetational zones in the dunes, at equivalent distances from the shoreline. This is some indication that the interactions amongst its species govern the outcome, especially when this leads to a uniform community in the undisturbed areas well away from the beach (Plate 5.10). Where there is less predictability and more frequent disturbance the early stages of most communities are the least stable.

Disturbance resulting from herbivory can also affect the eventual outcome of a succession. Herbivores that feed on early-successional species can delay the successional process and slow colonization by other species. If, on the other hand, the plant being grazed inhibits succession, the effect of herbivore may be to allow other species to colonize. Later in the sequence, herbivores have the power to either halt or even reverse a succession. Catherine Bach showed this in her investigation of the role of the flea beetle (*Altica subplicata*) in dune succession on the shores of the American Great Lakes. *Altica* feeds on the dune willow (*Salix cordata*). Bach's

long-term experiments compared dune development with and without the activity of *Altica*. Not surprisingly, the density of the willow declined in the presence of the beetle, with a corresponding increase in the density of herbaceous plants. Herbivory by *Altica* seems to facilitate colonization of the dune by plants otherwise inhibited by the willow. This three-way interaction shows how both facilitation and inhibition can result from the activity of a single species.

Few ecologists today would argue that succession can only have one outcome—a particular climax community for a particular climate. Most now recognize that chance and history and different starting positions lead to different results. We can, however, see recognizable and repeating patterns that allow us to identify broad community types, invariably shaped, if not defined, by the interactions between their species.

5.4 Communities, change, and conservation

Clearly, some communities have been displaced a long way from their natural state. Today most Mediterranean-type communities are plagioclimaxes that have incorporated varying frequencies of disturbance within their limited cycle of change.

The pressure of our numbers and activity has accelerated the pace of disturbance, hastening the loss of plant cover and prompting widespread soil erosion and degradation in these regions. Just as Plato described habitat degradation occurring in Ancient Greece, our changes to the Mediterranean basin today (Box 5.6) may mean we too will witness paradise lost. Despite our long history together, the pressures on the landscape, from tourism, agriculture and industry may mean that the scale of human disturbance is just too great for our partnership with this landscape to continue in this form (Section 6.6, Box 7.1). Along with the threat of increased aridity (Section 8.3), some are predicting a major wave of extinctions in the near future, as many of the endemic plants of these regions are lost (Section 9.5, Box 5.2).

Human disturbance has taken a different form in recent years. The impact of tourism and its associated development has resulted in the loss or damage of 75 per cent of Mediterranean sand dunes in the last 30 years. The very act of visiting a place involves the use of its resources, adds effluents, and demands facilities which, taken altogether, can add up to a potent form of disturbance. Visitor pressure, simply walking on the dune, can damage its fragile and early stages of its colonization by marram. Recent research has shown that a mere 500 pedestrian passes on a dune can have a significant, albeit temporary, effect. More intensive visitor pressure can result in permanent change in the vegetation.

Through our travels we have connected the five regions to each other, shuffling species between them. Not only do we now cultivate the vine in all Mediterranean communities, we have also introduced some less desirable species into each region. As we saw earlier, *Carpobrotus* from South Africa has become an invasive weed in the Mediterranean Basin and California. Another introduced species displacing

BOX 5.6   The North–South divide

Jacques Blondel and James Aronson highlight another way in which the Mediterrenean basin represents a microcosm of the larger world—the clear demarcation of wealth, population growth and economic development between its northern and southern halves . . . and the impact these have on its ecology.

Of the northern shore they say:

So a gloomy dichotomy emerges: far from the coast there are deserted fields, orchards and pastures, progressively encroached upon by shrublands, and increasingly dense, unproductive and ill managed woodlands. Along the coasts, in the densely urbanized and homogenized industrial zones that continue their inexorable sprawl, all ecological and cultural contact with the Mediterranean past is abandoned and lost.

And of the south:

A visit to any mountainous areas of North Africa today reveals that demographic pressure, combined with a highly conservative rural economy still largely disconnected from outside markets, results in very low crop yields and overall productivity. This leads to the all-too-familiar cycle of increasing ploughing and grazing areas followed by soil erosion, and then new clearing elsewhere.

On its northern fringe, much agricultural land has been abandoned and a scrubby woodland developed. Without the moderate grazing pressure from large mammals constraining the more competitive species, the land loses much of its floral and faunal diversity under a scrub of highly competitive sclerophyllous plants. In North Africa, by contrast, the area under cultivation has expanded with the rapid growth of its population. The attempt to feed the people using poor and unproductive soils has meant greater encroachment into the upland

forest areas, leaving the natural woodlands confined to remote and inaccessible pockets.

Today the resident population of the four major states on the Northern shore is not growing, whereas the African countries have an annual growth rate of 2 per cent, amongst the highest in the world. Typically these people have one-sixth of the income of those on the European side. In the starkest of contrasts, the basin is the most popular tourist destination in the world, receiving somewhere between 200 and 250 million visits each year, most of whom stay on the coast.

The pressure of numbers, either from the resident population on the southern shore or the migratory population that comes to rest on the northern shore, place major demands on the land. On its northern rim, agriculture, other than viticulture, has been considerably reduced so that most of the economic activity is concentrated in the coastal regions or larger cities—in some places (e.g. Mallorca) half of the coastal fringe has been lost under tourist development. In contrast, the expansion of agriculture on the southern rim of the basin has caused widespread environmental degradation, especially soil erosion.

The two sides do share one thing in common—an increasing shortage of water—a problem they share with the other Mediterranean climatic regions. Water shortage (and waste water treatment) is now a major issue in Southern California and Mexico, where respecting the rights of neighbouring states in both water extraction and purification are critical issues. Water rights are also centre stage in the political machinations of the Middle East. Each region has their own peculiar ecological problems which are likely to get worse if the predicted reductions in rainfall follow with the effects of global warming (Section 8.3).

---

native species is *Eucalyptus* from Australia (Plate 5.11). Adapted to the nutrient-poor soils of the mallee, its fast growing growth habit makes it a formidable competitor as it scavenges nutrients from richer Mediterranean soils. The result is an impoverished ground flora within *Eucalyptus* plantations, whilst the tree itself is unpalatable to the resident invertebrate community. Its high content of essential oils both sours soil and increases the occurrence of fires. As a consequence, communities change both from an increased frequency of disturbance, and from changed species interactions.

We are at least getting better at following these changes, if only because we have a better understanding of how species combine to form common assemblages. Using the power of computers, massive databases have been analysed to identify plant species likely to occur together. The modern study of phytosociology (Box 5.7) uses these programs to create a classification of assemblages. The National Vegetation Classification Scheme (NVC) in Britain is used to match a community, electronically, with the database and to decide its possible origins, and past,

BOX 5.7 Classifying communities

Using plant species to classify plant communities has its origins in the eighteenth Century. At that time scientists first documented vegetation patterns and began to recognize this as evidence of interactions between the plants themselves and with the larger environment. In 1825, Dureau de la Malle carried out a series of felling and regeneration experiments in his woodlands in Normandy and concluded that plant species live socially and that this was . . . "a condition essential to their conservation and to their development".

In the early years of the twentieth Century, the study of plant associations, **phytosociology**, was a peculiarly European pursuit, pioneered by research groups working in Zurich in Switzerland and Montpellier in France and a rival, but nonetheless related, Scandinavian system in Upsalla. Whereas the Zurich–Montpellier approach was descriptive, its rival was largely numerical, though both classified communities according to predictable species associations. More recent classification schemes include Britain's National Vegetation Classification scheme (NVC). Developed by a team led by John Rodwell and Andrew Malloch, this classifies

Britain's vegetation into 300 distinct communities with around 750 sub-communities.

The NVC serves to standardize terminology and survey techniques, allowing comparison between sites at both local and national levels. Each community is given an alpha-numeric code, such as W for woodland and H for heathland. Certain categories are identified by two letters such as CG for calcicolous grassland (grasslands of chalk and limestone) and MG for mesotrophic grasslands (grasslands of neutral soils). The communities are numbered and any sub-community is distinguished by a suffix. The codes can be used on maps or within diagrams to illustrate the nature and extent of a community (Figure 5.15).

Sites can also be included in **geographical information systems** (GIS) which use aerial and satellite imaging to map large areas with a computer. Community descriptions can also detail threats and community transitions which predict future successional trends, valuable in conservation management. Similarly, it can provide information on a site's past communities, particularly useful in projects that aim to restore lost or damaged habitats.

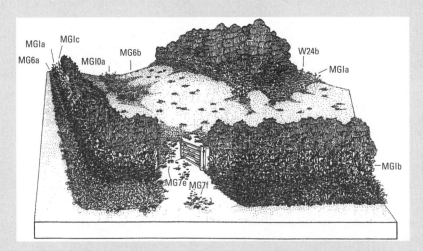

FIG. 5.15 An illustrated map of a mesotrophic grassland and associated communities. Here, the basic community MG1 (dominated by the grass *Arrhenatherum elatius*) occurs as three sub-communities. One of these (sub-community (a) is characterized by red fescue (*Festuca rubra*), and the second (MG1b) is a rougher sub-community containing stinging nettle (*Urtica dioica*). A third sub-community, MG1c is associated with the wetter conditions along the ditch and is typified by meadowsweet (*Filipendula ulmaria*). The diagram shows four other community types, including a scrub community (W24), classified along with the woodlands. A visual representation such as this is useful in showing how communities occur together and grade into one another within the landscape.

present and future management. The scheme is also used in the restoration of lost or damaged sites where it provides information on the characteristic species of an area.

Similar techniques were used by Janet Franklin in a recent analysis of Californian chaparral. Aware that conservation managers needed more detailed maps, she developed a system that could predict species occurrence. The existing vegetation classification (CALVEG) grouped the region's vegetation into 17 categories. In order to predict the species found at particular locations Franklin analysed data from 900 sample plots and incorporated the species assemblages into a database. A computer model was used to predict the occurrence of eight principal chaparral species (such as *Adenostoma* and *Arctostaphylos*).

Understanding communities in terms of their species interactions, and perhaps even generating universal rules of assembly, are invaluable in our efforts to conserve and re-create them. Sometimes, with succession, it may be enough to create the conditions and then stand back to allow the species to assemble themselves; in other cases, we may need to disturb them on a regular basis to create the desired community. As we seek to protect Mediterranean-type communities we will inevitably learn more of how they are put together. Perhaps, for the time being, it is enough to recognize that the concept of the community is a functional unit, rather like the species concept (Section 2.2), more of a device to help ecologists than an ecological reality. Nature, once again, does not come in neat parcels.

SUMMARY

A species assemblage or community represents an association of species adapted to the abiotic conditions of a region, especially climate, but which also reflects their interactions with each other. The Mediterranean-type community is found in five regions of the world sharing a climate very similar to that of the Mediterranean basin. Some regions also share similar topographies and geologies, but they differ in the species pools from which they have been colonized since the ice ages. Disturbance, in particular fire, plays an important role in these communities and, as a result, many sclerophyllous plants, typical of Mediterranean habitats, are fire-adapted.

Chile and California have similar plant communities, most probably because of their coastal mountain ranges and high latitudinal extent. A second group, South Africa and south-western Australia, have flatter landscapes and more ancient, more-impoverished soils. The Mediterranean basin itself has a wider range of habitats, because of its greater area and dominant geology. All show similar plant physiognomies and some equivalence in their animal communities, with high species endemism in all five regions. Even so, the details of their differences confounds attempts to find common features by which these communities are organized. Whilst there are indications, especially amongst the bird community, that some patterns are repeated in each habitat, well-defined assembly rules elude ecologists.

Communities change by a succession in which species colonize and go locally extinct. In primary succession, a community starts without previous occupiers and organic matter. Secondary succession occurs in disturbed areas re-colonizing from the remains of the previous community. Autogenic succession is where the community composition is largely determined by species interactions; allogenic succession is the result of factors from outside the community.

Species interactions through processes such as facilitation, inhibition, and tolerance mean the successional sequence generally moves from rapidly dispersed and quick-reproducing pioneer species to longer-lived competitive species. When species turnover is at a minimum the community is said to be at climax. Disturbance, and its frequency, can also determine the development of the community in an allogenic succession.

Despite having developed alongside humans for much of their history, the Mediterranean regions are today threatened by excessive disruption from tourism, increasing industrialization and changes in land management practices. We have also introduced non-native species that can shift community structure in some areas, threatening endemic species. Conservation depends on an ability to recognize and describe communities, along with an understanding of processes which control them.

FURTHER READING

Archibold, O. W. 1995. *Ecology of World Vegetation*. Chapman & Hall, London. (A comprehensive account of the major biomes.)

Blondel, J. and Aronson, J. 1999. *Biology and Wildlife of the Mediterranean Region*. Oxford University Press, Oxford. (An accessible review of the ecology and environmental issues of the Mediterranean Basin.)

Polunin, O. and Smythies, B. E. 1973. *Flowers of South-West Europe. A Field Guide*. Oxford University Press, Oxford. (A flora of the region, with detailed descriptions of its major ecosystems.)

WEB PAGES

A good general-purpose directory which includes links to sources on succession and other ecological principles covered here is
www.biologybrowser.org

EXERCISES

1 State three abiotic factors that govern the distribution of Mediterranean-type ecosystems.

2 Select the best answer to complete the following statement.

Sclerophylly has been suggested to be an adaptation to

(a) long summer drought,

(b) intense herbivory,

(c) nutrient-poor soils,

(d) all three.

3. Arrange the following successional stages of an idealized Mediterranean sand dune in their correct chronological sequence (youngest to oldest).

Stage	Characteristic vegetation
(a) Dune scrub	Shrubs such as juniper (*Juniperus*) and mastic (*Pistacia*)
(b) Dune woodland	Trees such as oak (*Quercus*) and pine (*Pinus*)

| (c) Yellow dune | Herbaceous plants such as marram (*Ammophila*) and sea medick (*Medicago marina*) |
| (d) Dune garrigue | Sclerophyllous, chamaephytes such as *Lavandula* |

4 Complete the following table and then decide which pair of regions you would expect to have the most similar communities on the basis of their shared abiotic features:

	Region				
	California	Chile	Mediterranean basin	South Africa	SW Australia
Soil type (fertile or infertile)					
Frequency of fire (high or low)					
Topography (rugged or low)					
Climate type (oceanic or continental)					
Scale of human disturbance (large or small)					

5 Select the best answer to complete the following statement.

In autogenic succession

(a) one species blocks colonization of another;

(b) the successional process is driven by interactions within the community;

(c) one species facilitates colonization by another;

(d) the successional process is driven by interactions outside the community.

6 Insert the appropriate words to complete the paragraph below (use the list below; some words may be used more than once and some not at all).

Opportunist ____, species are good colonizers where conditions such as soil and climate are ____. These species tend to be ____. Colonists might either ____ other species colonizing the site or they may inhibit them. Some species take a long time to become ____ even though they may have been present during the initial colonization process—these are known as ____ and tend to be ____. Colonizers of ____ environments (such as arctic alpine and ____) may share some of the tolerators' features and need to be slow-growing, ____ -persistors in order to survive.

adversity	hostile	r-selected
established	K-selected	short-lived
facilitate	long-lived	tolerators
favourable	late-successional	tundra

7 Explain why a plant like marram (*Ammophila arenaria*) is such a good pioneer species of sand dune ecosystems.

8 List and describe the three models that operate in successional processes.

9 Match the following plant life-forms with their correct attributes (listed below as (i)–(v)).

(a) Chamaephyte

(b) Cryptophyte

(c) Hemicryptophyte

(d) Phanaerophyte

(e) Therophyte

(i)   Has perennating buds 25 cm or more above the soil surface.

(ii) Has perennating buds within 25 cm of the soil surface.

(iii) Is an annual lacking perennating buds, surviving the unfavourable season as seed.

(iv) Has underground perennating buds.

(v)   Has perennating buds at the soil surface.

Tutorial/seminar topics

10 What are the ecological consequences of the large seasonal change in population size, associated with tourism? What features of the ecology of the Mediterranean basin make it particularly susceptible to these pressures?

11 Discuss the ways in which you might manipulate a dune ecosystem to examine whether its succession was primarily autogenic or allogenic.

12 Consider the possible reasons why the combination of a highly predictable abiotic environment with low nutrient levels might lead to greater niche differentiation.

6 SYSTEMS

'All flesh is grass'.

Isaiah 40:6

Sheep grazing maquis in Portugal.

SYSTEMS

On this point we have to differ. For although Isaiah is correct in the sentiment, he is wrong in the detail. Certainly, most animals on the planet ultimately depend on the photosynthetic plants for their nourishment, even if they are not actually grasses. However, it seems there was a time when there were no plants but there were consumers, simple single-celled animals that survived by scavenging organic molecules or eating other cells. At its outset, life on the planet depended on energy captured from chemical reactions.

Indeed, we can still find communities that depend on such energy today. At the bottom of deep and dark oceanic trenches, where volcanic vents supply both heat and reduced sulfur compounds, entire ecosystems have been developed around very primitive life-forms which use this chemical energy to build long-chained organic molecules (Box 6.1). Something similar to these oxygen-poor conditions must have prevailed on the young Earth since organisms able to use radiant energy to split water and to generate oxygen first appeared 3 billion years ago. Later, as some groups entered into symbiotic associations with cells that lacked their own energy-fixing capacity, the eukaryotic precursor of all major photosynthetic organisms evolved.

The success of this association was to have profound consequences for the planet, leading to a major shift in the chemistry of our atmosphere. Then the abundant energy in sunlight could be used to fuel the synthesis of the building blocks of cells, dwarfing the meagre supply coming from the earth. With energy fixed by photosynthesis, complex communities developed, adding consumers that used the energy locked in the tissues of other organisms—herbivores, carnivores and also decomposers eating their non-living remains. In Chapter 5 we saw how species form assemblages and structure themselves into communities. Here we look at how energy flows through these communities and how this drives community processes such as nutrient flow and species interactions. By analysing how the communities function in relation to their abiotic environment, we describe their properties as ecosystems.

We begin this chapter by examining energy fixation and the innovation of photosynthesis. We then go on to consider how energy moves from one species to another, along food chains and in food webs and consider what this tells us about community organization. Finally, we look at the oldest profession, agriculture, as a means of securing energy supplies for ourselves.

BOX 6.1 The extremophiles

Plants are not the only primary producers on Earth. Indeed, there was a time when life on the Earth's surface was so hazardous that life fuelled on sunlight was not possible. Descendants of these early life-forms are collectively known as **extremophiles**—literally, lovers of extremes—and live in a range of harsh environments ranging from ice to boiling water and from salt to battery acid.

The first extremophiles to be discovered were thermophiles and hyperthermophiles—primitive bacteria living in the hot springs and thermal vents associated with terrestrial and marine volcanic activity (Table 6.1). These bacteria obtain energy by oxidizing the hydrogen sulfide dissolved in the volcanic spring waters. Other hyperthermophilic sulfur bacteria are known from 'black-smokers'—underwater volcanoes that release molten magma and sulfurous gases deep within the ocean. They thrive in water close to boiling point and under immense pressure. These form the base of a food chain fuelled by sulfur rather than sunlight and a community of deep-sea animals, including worms, crustaceans, and fish have formed around them. One of the strangest is *Riftia pachyphila* a 1.5-m long tubeworm, which lives around the margins of hydrothermal vents and concentrates sulfur within its body fluids. Its body cavity is home to symbiotic colonies of sulfur bacteria which supply the worm's energy needs.

Elsewhere in the deep oceans, methanogens live on mounds of frozen methane hydride deep on the ocean floor. Other bacteria are found within the ice of the coldest place on Earth—Lake Vostock, deep beneath the Antarctic ice sheet. Salt lakes have colonies of halophiles: *Halobacter* and *Halococci* require salt concentrations three to four times that of seawater. These bacteria avoid dessication by keeping their internal sodium concentration either above or equal to that of their surroundings. Again, other organisms have adapted to the halophilic lifestyle, including fish and invertebrates. The unicellular photosynthetic alga *Dangearidinella altitrix* lives on the surface of the salt crystals.

In the hostile sands of the desert, archaebacteria, such as *Metallogenium* and *Pedomicrobium* live off the thin surface layer of iron and manganese known as 'desert varnish'. They obtain their energy by oxidizing iron and manganese, turning the surface of the desert into a rusty-coloured mixture of metal oxides. An even more extreme group of these 'rock-eating' bacteria—lithoautotrophs—are found deep below ground in rock fissures. For many years geologists thought that the bacteria-like shapes within rock cores were the result of contaminated drilling equipment. However, recent research has confirmed that these were neither imagined or accidental, but real. Recent findings by the Ocean Drilling Program have confirmed that archaebacteria play an important role in turning glassy volcanic basalt into a clay-like material (Figure 6.1).

Possibly the most extreme extremophile yet discovered is *Deinococcus radiodurans*. A radiophile, it can withstand gamma radiation at levels of 6 Mrad/h,

TABLE 6.1 Temperature limits for extremophiles occurring within different groups of organisms. (*Notice how Archaeobacteria are able to withstand greater temperatures than their non-chemoautotrophic counterparts. Also note the much lower temperature limits experienced by eukaryotes*).

Group	Upper temperature limit (°C)
Prokaryotes	
Archaeobacteria	
Sulfur-dependent bacteria	115
Methane-producing bacteria	110
Bacteria	
Photosynthetic bacteria	70–73
Heterotrophic bacteria	90
Eukaryotes	
Protozoa	56
Algae	55–56
Fungi	60–62
Plants	
Vascular plants	45
Mosses	50
Animals	
Crustaceans (ostracods)	49
Insects	45–50
Fish	49–50

BOX 6.1   Continued

which means it could survive life within a nuclear reactor! This bacteria has attracted a great deal of attention in the medical world as its survival is due to an
efficient DNA repair mechanism, of great interest to
those studying cancer. A recombinant strain of the
bacteria is used commercially in the cleaning of organic solvents such as toluene and trichloroethylene
from nuclear waste.

Enzymes isolated from extremophiles—
'extremozymes'— have revolutionized biotechnology
and are part of our everyday life. Biological washing
powders, 'stonewashed' jeans, contact lens cleaning
fluids, and many antibiotics are all products of
extremozyme technology. Extremophiles have helped
unlock the secrets of the gene, since the DNA polymerase enzyme used in genome sequencing and DNA
fingerprinting (Box 2.2) was first isolated from the ther-
mophilic sulfur bacteria *Thermus aquaticus*.

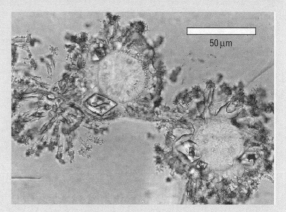

FIG. 6.1 Photomicrograph of basalt microbes at work
within volcanic rock deep below the ocean floor. Once
these were thought to be artefacts caused by contamination
but recent research has confirmed their identity.

6.1 █ Ecological energetics

Energy is the common currency of the living world.
The physical, chemical, and biological worlds are ultimately linked by the limits and rules that govern energy transformation and movement. Energy is stored
in chemical bonds and released slowly to drive all
biological activity. This maintenance of life is called
metabolism (Figure 4.1).

Energy is formally described as the capacity to do
work and is measured in various units. The internationally accepted unit is the **joule**—the amount
of energy in a force moving 1 kg through 1 m—
though you may be more familiar with the calorie
(cal), which is the amount of heat needed to raise
1 g of water by 1 °C (1 joule = 4.2 calories).

More important than the units of energy are the
first and second laws of thermodynamics which
describe its properties. These state that energy may
never be created or destroyed and that energy

transformations always lead to a reduction in usable energy. When energy is transformed from one
type (say, movement) to another (say, electrical energy) some is dissipated as heat. Any system, living
or otherwise, operates through a series of energy
transformations, so without further inputs these
heat losses mean usable energy must eventually run
down, or run out.

Energy at work is termed **kinetic energy**,
whereas stored energy is known as **potential
energy**. Besides synthesizing structural molecules to
build new cells or gametes, much of the activity of
living systems involves creating molecules to
store energy and drawing upon them to drive
metabolic processes. We detect this as the heat
lost through respiration. We can also measure the
energy fixed in tissues, which ecologists refer to as
production.

6.2 The producers

Energy enters the biosphere as sunlight, with each square metre of the Earth's surface receiving an average of 48 million kilojoules (kJ) of radiant energy per year. Without photosynthetic organisms, much of this energy, after heating the surface, would simply be re-radiated back to space.

Radiant energy is captured through the photosynthetic capacity of green plants, which are collectively termed **primary producers**. Because they can fix their own energy they are known as **autotrophs**, quite literally 'self-nourishers' (or acknowledging their primary energy source, **photoautotrophs**). The most primitive photoautotrophs are specialist bacteria, including the cyanobacteria which have a rudimentary photosynthetic apparatus. Before these evolved a very different source of energy drove ecological processes, the potential energy in reduced chemical compounds, released by **chemoautotrophs**. These primitive bacteria can still be found in very extreme environments where oxygen is low (Box 6.1).

In photosynthesis, the chloroplasts within plant leaves use radiant energy to split water molecules and combine the products with carbon dioxide to form glucose (or other simple sugars):

$$6CO_2 + 6H_2O \xrightarrow{\hspace{2cm}} C_6H_{12}O_6 + 6O_2$$

Carbon dioxide Water Radiant energy Glucose Oxygen

Glucose is then used to build more complex molecules either as structural components or as an energy source to fuel metabolic processes.

A key component of photosynthesis is a molecule called **chlorophyll**, a group of green pigments common to all plants (Figure 6.2) and contained in chloroplasts (Figure 6.3). Here, chlorophyll is arranged within a mass of tightly folded membranes, called thylakoids, to pack as much pigment into as small a volume as possible.

Chlorophyll appears green because it absorbs light at the red and blue ends of the spectrum and reflects only green (Figure 6.2). Leaves that are not green still have

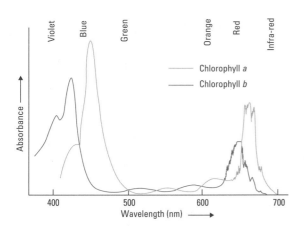

FIG. 6.2 Absorption spectra of the two forms of chlorophyll (*a* and *b*). The graph shows the pattern of light absorption of chlorophyll. Note the absorbance peaks at the red and blue ends of the spectrum: these are the wavelengths that are best at exciting electrons within the chlorophyll molecule. Green light is not absorbed and is reflected back giving plants their characteristic green colour.

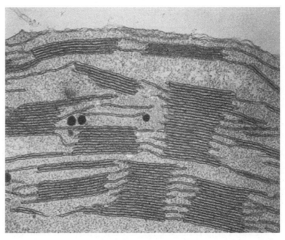

FIG. 6.3 Electron micrograph of the chloroplast, the cell organelle responsible for photosynthesis. It consists of a series of membranes (thylakoids) on which chlorophyll and its associated proteins are situated. Chloroplasts have their own DNA and this is thought to be evidence of their evolutionary past as free-living organisms which took up symbiotic residence within other cells.

chlorophyll but also have a range of other pigments. These accessory pigments are used by plants which grow in poor light conditions, in the shade of the forest floor or deep coastal waters. We see some of these in the autumn when the leaves turn red; then the chlorophyll is degraded to reveal the pigments it once masked.

Not all of the light hitting a plant is used in photosynthesis—most passes through the leaf and never encounters chlorophyll. Just 44 per cent is at a wavelength that can be absorbed by chlorophyll (Figure 6.2) and only 2 per cent of the radiant energy falling on a leaf becomes fixed in the sugars produced (though the warming effect of all the sunlight helps to increase photosynthetic activity). Despite this low efficiency, the productivity of autotrophs makes up 99 per cent of the living biomass on the planet, and it is this energy which supports the living components and processes of the biosphere.

The productivity of an autotroph is the rate at which it produces living material or **biomass**. This can be expressed either in terms of the amount of matter within a given area over time (kg/m^2 per annum) or its energy content (kJ/m^2 per annum). The full photosynthetic output of a primary producer is known as its **gross primary production** (GPP), but not all of this energy becomes fixed in the tissues. The largest proportion, perhaps more than 60 per cent, is used in the plant's own metabolism, eventually to be lost as heat. Only the energy that remains can be devoted to growth or the production of gametes, the production of biomass, and this is termed **net primary productivity** (NPP):

Net primary production (NPP)
= Gross primary production (GPP) − Respiration (R)

The biomass accumulated as NPP in an area over a period of time is referred to as the **standing crop,** and it differs widely both between species of plant and between ecosystems (Table 6.2).

Different plants allocate their biomass between growth and reproduction (gametes and reproductive structures) in different ways. They have different life history strategies (Section 3.4). Some have very short lifespans: thale cress (*Arabidopsis thaliana*) completes its life cycle in little more than a month. Others like the bristlecone pines (*Pinus aristata*) can live for several thousands of years. A plant with a short lifespan typically devotes much of its resources to a

single reproductive event with relatively little of its biomass used in non-reproductive growth, a strategy matched to the changeability of the habitat.

Annual plants complete their life cycle within a year (Section 4.4). Early on, energy is used to produce as many leaves as possible and these plants only shift their resources to reproduction towards the end of their life. Perhaps as much as 90 per cent of their energy is invested in flowers and seeds. Such plants are characteristically r-selected species, fast-growing opportunists. More long-lived **perennials**, which

TABLE 6.2 Global primary production

Ecosystem type	Mean net primary productivity (g/m^2/year)	Mean biomass (kg/m^2)
Continental		
Tropical rainforest	2200	45.0
Tropical seasonal forest	1600	35.0
Temperate evergreen forest	1300	35.0
Temperate deciduous forest	1200	30.0
Boreal forest	800	20.0
Woodland and shrubland	700	6.0
Savanna	900	4.0
Temperate grassland	600	1.60
Tundra and alpine	140	0.60
Desert and semi-desert scrub	90	0.70
Extreme desert, rock, sand, ice	3	0.02
Cultivated land	650	1.00
Swamp and marsh	2000	15.00
Lakes and streams	250	0.02
Mean continental	773	12.3
Marine		
Open ocean	125	0.003
Upwelling zones	500	0.02
Continental shelf	360	0.01
Algal beds and reefs	2500	2.0
Estuaries	1500	1.0
Mean marine	152	0.01
Grand total	333	3.6

survive over a number of seasons, allocate more energy to structural tissues or energy storage, allowing rapid growth at the beginning of a new season. Energy reserves often take the form of tubers, roots, or swollen underground stems, rich in starch. Perennials may delay reproduction until they have accumulated sufficient energy to fuel the process but then they may reproduce repeatedly. You may recognize this as a *K*-selected strategy, adopted by competitive species that dominate stable and predictable environments (Section 3.4).

Trees and shrubs go a stage further. As they grow, much of their energy goes into producing large amounts of woody tissue. When they are saplings, more than half of their biomass may be in the form of leaves, but this changes as they grow. A mature tree may consist of 95 per cent non-living woody tissue, with the live biomass confined to the shoots, roots, leaves, and a thin layer of living tissues beneath the protective bark, wrapped around a superstructure of dead wood. The wood serves as a scaffold to support the light-gathering surfaces of the crown and the nutrient gathering network in the roots.

A large tree represents a considerable accumulation of energy and this becomes obvious when wood burns and this energy is released as heat. Wood is primarily a water-conducting and structural tissue and the plant's investment is rewarded where water is abundant but there is severe competition for light and nutrients. Forests dominate the wetter parts of the world within each major climatic region (Section 8.1).

Even under optimum conditions, different plants have different capacities to fix radiant energy. Generally, plants adapted to habitats that have abundant nutrients have the highest photosynthetic efficiencies, including many of our agricultural crops. Photosynthesis also rises with temperature, though each species will have a range to which it is adapted and in which it is most productive.

Light is rarely limiting to photosynthesis in terrestrial ecosystems. Most plants have their highest efficiency at relatively low light levels and bright light can inhibit photosynthesis. However, light is quickly extinguished in deep waters, both by absorption and due to reflection by suspended matter. Only certain wavelengths can penetrate any distance and plants living at depth have to be able to collect energy from whatever light is available. The large kelps that live in

deep coastal waters have a range of pigments that absorb energy over those parts of the spectrum that penetrate furthest. Similarly, many terrestrial plants growing under deep shade may also have accessory pigments to collect energy over these residual wavelengths.

Plants can only invest in new tissues when the energy fixed in photosynthesis is greater than that used in respiration. If the two are in balance the plant is said to be at its **compensation point**. In aquatic systems, this is the depth below which photoautotrophs are unable to survive. In the same way, some species cannot persist in the deep shade of a forest.

Even so, many plants have to survive extended periods below their compensation point when they are burning more energy than they can fix. Most plants can go into debt for a while, living off energy reserves in the form of starch, but they may not live beyond their means indefinitely. One strategy is to reduce respiration costs by shedding photosynthetic surfaces that are costly to maintain. This happens annually in the temperate regions, where the autumnal leaf-fall is triggered by a decline in temperature, light quality and daylength, reliable signals of the forthcoming winter.

This is not the only possible response to low light and temperatures. Many evergreens are able to photosynthesize throughout the winter months, albeit at a lower level than during the summer. The ability to harvest light energy over an extended season, means they can match, and even beat, the productivity of deciduous trees limited by the short season when they have leaves (Figure 6.4).

Obtaining carbon dioxide, water and mineral nutrients necessary for life places plants in something of a dilemma. Carbon is captured when carbon dioxide is absorbed by the leaf, having entered it through a series of breathing pores (**stomata**). The loss of water through the stomata is part of the mechanism by which water and dissolved minerals are drawn up from the roots in a process known as **transpiration**. So a plant must keep its stomata open to meet its carbon, water, and mineral needs. Even when water is scarce they cannot simply shut down their transpiration stream since this would halt photosynthesis. As a consequence, plants are at risk of dessication when water is scarce.

Stressful conditions such as cold or heat can lead to leaf shedding and dormancy, shutting down metabolism to reduce energy consumption. Many plants of

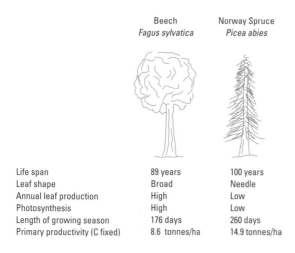

	Beech *Fagus sylvatica*	Norway Spruce *Picea abies*
Life span	89 years	100 years
Leaf shape	Broad	Needle
Annual leaf production	High	Low
Photosynthesis	High	Low
Length of growing season	176 days	260 days
Primary productivity (C fixed)	8.6 tonnes/ha	14.9 tonnes/ha

FIG. 6.4 Evergreen species sacrifice short-term productivity for the ability to photosynthesize for longer. The Norway spruce (*Picea abies*) photosynthesizes for almost half as long again as the beech tree (*Fagus sylvatica*) and, despite its low annual leaf production and photosynthetic rate, it is more effective over the long term.

mediterranean and arid areas are drought-deciduous, shedding their leaves when subjected to drought stress (Section 5.1). Black sage (*Salvia mellifera*) overcomes the drought stress of the Californian chaparral by shedding its lower, older leaves in favour of the younger leaves at the shoot tips. Other chamaephytes take this a stage further; the germander, *Teucrium polium*, is one of a number of species that produce different sized leaves according to the season. During the months of January–March it produces winter leaves, larger than the small leaves produced during the summer months (Figure 6.5). In this way, the plant can achieve year-round photosynthesis but reduce its water loss during the hot summer months by between 50 and 76 per cent.

Some plants avoid having to shed their leaves in response to extremes of heat or cold. The Norway spruce (Figure 6.4) reduces water loss by producing needle-like leaves with a very low surface area. This allows it to photosynthesize when water is frozen and unavailable. At the other extreme, cacti survive in hot, dry deserts using their swollen stems for water storage and photosynthesis and with leaves reduced to spines. Others produce leaves covered with downy leaf hairs that serve to trap a layer of still air that not only reduces water loss, it also insulates against

excessive cold or heat. Examples include the alpine edelweiss (*Leontopodium alpinum*) and members of the genus *Salvia* of hot mediterranean climates. Oils produced as secondary plant products can also be used in this way (Section 4.4); as the temperature rises, the oils volatilize forming a protective layer around the leaf. This increases the diffusion resistance from the leaf surface and so reduces water loss. As we have seen, such compounds have other benefits such as deterring grazers and regulating the spread of wildfire in fire-adapted species (**pyrophytes**) (Section 5.1).

One other strategy to cope with drought is to adapt the photosynthetic process itself, making its water-use more efficient. Some plants have developed a metabolic pathway (termed the *C4* **pathway**) which allows them to open their stomata night and day. These plants have a special enzyme—phosphoenol pyruvate (PEP) carboxylase, which enables carbon dioxide to combine with phosphoenol, already present in the plant. The product is a series of four-carbon compounds such as acetic, aspartic, malic, and oxaloacetic acids which act as a carbon store. When daylight comes, the four-carbon acids can be broken down and fed into the photosynthetic process. A number of plants associated with harsh or hot environments use this mechanism but others include crops such as sorghum (*Sorghum bicolor*), maize *(Zea mays)*, and sugar cane (*Saccharum officinale)*. Some species have a further modification, known as the **Crassulacean acid metabolism** (CAM), and are able to keep their stomata firmly shut during the heat of the day and open them at night when they can absorb carbon dioxide without the risk of losing too much water storing the carbon as malic acid.

The above adaptations—leaf shed, leaf hairs, secondary plant products and alternative metabolic pathways—all come at a metabolic cost and their use represents a trade-off between production and long-term survival. The water requirements of photosynthesis represent a limiting factor to plant growth and thus productivity is closely related to water availability. This explains why the most productive ecosystems are those with abundant moisture (Table 6.2).

The productivity of forests is evident from their complex structure. Unlike grasslands, or communities of low scrub, forests are multilayered with a tree canopy, shrub layer, herb layer and forest floor. Leaves do not need bright sunlight to be effective and light filtered or reflected from upper layers is sufficient for

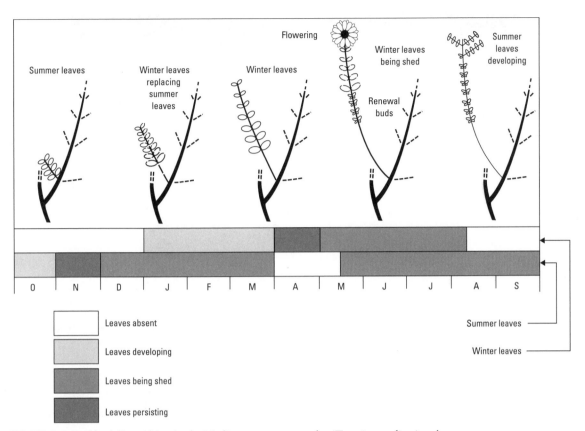

FIG. 6.5 Seasonal leaf dimorphism in the Mediterranean germander (*Teucrium polium*) a plant that produces different leaves for different seasons so as to minimize drought-stress.

those plants adapted to shade. A multi-layered community can develop if there is sufficient moisture to allow all of these photosynthetic surfaces to work.

We compare the complexity of the primary producer community of different ecosystems using the **leaf area index** (LAI); this is simply the leaf area as a proportion of the ground area it covers:

$$\text{Leaf area index} = \frac{\text{Total leaf area}}{\text{Area of ground}}$$

Forest ecosystems can have LAIs as high as 9, which means that light passes through nine layers of leaves before reaching the ground. Arid and semi-arid areas may have a LAI of less than 1. The LAI is therefore useful as an indication of both the structure and potential primary production of an ecosystem.

Algal beds and reefs have the highest rates of primary productivity known (Table 6.2). With an annual NPP of 2500 g/m^2, they can out-perform even

tropical rain forest. They too have a multilayered structure, with plants of different sizes, though the primarily adaptations to the light at different depths are the accessory pigments, obvious in the red and brown seaweeds. Swamps and marshes too are highly productive, both because of their shallow water and abundant nutrients.

Grasslands tend to have LAIs somewhere between forests and desert and with an intermediate NPP of 600–900 g/m^2 each year. They have none of the vertical construction of a forest and this is reflected in their standing crop (1.6–4 compared with 20–45 kg/m^2). These differences reflect the allocation strategies of the grassland plants, adapted to seasonal rainfall, periodic fires and grazing. Grasses are perennials that invest little in support tissues, but instead sprout leaves that will function as long as water is abundant. When the dry season brings primary production to a close, the grass withdraws

TABLE 6.3 Photosynthetic efficiencies and growth rates of crops

Crop	Country	Crop growth (g/m²/day)	Total radiation (J/cm²/day)	Light conversion efficiency
Maize *Zea mays**	USA	52	2090	9.8
Millet *Pennisetum typhoides**	Australia	54	2134	9.5
Sugar beet *Beta vulgaris*	UK	31	1230	9.5
Millet *Pennisetum purpureum**	El Salvador	39	1674	9.3
Sugar cane *Saccharinum* spp.*	Hawai'i	37	1678	8.4
Tall fescue *Festuca arundinacea*	UK	43	2201	7.8

*C4 plants.

resources from its leaves. Like other hemicrypto-phytes (Figure 5.8), a grass protects its growing tissues by keeping them close to the ground, away from grazers and flash fires. New leaves will sprout when water is next available. This is a highly success-ful strategy and one which also works well in moister areas, if there is intense grazing pressure. It is the pressure of grazing, primarily by our livestock, that has led to much of the northern temperate forests being replaced by pasture.

Cultivated land appears to have a low net productiv-ity (650 g/m² per annum), but this merely reflects the highly seasonal nature of agricultural activity. Such land is in use for only part of the year when it supports one or two crops; at other times it is being prepared or waiting the next growing season. However, when they are actively growing, crops are among the most

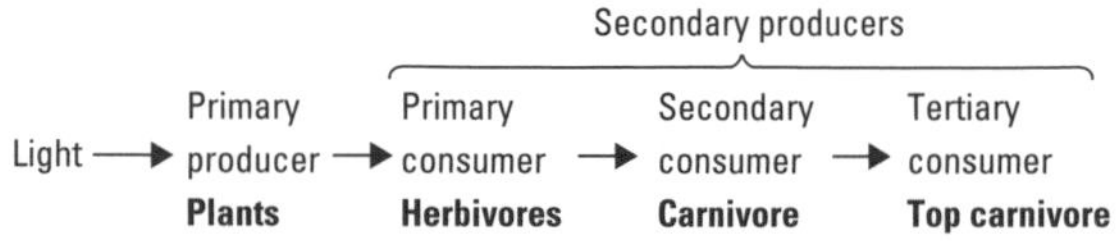

FIG. 6.6 A generalized food chain—the pathway of energy through a simple food chain. The organisms can be divided up into primary producers, which fix radiant energy and secondary producers which consume this energy second-, third-, and fourth-hand.

efficient primary producers (Table 6.3). Photosynthetic efficiencies close to 10 per cent are achieved in some cases, primarily because we remove some of the checks on photosynthesis, by adding water and nutrients. Notice, too, that many crops are C4 plants whose photosynthetic processes are adapted to drier conditions.

6.3 Links in the chain

A **food chain** describes the route by which energy passes through a community and the feeding rela-tions between some of its species. When they are con-sumed, the energy fixed by primary producers passes to the consumers or **heterotrophs** (literally 'nour-ished by others'). Consumers are collectively called **secondary producers** and form the food chain as a series of **trophic levels** (Figure 6.6).

Primary consumers are **herbivores**, feeding on plants. Although vegetation is abundant, plant mate-rial often represents a poor quality food which re-quires a considerable investment to digest it, including a long digestive tract. This means the assimilation efficiencies of herbivores are generally low (Box 6.2). Some herbivores concentrate on parts of the plant which are more nutritious or more readily

BOX 6.2 Energy efficiency

Because it involves a transformation, any movement of energy from one organism to another inevitably leads to some energy loss. Of the energy that is assimilated, much is used in the metabolic processes, the cost of building, maintaining and degrading the cell. From that assimilated (A), some energy goes to respiration (R) and some to production (P). Production can be divided into that used in growth (Pg) and in reproduction (Pr)—Figure 4.1.

We can work out the proportion of energy in the diet taken up by a consumer simply as its assimilation efficiency:

$$\text{Assimilation efficiency} = \frac{\text{Energy assimilated (A)}}{\text{Energy consumed (C)}}$$

(An equivalent equation derives photosynthetic efficiency by making the divisor the radiant energy received at the leaf surface).

A simple extension of this allows us to work out how efficiently energy consumed is converted into new tissues:

$$\text{Production efficiency} = \frac{\text{Energy fixed in tissues (P)}}{\text{Energy consumed (C)}}$$

The proportion of energy assimilated that is converted into tissues is the growth efficiency:

$$\text{Growth efficiency} = \frac{\text{Energy fixed in tissues (P)}}{\text{Energy assimilated (A)}}$$

Ecological efficiencies vary amongst animals according to their metabolic costs. Warm-blooded animals such as birds and mammals—**endotherms**—have high metabolic costs and can spend over 90% of their energy income in maintaining their body temperature. **Ectotherms**, on the other hand, rely primarily on external heat sources and do not have these costs. They can devote more of their energy to production (Table 6.4).

In general, organisms further along a food chain have higher assimilation efficiencies, largely as a result of the quality of their diet. Herbivores may have a plentiful supply of plant material, but up to one-third of this may be cellulose, which they are unable to digest without the help of bacteria in their gut (Box 6.5). Their assimilation efficiencies are low, typically around 10% and herbivores therefore need to consume large amounts of vegetation to meet their energy demands.

Carnivores, on the other hand, receive their energy in more usable form, as proteins and fats that are both richer in energy and more readily digested. They have relatively high assimilation efficiencies, as much as 90% in some exceptional cases. This efficiency and the high nutritive value of their food mean that carnivores, compared to herbivores, need to eat less and eat less often.

Production efficiency, which takes into account the energy used in respiration, also varies with trophic position (Table 6.4). Within a group sharing a similar metabolism, say the terrestrial invertebrates, herbivores again have the lowest efficiency (21–39%) and carnivores the highest (27–55%). Across groups, ectotherms have production efficiencies of between 10 and 55% and endotherms of just 1–3%.

TABLE 6.4 Production efficiencies of different animal groups

Group	Production efficiency (%)
Endotherms	
Insectivores	0.9
Birds	1.3
Small mammals	1.5
Other mammals	3.1
Ectotherms	
Fish and social insects	9.8
Non-insect invertebrates	25.0
Non-social insects	40.7
Non-insect invertebrates	
Herbivores	20.8
Carnivores	27.6
Detritivores	36.2
Non-social insects	
Herbivores	38.8
Detritivore	47.0
Carnivores	55.6

These figures might seem to argue for being a carnivore and never being a herbivore. However, most of the biomass of the planet is vegetation and there is consequently more energy available to herbivores than carnivores. The figures also suggest that being an ectotherm is a better strategy when energy is in short supply and this is probably true.

Ecological efficiencies also make it obvious why energy decreases so rapidly along a food chain. A carnivore three or four steps away from the primary producers of an ecosystem might have one ten-thousandth of energy originally fixed by the plant. Figure 6.12 shows how little energy becomes fixed in each successive trophic level of a grassland food chain, so that a weasel fixes a mere 0.0026% of net primary productivity into its own tissues. This helps to explain why there are so few weasels and why everybody cannot be a carnivore.

digested than others, such as new shoots or buds. Seed-eaters, for example consume a food rich in stored carbohydrates and oils. They benefit from an easily digested food with high levels of nitrogen which otherwise would have fuelled the germination of the seed. Other herbivores have found ways of unlocking the energy contained within the indigestible cellulose that constitutes the major part of a plant's biomass (Box 6.5).

Carnivores face a different set of challenges. As secondary or tertiary consumers, they live on the energy fixed in the tissues of herbivores or other animals. Flesh is primarily protein and fat, high-energy compounds that are readily degraded and, in the case of protein, are also rich in key nutrients. Carnivores have a relatively short digestive tract and can digest their food more readily. However, they must still meet the costs of catching and killing their prey—costs that are minimal for a herbivore. A variety of adaptations to make predation efficient have arisen, ranging from the spider's web to the claw of a cheetah (Section 4.4).

Some feeding strategies do not locate a consumer on just one trophic level. **Omnivores**, feeding on both plants and animals, and perhaps herbivores and carnivores, straddle several trophic levels. Consequently, the energy flow through such a food chain does not follow a simple progression of trophic levels. These patterns become even more complex where omnivores scavenge dead animals or plants.

A scavenger diverts energy heading for a decomposer food chain back into a chain based primarily on herbivores. We can distinguish two basic routes for energy moving through ecosystems—from herbivores to carnivores (termed a **grazing food chain**), and from decomposers to carnivores (a **decomposer food chain**) (Box 6.3). However, these two chains are connected every time a blackbird pulls a worm from the soil, when any decomposer is eaten by a consumer from the grazing food chain. The amount of energy moving down each pathway depends upon the ecosystem; in most cases, a large proportion of the primary production passes into the decomposer route. On average, only 10 per cent of net primary productivity in terrestrial ecosystems finds its way into herbivorous consumers. The rest, 103.5 billion tonnes globally, fuels the decomposer chain. This makes the decomposer chain the most significant route by which energy passes to the rest of the system (Box 6.3). In contrast, a large proportion of the primary production represented by the phytoplankton is consumed by herbivores, thereby directing energy into the grazing food chain of some aquatic systems.

Chain length

Figure 6.12 shows a simple food chain for a grassland ecosystem and the amounts of energy passing through a herbivore (a mouse) to one of its predators (a weasel).

Notice how little energy actually becomes fixed in the tissues of the carnivore. The inefficiency of energy transfer (Box 6.2) and the inevitable losses with each transformation means little is

BOX 6.3 The reducers

Consider a fallen leaf. Having escaped being eaten, it lies, along with other vegetation and dead organic matter, within the litter layer. What happens to the energy locked within its tissues? How are nutrients and energy recycled back into the system?

The breakdown of the litter layer proceeds through several stages and through a series of physical, chemical, and biological processes. The first is a **leaching** of nutrients—salts, sugars and amino acids—by water, which can lead to as much as 30% loss of biomass. The next stage involves the fragmentation into **detritus** through the feeding activity of **detritivores**, such as earthworms, millipedes and woodlice. On average, these animals are able to assimilate a small fraction of the energy and nutrients they consume and the rest becomes a resource for the microbial and fungal decomposers (Figure 6.7). These are often added from the gut flora of the detritivores, so the leaf's passage through their digestive tract accelerates its decomposition. The bacteria and fungi release enzymes that breakdown complex carbohydrates and other macromolecules, further mobilizing the nutrients in the remains. Added to the soil, these become available to primary producers.

The speed of this process varies according to the quality of the material being decomposed. Decomposition is most rapid in detritus which has a high content of simple sugars and slower in material with a high lignin content. Mike Swift and his colleagues measured the rate of breakdown of different components of straw (Figure 6.8) confirming that simplest compounds are the fastest to be lost. Meanwhile, complex carbohydrates and compounds high in tannins and lignin decompose more slowly as they are broken down by specialist fungi and bacteria.

Environmental conditions also govern the speed of decomposition, fastest in warm, wet conditions where there is abundant oxygen. This is why nutrient turnover is quick in tropical rainforests and so slow under the cold, waterlogged conditions of the tundra (Section 8.2).

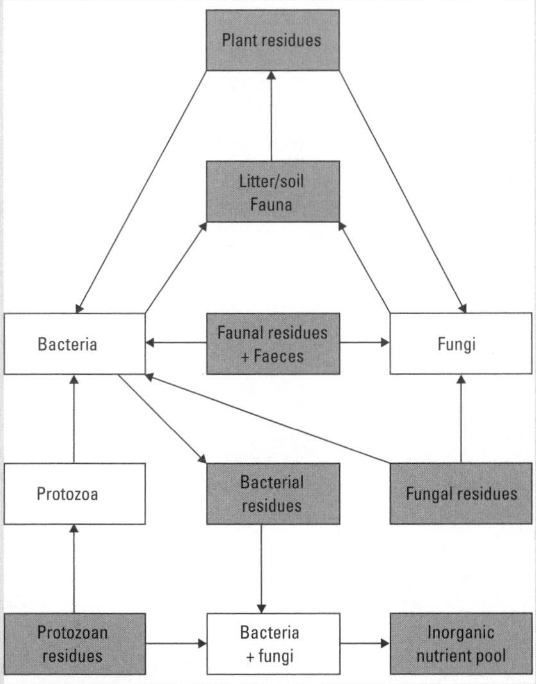

FIG. 6.7 A generalized diagram of the flow of material and energy from detritus to a range of detritivores and decomposers.

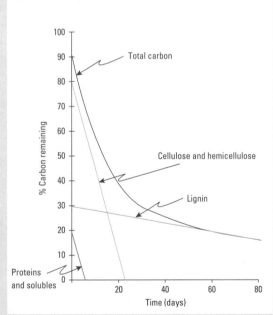

FIG. 6.8 The relative rates of decomposition of carbon compounds within straw left on the soil surface. Here, it can be seen that some compounds break down faster than others.

BOX 6.3 Continued

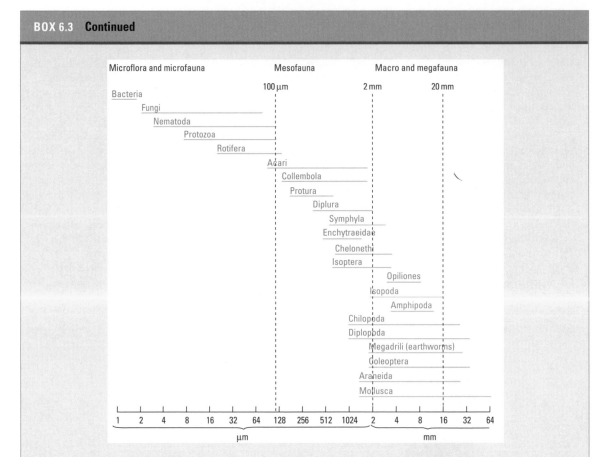

FIG. 6.9 Reducers great and small: soil decomposers classified according to width.

The reducers might be inconspicuous but they are far from insignificant and account for a staggeringly large proportion of global biomass (Figure 6.9). A single gram of temperate woodland soil may contain as many as 6 million bacteria and 3000 m of fungal threads (hyphae); a square metre of woodland floor may support a further 200 million decomposers ranging from protozoa through to molluscs.

Worms process detritus at a rate of between 50 and 170 tonnes per hectare each year. Soils that lose their worm population, through pollution or mismanagement, suffer a correspondingly drastic decline in their structure and fertility. Returning them to a productive state frequently involves the use of nutrients and organic matter to re-establish the reducer community and bring the soil back to life.

Where nutrients are short and invertebrate detritivores are few, fungi play an important role in the degradation of dead organic matter (Figure 6.10). Their hyphae will degrade organic matter with a low nitrogen content. It is only when their caps appear above the litter or sprout from a log that we get some idea of their number and variety (Figure 6.10).

The soil ecosystem has its own network of organisms that interact with each other to form a complex community which is built upon dead organic matter (Figure 6.11). But the soil is not the only place we find reducers—some specialized decomposers live within the digestive tract of other organisms where they break down undigested material, an arrangement that works well both for them and their ruminant hosts (Box 6.5).

BOX 6.3   Continued

FIG. 6.10 The familiar sight of the fruiting bodies of fungi. Underground is a large network of fungal hyphae which live by breaking down organic matter.

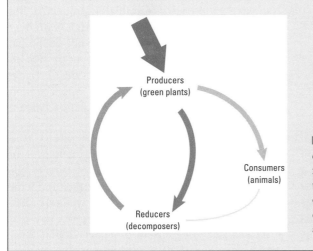

FIG. 6.11 Decomposers. In most ecosystems the bulk of the energy fixed by primary producers pass to the decomposer community. This not only fuels the productivity of the decomposers and detritivores, but also drives the recycling of nutrients.

available to the higher trophic levels. At each level, energy is expended in respiration and only a fraction of the energy consumed becomes fixed in the tissues. A plot of the energy content of the trophic levels in sequence produces a pyramid (Figures 6.13 and 6.14). In our example, the NPP of the primary producer (200 million kJ) is reduced to just 546 kJ of weasel.

This chain has just three links. Long food chains are rare in nature and rarely do they extend beyond

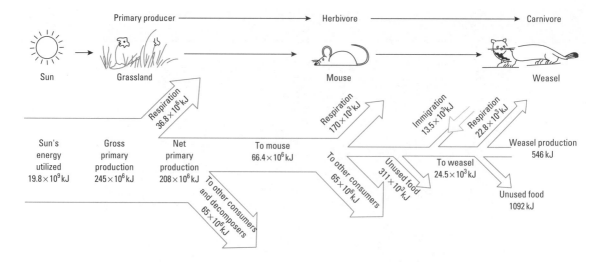

FIG. 6.12 The loss of energy as it passes along a simple food chain (in this case a North American grassland). Energy values are in kilojoules (kJ).

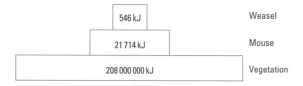

FIG. 6.13 Pyramid of energy (not to scale), containing the productivity data from the food chain shown in Figure 6.12. Note the rapid decrease in the amount of energy available at higher trophic levels.

five links. The dramatic decline in energy at each trophic level was thought to limit the number of links, with too little energy remaining to support a fourth or fifth trophic level. However, if this was the case, we would expect food chains dominated by the more energy-efficient ectotherms (such as the insects) to be longer than those with endotherms (such as the mammals—Box 6.2). In fact, the evidence from a number of studies suggests that the energy efficiency of the organism makes little difference to chain length.

Perhaps, then, chain length is controlled by primary productivity, the amount of energy entering the ecosystem. Then we should expect food chains to be longer in communities with higher primary productivity. Stuart Pimm, John Lawton, and Joel Cohen compared food chains from highly productive biomes, (such as tropical rain forest) with low productivity biomes (such as tundra) and found no difference—each with an average of four trophic levels. Pimm and Kitching also tried boosting the energy input (by adding leaf litter) to some simple communities, but their experimental manipulation failed to extend chain length.

It seems we need to look more closely at other features of food chains. Some patterns tend to recur, sometimes for fairly obvious reasons. One is the tendency for consumers to become larger and fewer with each link of the chain (Table 6.5). Large carnivores require larger territories from which to collect sufficient prey. Organisms at higher trophic levels tend be longer lived, delaying reproduction until later in their life cycle.

Many of these characteristics result from an increase in size, and this could be why some food chains are not longer. As Paul Colinvaux has suggested, it may simply be that a predator would

TABLE 6.5 General trends along food chains

Fewer species
Each trophic level of the food chain generally supports fewer species than the preceding one

Larger body size
Moving along the food chain species tend to display an increase in body size as predators are generally larger than their prey

Lower population numbers
Higher energy costs of species nearer the top of the food chain mean that ecosystems can support relatively few of them

Longer lived
Larger species tend to have longer lifespans (e.g. contrast a butterfly with an eagle)

Lower reproductive rates
Longer-lived species tend to mature more slowly, delaying reproduction until later on in their life history

Increased home range
Larger organisms have to cover larger areas to meet their energy requirements in terms of prey. They therefore have bigger territories.

Higher powers of dispersal
In order to cover large territories an organisms has to be able to move across a wide area

Increased searching ability
Looking for food within a large area requires the ability to find and recognize food species

Increased behavioural complexity
Large territories, long lifespan, and lower densities mean that species near the top of food chains have more complex behaviour patterns with regard to feeding and reproduction

Higher metabolic cost
Increased size brings with it an increased energy requirement for the maintenance, growth, and reproduction of the organism

Require food of higher calorific value
Food must have a higher energy value so that it can fuel higher the metabolic rates of species nearer the top of the food chain

Reduced feeding specialization
At higher trophic levels, there is a tendency for species to be more generalist in their feeding strategy; this also includes omnivory

Greater assimilation efficiency
Species that are further along the food chain have increased efficiency in the use of the food they ingest.

have to be so large to feed on the big carnivores at the end of most food chains that it would be unlikely to secure enough energy to maintain itself or a population.

Predators also compete with each other for prey. Collectively, they may well hold herbivore numbers in check, so that despite the abundant vegetation in most ecosystems, herbivore numbers are capped by the predators that consume them. The predators themselves are similarly limited by the numbers of those predating or parasitizing them (Section 4.4). Together it is the interactions between trophic levels that serves to limit and organize the whole system (Section 9.4).

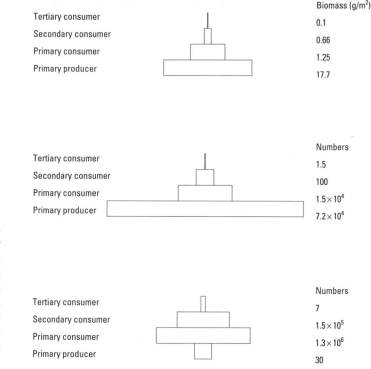

FIG. 6.14 Some other ecological pyramids. Plots of (a) biomass or (b) numbers of individuals at each trophic level both show a pyramidal shape for successive trophic levels within an ecosystem. However, numbers can also be misleading; (c) a single tree may support a large number of herbivores, producing an inverted pyramid of numbers.

6.4 The web

If a food chain follows one possible route by which energy flows through an ecosystem, a food web maps all possible routes. In a food web the significant food chains of a community are linked together as a network.

Figure 6.15 shows a food web for the community associated with oak trees (*Quercus robur*) in an Oxfordshire woodland. This web represents an aggregation of several food chains, some of which overlap. Most are only four links long, while others are even shorter. Note that the birds and mammals within this ecosystem have tendency to be omnivorous. Titmice (birds of the family Paridae) feed at several trophic levels, and both blue and great tits feed on seeds (primary producers), winter moths (herbivores), beetles and spiders (carnivores). Mice and

voles are also omnivorous, eating both insects and plant material.

Webs provide us with a more complete description of the energy pathways in an ecosystem. They allow us to locate organisms that do not fit neatly into one trophic level. We can include detritivores and decomposers, so uniting grazer and decomposer food chains. It is also possible to include parasites and hyperparasites, higher-level consumers ignored in many food chains. However, like any map, our representation of a web is inevitably a simplification of reality, with a somewhat arbitrary boundaries around the system. Not all the trophic connections may be identified, if only because of the considerable amount of work needed to map every pathway and often species are simply lumped together into trophic or 'feeding' groups.

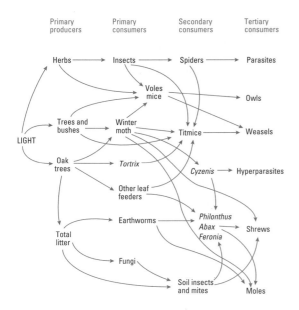

Primary producers | Primary consumers | Secondary consumers | Tertiary consumers

FIG. 6.15 A food web as a series of interconnected food chains. In this case, the web shows connections between animals associated with oak trees in Wytham Wood, Oxfordshire.

Early studies of food webs were limited by the scale and quality of observations. The so-called 'rare feeding events' were readily incorporated into a web as if they were everyday occurrences. The more detailed analyses of later studies, analysing gut contents and faeces provided more precise profiles of the dietary habits of a species and today these are being extended by DNA fingerprinting (Box 2.2). This technique has the potential to revolutionize food web research as it can positively identify dietary species, so confirming links within food webs. Simon Jarman and his colleagues used the method to identify which consumers fed on krill in marine food webs and found them the diet of a number of vertebrate predators including Adelie penguin (*Pygoscelis adeliae*) and the pygmy blue whale (*Balaenoptera musculus brevicauda*).

Despite this, webs are still only snapshots of nature and freeze a dynamic community into a static diagram. Species come and go with time, and some even change their feeding position during their life cycle. Most webs ignore this and few give any measure of the strength of the connections, distinguishing

between rare feeding events and very strong trophic links. Nevertheless, some patterns recur. For example, there is evidence that the number of predator to prey species is relatively constant and the proportion of consumers on each trophic level seems to be independent of the total number of species in the web. The number of linkages is typically twice the number of species and, as we have seen, food chain length appears to be independent of the web. Food chains are generally shorter in small or frequently disturbed habitats and in ecosystems with little vertical development (a grassland, say, compared to a forest). However these differences are relatively small and, as Pimm and Kitching discovered, the overall productivity of the ecosystem makes little difference to chain length.

So what causes these patterns? Computer models that incorporate the effects of population sizes and the intensity of the interactions between species indicate that webs may be limited by the number of species they can contain. Webs are then not easily invaded by a new species. A predator that attempts to invade a well-developed community must gain access to a prey species, and thus compete with established predators. Consequently, interspecific competition must increase (Section 4.3) and become more intense for more species at that trophic level. Similarly, interactions between levels will constrain the structure of a web; in their study of 62 published food webs, Frederic Briand and Joel Cohen found evidence that the number of predator species was a better predictor of the number of prey species rather than the other way around.

Perhaps these are 'rules of assembly' that govern how a food web can be constructed (Box 5.3). As with many ecological concepts food webs are an artificial construction to aid our understanding, helping us to describe important ecological processes, but which may fail to capture the fluidity of the real world. Even so, they point to constraint and regulation amongst species interactions and help us to both manipulate these processes to our benefit and to quantify them. In particular, they can be used as ecological road maps by which we can track the route of pollutants such as pesticide residues, heavy metals and radioactive isotopes, through the ecosystem (Box 6.4).

BOX 6.4   Bioconcentration, bioaccumulation and biomagnification

For many people, it is inevitable that pollutants entering ecosystems will move along food chains, becoming more highly concentrated as one species consumes another, to eventually poison the species at the end of the line. An unnatural and toxic substance that does not decompose will simply follow the natural trophic routes through an ecosystem, accumulating at each step along the way.

The classic example, and the first pollutant that appeared to show this behaviour, was dichlorodiphenyltrichloroethane (DDT). Following its widespread use as an insecticide in the 1950s, DDT was found to have caused the population collapse of some predatory birds, by reducing the number of eggs they hatched. The realization that spraying insects killed animals two or three trophic levels further along sparked off the environmental movement in the West and began the tradition that pollutants were 'biomagnified' along food chains.

When we look at this process in more detail, we see the picture is not so simple. **Biomagnification** occurs if the concentration of a pollutant is higher in a consumer (or a plant) than in its diet (or the soil). We can thus measure the **concentration factor** (CF):

$$Cf = \frac{\text{The concentration of the pollutant in the consumer}}{\text{The concentration of the pollutant in the diet}}$$

When CF > 1, we have biomagnification; if we include both the diet and water as a source for aquatic organisms then we have **bioaccumulation**; **bioconcentration** refers specifically to uptake from water alone. To have a CF greater than 1 the consumer's rate of uptake must exceed its rate of loss.

This equation is equivalent to the assimilation efficiency used to measure energy transfer in Box 6.2. Because it is passed on in the diet, a trophically mobile pollutant will trace the energy pathways within a food web.

These are very simple measures, so how could they be misleading? First, all of these terms are time dependent: levels in the diet change continually, as do tissue levels in the predator. For this reason a concentration factor only has meaning when the two terms are in balance with each other. To be in equilibrium with its diet, a consumer must change in step with dietary concentrations. Rarely are these conditions properly considered in collecting specimens from the field.

Next, should we measure the whole body concentration or just those tissues likely to become part of the diet

of the next trophic level—say in the fat and muscle? Often, much of the pollutant burden will remain in the leftovers—in the bones, shell, or hair—and so is lost from the food chain we are mapping. Comparing whole-body concentrations between different species can therefore be misleading. In the same way, a comparison of a single prey species with a predator that catches several prey species is also too simplistic; a food chain is only one subset of possible routes through a food web. Indeed, many predators are omnivores and feed on more than one trophic level. Demonstrating bioconcentration means placing a species in its appropriate trophic position and that requires detailed knowledge of the larger food web.

Neither is the pollutant necessarily transferred, unaltered, along the food chain. This can be the case for elemental pollutants, such as toxic metals, but many organic compounds are degraded by the metabolism of plants or animals. Even DDT, which is highly insoluble in water, can be degraded at a very slow rate, and its concentration falls as it is broken down in some animal tissues. Another factor is body size; large animals tend to eat more and consume a greater mass of a pollutant, but lose it more slowly because they have slower rate of metabolism per unit of body weight. Thus, the larger animals found at the end of food chains may have a higher concentration simply because they are bigger, not necessarily because of their trophic position.

Once we begin to address these questions (and consider whether they have been addressed in previous studies), we realize that biomagnification does not occur for every pollutant, for every top predator. Take, for example, the polar bear. Susan Polischuk and her colleagues studied the concentrations of various organochlorine (OC) pollutants in these top predators and showed that DDT (and its derivatives) and PCBs had very different dynamics.

Organochlorines have an affinity for the lipid tissues the bears lay down when they are well-fed. During the winter, when they hunt from the ice, the bears of the western Hudson Bay feed on ringed seal (*Phoca hispida*) often preferentially eating the blubber of their prey. A well-fed bear can have 50% of its body mass as lipid, though by the end of the summer, after fasting on land, this may fall to 10%.

Polischuk and her colleagues measured OC levels in the fat, blood and milk before and after the bears under went their summer fast. The fast lasted an average of 56 days, during which time the bears do not defaecate, so pollutants could not be lost via this route. Their fat reserves are used to maintain the bear through

BOX 6.4   Continued

the fast so if a pollutant is not broken down by the bear's metabolism we would expect its concentration to rise as the bear gets thinner. The researchers found that PCB concentrations did rise, but the DDTs declined; DDT was metabolized but the PCBs were not. John Kucklick and his colleagues have found the same pattern in Alaskan polar bears, where again PCBs are always retained, but at very different levels in different populations (Figure 6.16).

The value of this study is in the steps it took to make valid comparisons, by separating out the effect of gender, by comparing the same bear before and after fasts, and looking specifically at the lipid tissues. Indeed, by comparing concentrations of OCs in the blood and milk with the fat tissues, it suggested that the different tissue fractions were close to equilibrium with each other on both sampling occasions. PCBs are probably lost through subsequent excretion, and in the case of nursing mothers, with the milk. Because these compounds are not metabolized, milk concentrations rise during the summer fast, at the time the cubs are becoming stressed by a lack of food. At this crucial stage, the cubs are being fed increasing concentrations of a pollutant that may affect their development (see First Words).

Interestingly, polar bears have a great proficiency in metabolizing DDT and this is one reason why the polar bear, a top predator, does not show biomagnification for DDT. Bioconcentration is far from inevitable for a pollutant and concentration factors can show considerable variation both between populations and between species.

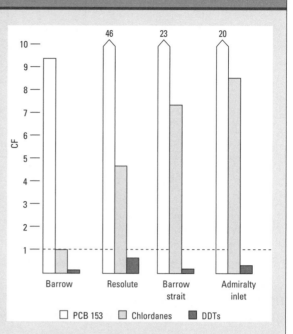

FIG. 6.16 The concentration factors (CF) of organochlorine pollutants in the lipid tissues of polar bears from four populations in Alaska, compared to the blubber of its main prey species, the ringed seal. DDT and its derivatives never achieve a concentration factor of 1 whereas PCB 153 is always highly biomagnified. Yet, as with other OCs (such as chlordanes), there is little consistency in CFs between neighbouring populations.

6.5 Working the system

Some of the shortest food chains are the ones we use ourselves. As omnivores, agriculture allows us to feed as both herbivore, consuming our crops directly, or as carnivores feeding on our herbivorous livestock (Figure 6.17).

From the earliest times, human beings responded to the patterns of productivity in the world around them, but a capacity to manipulate primary and secondary productivity marked the outset of civilization. Jacob Bronowski maintained that the change

from nomad to **village agriculture** was the largest single step in the *Ascent of Man*. Since then, our ability to work the system has been a key factor in our success as a species.

Humankind started out as **hunter-gatherers**—relying on what could be collected or hunted down (Figure 6.18). This way of life is still practised today by many peoples in tropical and sub-tropical areas, among them the bushmen of the Kalahari and the Australian aboriginals.

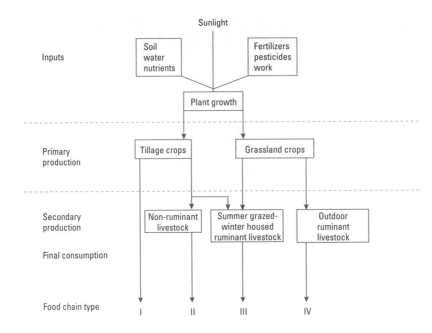

FIG. 6.17 A 'web' of agricultural food chains, ranging from (I) humans directly consuming crops and (II) intensive livestock rearing to (III and IV) the grazing of ruminants.

FIG. 6.18 A checklist of agricultural systems. The five principal agricultural systems are ranked in terms of their degree of intensity and the amount of mobility in the societies in which they occur. These range from the highly mobile, low-intensity lifestyle of hunter-gatherers through to the highly settled, intensive practices of modern agriculture.

Although it can never support a large population, certainly in semi-arid areas, hunter-gathering still represents a viable and sustainable feeding strategy. Energetically, at least, it provides a relatively high return on the energy and effort invested—about 10 times as much. By contrast, the epitome of intensive western agriculture, poultry production, returns just half of the energy put in.

Nomadic-pastoralists are found over a much wider range of habitats, primarily because they follow or lead herds of semi-domesticated or domesticated grazers searching for suitable forage. The Lapps of northern Europe follow the movements of reindeer to exploit the brief summer of the tundra. The Bakhtiari of Iran and Iraq herd their domesticated sheep and goats over mountain ranges to avoid summer droughts and to find sufficient grazing. Herdsmen of the Alps, the northern Italian *alpeggios*, Spanish *dehesas* and *montados* of northern Portugal also herd their stock from one place to another according to the season.

This seasonal movement of livestock is termed **transhumance** and, like hunter-gathering, is a long and ancient tradition which is practised by some very ancient cultures. It is also has more recent equivalents, such as the ranchers of the Rocky Mountains, who exploit the high mountain pastures that briefly flourish after the winter snows have melted. Alpine pastures have been traditionally managed in this way, with herdsfolk moving their stock (and their homes) to the higher slopes during the summer months, returning to the valleys for the winter. The practice requires a highly integrated human society where those who herd stock are supported by others in the valley who cultivate crops and make hay for use in the winter months.

Cultivation of crops probably developed alongside the domestication of the first agricultural animals and almost certainly followed the evolution of a series of heavy-seeded grasses (Box 2.5). Indeed, the sowing, harvesting, threshing, storing and trading of that seed prompted the settled existence and societies that dominate most cultures today. As agriculture has developed and we have learnt to play the energetics of ecosystems, so our relations with the living world have become increasingly sophisticated, if somewhat strained.

Grazing and the cultivation of plants dominate world agriculture. Of the 36 per cent of the Earth's land surface in agricultural production, 11 per cent is used for cultivation and the remaining 25 per cent is given over to grazing. At the small scale, the most established form of animal husbandry is the **infield–outfield** system, characteristic of highly settled village communities. Generally, it operates on a roughly 10 : 1 ratio of outfield to infield. The outfield is relatively unproductive land that is unfenced, and used for grazing by domesticated livestock. The infield surrounding the settlement intensively cultivated, growing both food for the villagers and winter fodder, but is primarily used as meadowland for high quality grazing.

Using this system, farmers can obtain a return from the outfield, exploiting its productivity and nutrients, whilst making minimal investment in its management. In contrast, considerable effort, principally cutting or mowing, is used to manage the infield grasslands. While the livestock are grazing the extensive lands of the outfield, the hay meadows are allowed to grow unhindered for several weeks. After their grasses have set seed, the standing crop is cut for hay. As the vegetation dries, its seeds fall to the ground, providing new seedlings for the grassland sward. After the hay has been collected, livestock may be allowed to graze the infield (aftermath grazing) and are then fed on the hay during the winter.

Meadows (and other grazed grasslands) can have very rich floras—a square metre of chalk grassland can support more than 40 species of flowering plants. This diversity follows from the frequency of disturbance and release of nutrients that the grazers facilitate. Grazers remove some of the biomass of the faster growing grasses, recycling its nutrient content in their faeces and urine.

New methods of intensifying grass production are bringing about changes in these meadows. Old pastures are oversown with vigorous agricultural grasses, such as perennial rye grass (*Lolium perenne*), supported by the use of inorganic fertilizers. These out-compete the other species and wild flowers that do not respond well to the increased nutrients are lost (Section 4.3). Even greater changes follow when the meadow is used for silage production. By mowing several times during the year, the farmer collects the net production of the meadow and stimulates further growth of its grasses. The clippings are then fermented to break down part of their cellulose, increase the nutritional value of the fodder and reducing its bulk. Unfortunately, frequent cropping means that the grassland and its wild flowers get little chance to set seed and, inevitably, floral diversity declines. In Britain alone, 95 per cent of flower-rich hay meadows were lost during the last 50 years (Plate 6.1), endangering a large number of wild flower species. Many temperate countries have suffered similar losses and now have programmes to promote traditional grassland management.

Why grow animals?

The metabolic costs and variable efficiencies of each trophic level mean that energy is lost with each transfer along a food chain (Boxes 4.1 and 6.2). Why then do we not dispense with grazing land and its animals, and instead grow crops we could consume ourselves? Surely, we would then recover more of the energy entering the system.

One simple answer is that, as omnivores, we lack the physiology to assimilate much of the energy available in most plant tissues. Instead, we use the capacity of herbivores to release this energy and convert into a form we can assimilate by consuming their tissues (Figure 6.17). Grazers represent an energy-efficient means of harvesting primary productivity. They range over many hectares of land, concentrating part of its energy in their tissue in form we can digest. This is particularly true of ruminants, which, with their symbiotic microorganisms, are able to digest cellulose, the major product of primary production and which is otherwise unavailable to us (Box 6.5).

BOX 6.5   Rumination

Some herbivores have a symbiotic relationship that allows them to recover energy from cellulose which would otherwise be excreted. All animals have communities of bacteria in their gut, but ruminants—sheep, cattle, deer, antelope, gazelles and giraffe—have developed this association to exploit the degradative powers of cellulose degraders. They are fermentation vessels on legs, operating a highly controlled continuous reaction is an anaerobic ecosystem within their gut.

The ruminant stomach is divided into four compartments and contains between 100 and 1000 million bacteria per millilitre and roughly half that number of protozoa which live off the bacteria. That is, they form a decomposer food web within the host.

The animal maintains a steady environment through its own body heat, keeping the pH of the reaction vessel constant (by producing copious amounts of alkaline saliva) and by providing a supply of vegetation for the bacteria to work on. After it has been pre-processed by chewing, food is passed into the rumen for fermentation.

From here it is regurgitated at regular intervals to be re-processed by 'chewing the cud'. In this way, the ruminant breaks the vegetation into progressively smaller fragments, providing the bacteria with a greater surface area on which to work.

The microbes can be grouped into guilds that specialize in attacking particular components. Cellulolytic bacteria break down cellulose, whilst the amylolytic bacteria work on starch. Some are able to deal with both. The main products of the fermentation are volatile fatty acids, such as lactic, butyric, propionic and acetic acids which the host is able to absorb across the rumen wall into the bloodstream. As with all anaerobic fermentation, the process produces large amounts of carbon dioxide and methane, which the ruminant 'vents'.

Proteins within the food are broken down into their constituent amino acids in the final chamber—the abomasum. This differs from the others since it is highly acidic and that means death to microbes that pass into it. Eventually, the bacteria themselves become part of the

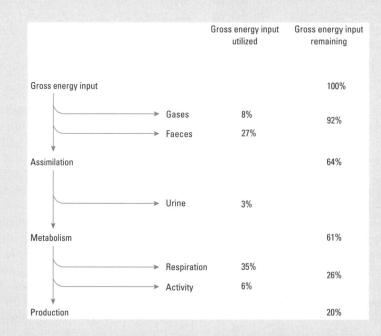

FIG. 6.19 Energy budget of a ruminant, charting the fate of energy taken in as food (gross energy intake). Losses occur at various stages in the process, with the largest being due to metabolic costs. The last column shows how much of the original energy is left at each stage. Eventually, 20% becomes fixed in the cow's tissues.

BOX 6.5   **Continued**

diet of the ruminant and represent an important part of its protein intake.

The benefits of carrying and maintaining a large bacterial culture in an enlarged stomach are considerable (Figure 6.19). Forage, such as hay can contain as much as 30% cellulose and being able unlock an energy source denied to most other herbivores offers a key advantage to the ruminant. The partners in the mutualistic association, the bacteria, benefit from the benign conditions in most of the animal's stomach.

There is a third, more recent, beneficiary from in this relationship—the many people for whom ruminants are the mainstay of their agriculture. Compared to pigs and poultry, which are able to convert 14 and 12%, respectively, of their gross energy intake into production, ruminants are much more efficient, at around 20%. Cattle benefit from our protection and husbandry and that has enabled them to become one of the most populous mammals on the Earth. We, in turn, benefit from their capacity to convert low grade plant material into protein and fat.

There are also sound ecological reasons why cropping is often not the most appropriate system. Much of today's pasture became grazed because it could not be used for anything else. Outfields and rangelands are frequently rough, stony places, with steep hills, deep valleys and poor soils. This makes them impractical and costly to cultivate and to maintain, so it is no accident of history that these areas have escaped cultivation.

Despite their limited structural complexity, grassland communities can be highly productive. Alpine meadows, for example, can produce as much as 1200 g/m^2 a year and this rises to 1500 g/m^2 for prairies. This productivity is almost double that of crops grown for our consumption or for fodder (Table 6.2). Because our crops are in the soil for only part of the year they cannot match the primary production of permanent pasture.

The soil beneath the pasture is also better able to maintain its structure and decomposer community. With minimum disturbance, the grass sward provides year-round protection from severe weather and its roots bind topsoil preventing erosion. Faeces left by grazers and the remains of the overlying vegetation raise its organic content, enabling the soil to hold water in its upper layers where it is of most use to plant life. Again, compared with other crops, grass has a more efficient water budget and there are a large range of grass species able to grow under all but the most extreme rainfall regimes. In contrast, the water requirements of crops can be considerable (Table 6.6) and may require regular irrigation. In

TABLE 6.6 Water requirements of crops. The amount of water (in grams) required to produce 1 g of dry weight of foodstuff.

Rice	710
Potato	636
Oats	597
Wheat	513
Maize	368

these situations, grasslands may be the only viable option.

Getting the grazing right

In many areas, it is the grazing activity of the livestock which maintains pasture and prevents it from becoming scrub or woodland. In such cases, the grazers are acting as keystone species (Section 3.5). However, as we saw earlier with the browsers of the African savannah, above a certain level, their feeding can damage the system's capacity to recover.

Grazers are selective and will continue to graze their preferred food, even when it becomes scarce. On the upland pastures of northern Britain, hill-farming has always been a tough existence for farmer and stock alike. To keep sheep-farming viable (and prevent rural depopulation and loss of Britain's open moorland) financial incentives were used to support upland grazing. Unfortunately, the incentives took the form of 'headage payments' whereby farmers were paid a subsidy for every sheep they had. Not suprisingly, this led to widespread overstocking of hill country and considerable environmental degradation.

Sheep prefer bent and fescue grasses (*Agrostis* and *Festuca* species) and these were the first species to decline with overgrazing. Mat grass (*Nardus stricta*), on the other hand, is tough and unpalatable, and tends to be avoided by the sheep. Left ungrazed, *Nardus* rapidly spread over large areas where the other grasses were now absent. The deterioration in the grazing quality of the hill pastures was not only a concern to the farming community but conservationists too. Hungry sheep headed for moorland and started to overgraze heather. Because heather plays a key role in the moorland ecosystem, the loss of such an important primary producer had severe implications for the animals it supported further along the food chain. Once the problems were identified, measures were taken to adjust stocking densities to suit the carrying capacity of each area rewarding farmers who stocked their land to an optimum density.

In some cases, undergrazing can be as much of a problem as overgrazing. As we saw earlier, moderate grazing controls and rejuvenates the plants within the sward, raising their productivity. It also helps to accelerate nutrient cycling within the ecosystem, promoting an active decomposer food web. Small patches of bare ground, dung heaps and other localized disturbance provide conditions for young seedlings to become established.

In the absence of grazing, changes start to occur. There is an increase in the standing crop, a gradual accumulation of biomass, and the community may slowly start to revert to scrub and, possibly, woodland. Traditional agriculture therefore has techniques to compensate for low or no grazing. Most of these involve some form of biomass removal, principally cutting and burning. Farmers in the American mid-west now recognize that wildfire plays a key role in the maintenance of the prairie ecosystem and incorporate it into their grassland management. In this way, they create localized disturbances to maintain its high floral diversity, preventing competitive species from becoming dominant. Mediterranean type environments also depend on disturbance in the

form of grazing and fire (Section 5.1). Rural depopulation in the Mediterranean has resulted in a cessation of grazing so that much of the upland areas is reverting to scrub (Box 5.6).

Added energy

Agriculture uses energy to manipulate food chains and food webs. We try to prevent some species (weeds and pests) from using the energy entering the system and ensure that most goes to those we are cultivating. We make an investment in time and effort, which can be measured by the energy applied to the system. This is termed its **energy subsidy**, and refers to everything from one person turning the soil with a hoe, to the waves of combine harvesters that pass over our wheatfields (Figure 6.20).

We use energy to remove competition from non-crop plants by tilling the soil or by the application of herbicides. Fossil fuels are used to manufacture and apply fertilizers, enabling us to side-step nutrient limitations on crop productivity. The energy we use to manufacture agricultural equipment or the fuel used in farm machinery, all represent indirect costs, subsidizing our cultivation.

These inputs vary with farming practices, as does the amount of energy in the form of human labour.

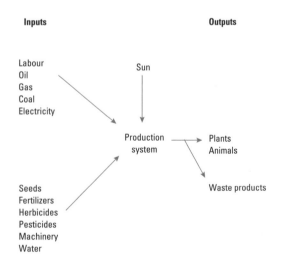

FIG. 6.20 Energy subsidies can be direct energy inputs in the form of fuel, or indirect inputs from materials (e.g. fertilizers and pesticides).

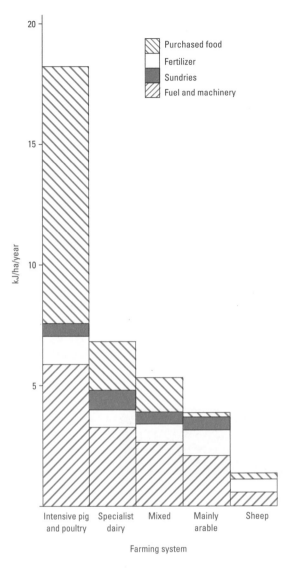

FIG. 6.21 An energy subsidy breakdown for selected agricultural systems. The graph illustrates the main energy inputs for five major agricultural systems, from intensive 'factory farming' of livestock, through to extensive grazing of sheep.

TABLE 6.7 Energy inputs and protein yields of four major agricultural systems

Agricultural system	Total energy input (10^6 kJ/ha)	Protein output (kg/ha)
Hill farming (sheep)	0.6	1–1.5
Mixed farming	12–15	500
Intensive crop production	15–20	2000
Intensive animal production	40	300

An African pastoralist, for example, will expend 21 420 kJ per hectare per year—43 per cent of the energy costs of an American maize farmer. Figure 6.21 shows how types of energy input vary with different farming systems. Feedstuffs, for example, are a principal source of energy for intensive livestock production. In Table 6.7, we see the effects of energy input on the protein output of four agricultural systems, ranging from extensive farming through to intensive crop and animal production. (Note that the link between protein output and energy input is problematic since not all organisms will produce protein in the same proportion to their energy intake.)

The returns are greatest in intensive crop production, where the food chain is shortest (chain I in Figure 6.17). The three other systems (chains II–IV) involve a third link—the livestock—and this leads to a reduction to the energy available for consumption.

Intensive animal production is costly, but it is a reliable and convenient means of producing protein. Not only is energy added with the food supply, animals reared indoors often require a highly controlled internal environment and fuel is required to maintain lighting, heating and ventilation. Western dairy farming also demands a large energy subsidy, both in terms of handling the animals and the milk they produce.

Extensive grazing, in the form of rangeland and pastoralism, remains the most widespread means of producing animal protein. Whilst protein returns are comparatively small, grazers can go where no plough can and are able to capture the productivity of land which might otherwise be agriculturally unproductive. Sheep farming incurs the lowest costs since the bulk of their grazing is confined to the extensive outfield or unmanaged pasture. These animals only need an occasional food supplement and the principal costs involve the management of periodic grazing within the infield entailing a small input of human labour.

The energy subsidy buys us out of some of the constraints which limit productivity, but not all of them. Eventually the costs start to outweigh the benefits and either returns on the investment decline or deterioration in environmental quality follows. A dramatic example of this can be seen in southern Europe where the intensification of olive oil production has resulted in widespread environmental change. This has been further fuelled by financial incentives that failed to take the needs of the environment into account (Box 6.6).

Helen Caraveli recently used changes in the energy subsidy of European agriculture (in the form of tractors and fertilizers) as a measure of intensification in Mediterranean farming (Table 6.8). She was able to show that over a 15-year period Mediterranean countries had intensified at a disproportionate rate when compared with their northern Europe counterparts. During this time, tractor uses in the Mediterranean increased by 70 per cent (compared with a 7 per cent decline in northern Europe). Given the small size of average Mediterranean farms and the nature of their soils and landscape, the potential for adverse environmental impact is considerable.

It is not only the cost of the energy that is important, but the way we use it. For example, phosphate fertilizers allow us to by-pass the limits on productivity imposed by the low availability of phosphates in most soils (Section 7.1). These fertilizers are manufactured from phosphate-rich rocks at considerable energy cost. However, a significant proportion of that applied does not fertilise the crop but is lost with run-off, leading to the enrichment of soils and aquatic systems (Box 7.3). This can lead to dramatic changes in the species composition of their communities.

In a different way, simply adding more water to an ecosystem can cause long-term damage. Pumping groundwater, to irrigate crops more suited to wetter climates, may deplete underground reserves and, without due care, it may also disrupt the nutrient balance of the soil (Box 7.1). This is a problem especially in arid areas (Section 7.1), where high levels of evaporation cause salts to accumulate in the surface layers. Around one-third of all irrigated land in India has been degraded as a result of salinization.

Intensive cultivation can also damage the soil by altering its structure and disrupting the soil biota (Table 6.9). Frequent ploughing promotes the rapid

TABLE 6.8 Agricultural intensification within the Europe between 1980 and 1995 as defined by energy subsidy in the form of tractors and fertilizer consumption

Country	Average farm size (ha)	Number of tractors per km² of agricultural land		Total consumption of fertilizers (kg/ha)	
		1980	1995	1980	1995
Northern countries					
Denmark	37.1	7.1	6.5	236	196
France	35.1	7.8	6.7	297	242
Germany	28.1	12.9	10.8	413	241
Netherlands	16.8	21.7	19.9	825	564
United Kingdom	67.3	7.3	8.4	294	502
Mediterranean countries					
Greece	4.3	3.6	7.5	134	154
Italy	5.9	8.6	13.7	170	176
Spain	17.9	2.6	4.0	81	92
Portugal	8.1	2.7	4.9	83	84

Table 6.9 Biological functions of soil organisms and management practices that affect them

Biological function	Functional group	Management practices affecting them
Residue breakdown/ decomposition	Residue-borne microorganisms, meso/macrofauna	Burning, soil tillage, pesticide applications
Carbon sequestration	Microbial biomass (particularly fungi), macrofauna	Burning, shortening fallow period after slash and burn, soil tillage
Nitrogen fixation	Free and symbiotic nitrogen-fixers	Reduction in crop diversity
Organic matter/ redistribution	Roots, mycorrhizas, soil macrofauna	Reduction in crop diversity, soil tillage fertilization
Nutrient cycling, mineralization/ immobilization	Soil microorganisms, soil microfauna	Soil tillage, irrigation, fertilization, pesticide applications, burning
Soil aggregation	Roots, fungal hyphae, soil macrofauna, soil mesofauna	Soil tillage, burning, reduction in crop diversity, irrigation
Population control	Predators/grazers, parasites pathogens	Fertilization, pesticide application, reduction in diversity, soil tillage

breakdown (oxidation) of its organic matter. When this happens, populations of decomposers and detritivores fall dramatically and can even crash, effectively decoupling the producer and reducer food chains. Without the soil-processing activity of decomposers, soil texture and its capacity to hold nutrients and water deteriorates. Without substantial plant cover, soils are more mobile and without a significant organic fraction soil particles do not bind to each other. As a consequence, vast amounts of soil are blown or washed away; some estimates suggest that the loss of topsoil amounts to about 1 per cent of the world's cropland each year. In the Andalucian region of Spain alone, an estimated 80 million tonnes of topsoil are lost each year as a result of intensive olive production (Box 6.6).

Around 80 per cent of soil erosion is directly attributable to human activity and cultivation practices which are unsympathetic to local conditions. Farmers try to achieve the highest levels of productivity, whilst also seeking to sustain this production over the long term (a principle we have met before in Section 3.3). Many traditional practices have evolved in response to the local conditions of climate, slope, geology and soils. Techniques such as shifting cultivation are those best suited to habitats where nutrient supply and soil need time to recover. Without sympathetic cultivation a good quality soil can be rapidly reduced to little more than a poorly structured dust, lost with the first rains or high wind.

Agricultural productivity is so central to our society that it has always been a focus for technology and innovation. Following the dramatic successes raising productivity with cheap artificial fertilizers in the 1940s and the 1950s, high hopes were held for solving food shortages. In the 1960s and the 1970s improvements in farm management techniques, plant and animal breeding programmes, pest and disease control all seemed to promise a way of feeding a rapidly growing world population.

The **green revolution** as it became known, hinged on the development of new varieties (cultivars) of crop plants. These were supported by a range of agrochemicals, pesticides and fertilisers and, in some places, new irrigation schemes. Cultivars were produced that allowed for easier mechanical harvesting and most importantly, increased productivity.

BOX 6.6   Mounts of olives

Olives are synonymous with the Mediterranean. They are inextricably linked with its landscape, peoples, history and culture. However, this icon of the Mediterranean and the environment that produces it are now under threat from agricultural intensification.

The olive-producing countries of the Mediterranean are so good at producing olive oil that they have become victims of their own success. Although olives are grown commercially in North Africa, the Middle East, California, Australia and Argentina, the European Union (EU) remains the world's major producer of olive oil. Together, Spain, Italy, Greece and Portugal dominate the market, producing around 80% of the world's olive oil from an area of 5 million hectares (Table 6.10).

However, all is not well in the olive groves. The farms of southern Europe are predominantly small, traditional family affairs (Table 6.8), but the 1960s and the 1970s saw an exodus of young people from mountains and in-land villages to the more prosperous coastal plains, to work especially in the tourist industry. This left an ageing population that was left to contend with the vagaries of farming where several good harvests might be followed by a disastrous one.

The EU subsidizes the farming industry through its Common Agricultural Policy (CAP). These provide a guaranteed price for goods regardless of the market value and also give welfare assistance to poorer farmers though the European Social Fund. Whilst the motives are laudable, helping stem the tide of rural depopulation, their less desirable side-effect is to promote intensification. The subsidies for olive oil are linked to

(a)

(b)

(c)

TABLE 6.10   Olive oil production within the European Union (1997 figures)

Country	Olive area (ha)	Approximate percentage of the world output
Spain	2 000 000	28
Italy	1 100 000	24
Greece	900 000	16
Portugal	340 000	2
France	40 000	>0.1
Total EU	4 380 000	70

FIG. 6.22 The changing face of olive production. (a) Traditional olive groves with their old trees, terraces and rich biodiversity. (b) Intensive olive production—young trees grown in serried ranks in almost clinically tidy conditions. (c) Casualties of change—ancient trees grubbed up to make way for intensification.

BOX 6.6   **Continued**

Table 6.11 Comparison of three types of olive production

	Traditional	Intensive	Semi-intensive
Tree characteristics	Big and old	Smaller and younger due to replanting	Dwarf varieties replanted regularly
Tree density	80–150 per ha	150–200 per ha	200–400 per ha
Terraces with supporting walls	Common	Occasional	Rare
Understorey weed control	Harrowed occasionally	Harrowed/cut repeatedly	Controlled with herbicides
Grazing	Rare or common	Rare	No
Chemical inputs	Very low	High	High
Irrigation	Uncommon	Increasing	Common
Harvesting method	By hand	By hand or mechanical	Mechanical
Typical yield	1200 kg/ha	2200 kg/ha	5500 kg/ha
Consistency of annual yield	Very low	Low	High
Soil erosion	Usually low	Often very high	Medium
Biodiversity	High	Low	Very low
Landscape value	High	Low	Low
Other environmental impacts	Fire prevention in marginal areas	Pesticide pollution, irrigation reservoirs	Pesticide pollution, irrigation reservoirs
Subsidy per ha (Euros)	200	800	1500

production performance and the more farmers can produce the more money they can claim. An intensive farm can produce 10–20 times more olives than a traditional olive grove (Table 6.11).

Traditionally, olives were harvested from ancient, large-canopied trees, many of which are over 500 years old. These low-input plantations have scattered trees and are typically planted on terraces banked by stone walls. Management here is minimal with little or no chemical inputs and harvesting is by hand. Traditional olive groves tend to be multipurpose areas—grazed by sheep or cattle and occasionally ploughed for the small-scale production of crops beneath the trees.

Needless to say, the productivity of this land can be increased quite easily with inputs such as fertilizers, pesticides, and irrigation. In these groves there is no room for the grazers or crops, as extra trees are planted to boost olive production. Many farmers have gone further, uprooting the ancient trees and replacing them with closely spaced modern varieties in a monoculture similar to arable farming. Tree densities are increased five-fold with a working life of just 25 years, after which they are ripped out and replaced with new stock.

All this has led to a profound change in the landscape and ecology of olive-producing areas. Terraced groves are increasingly being replaced with plantations filled with hundreds of little trees (Figure 6.22). Meanwhile old hill farms continue to be abandoned as their owners clear away natural areas in the valleys better suited to intensive olive production. The environmental costs are great including soil erosion and loss of biodiversity. The loss of older trees has resulted in fewer nesting sites birds such as little owl (*Athene noctua*), which need holes in the trunk. Closely spaced trees are also of little use to ground nesting and feeding birds like stone curlew (*Burhinus oedicnemus*), quail (*Coturnix*

BOX 6.6 Continued

coturnix) and partridge (*Alectoris rufa*). Plants too, are affected as the highly diverse wildflower communities of traditional olive groves are regarded as weeds in intensive plantations and are ploughed or sprayed to remove competition.

Currently, the CAP olive oil subsidy regime is under review. The proposed changes seek to shift the emphasis away from production and towards a more sustainable system. Incentives that reward good environmental practice such as reduced usage of fertilizers, pesticides, and irrigation may also be introduced and there may be bounty payments to farmers who retain ancient trees, maintain terraced groves, and manage wildlife habitats on their land.

High-yielding varieties were selected for particular characteristics, such as high leaf:stem and low carbon : nitrogen ratios. That is, they were bred to produce as many leaves and with as little stem as possible so as to pack protein into leaves, fruits or seeds. High-yielding varieties of rice can divert a massive 80 per cent of net production into their seeds, compared to the more normal 20 per cent.

However, the promise of many of these new strains has not always materialized. Crops that produced high yields in experimental trials failed to live up to expectations once out in the field. Sometimes the reason was blindingly obvious; varieties bred by developed countries to help feed developing countries, ended up being grown without Western technologies to support them. Local farmers found that the energy subsidy in terms of fuel, machinery, fertilizers and pesticides was too costly (Plate 6.2).

Important lessons can be learnt from past mistakes. Today's breeding programmes have to aim to fit crop varieties and cultivation techniques to the environment in which they will be grown. Applying ecological principles to agricultural systems enables us to regard agricultural land as a habitat like any other and to compare it with other, more natural communities (Figure 6.23). We can then choose technologies that are more suited to local economies and ecologies and which respect local traditions and practices—from established patterns of crop rotation to pest control methods.

Many people see land degradation, and its implications for food and fuel production in developing nations, as the major problem facing the globe over the next 50 years. Over the last 50 years, as much as

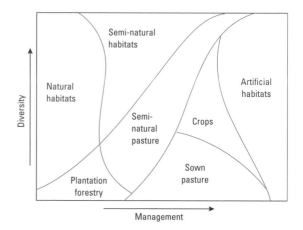

FIG. 6.23 A schematic template of diversity versus management for a range of habitat types. Management input increases as one moves from natural to artificial habitats. In many semi-natural ecosystems our intervention helps to maintain a high diversity though some forms of agriculture can produce very species-poor habitats.

9 million hectares of productive land has been severely degraded, mainly in over-populated areas (Box 5.6). Perhaps the most significant development at the United Nations Convention on Environment and Development, the Rio Convention of 1992, was Agenda 21, and its *nonbinding blueprint for sustainable development*. This, at least, recognizes that future economic development, both agricultural and industrial, needs to be sustainable. The ultimate test of sustainability will be the capacity of agricultural systems to maintain productivity without signs of increased species loss or environmental degradation.

SUMMARY

Primary producers convert sunlight into chemical energy through the process of photosynthesis. This involves absorbing light with the pigment chlorophyll and using the energy to fuel chemical reactions that convert carbon dioxide and water into sugars with oxygen as a by-product. The overall efficiency of the process is around 2 per cent. Some plants have modified the way in which they capture carbon dioxide to help conserve water in dry environments.

Much of the energy fixed in gross primary production (GPP) is expended in respiration. Net primary production (NPP) is the energy fixed in the tissues and available to higher trophic levels. Herbivores consume this primary production and begin a line of consumers that form a trophic or food chain. A food chain marks one pathway through which energy moves through a community and most are rarely more than four or five links long. This may be because they are limited by the inefficiency of the energy transfers from one trophic level to another, but evidence suggests that it is more closely tied to the structure of the larger food web. It seems that food webs show some consistency in their structure and organization as a result of species interactions, including energy transfers.

Human beings use a range of agricultural strategies to exploit primary production and the energy at different trophic levels. Some of the most ancient, from hunter-gathering to transhumance, are still practised in different parts of the world. Although agriculture seeks to improve the productivity of an ecosystem by applying energy subsidies (to remove competitors, supply nutrients, and so on) these only produce sustainable systems where they are sympathetic to local ecology. In the last 50 years, we have learnt valuable lessons from our attempts to manipulate the energetics of agricultural ecosystems which should inform future policy and practice.

FURTHER READING

Tivy, J. 1990. *Agricultural Ecology*. Longman, Harlow. (A classic text in the area of agricultural ecology—a helpful introduction into agricultural systems and issues such as the energy subsidy.)

Tivy, J. 1993. *Biogeography: A Study of Plants in the Ecosphere*, 3rd edn. Longman, Harlow. (A useful introduction to the subject which examines biogeography in its broadest sense (including agricultural ecosystems). This text considers ecological process such as producer and reducer food chains.)

Whittaker, R. H. 1975. *Communities and Ecosystems*, 2nd edn. Macmillan, New York. (A wide-ranging book whose title describes its content exactly. In it, Whittaker reports his original research on the productivity of ecosystems.)

Salisbury, F. B. and Ross, C. W. 1992. *Plant Physiology*, 4th edn. Wadsworth, Belmont, CA. (A standard text on all aspects of plant growth and development—this book is also particularly useful as an introduction to environmental plant physiology.)

WEB PAGES

www.unep.org
United Nations Environment Program—this has numerous internal and external links to sustainability issues, including the Earth Summits in 1992 and 2002.

www.panda.org/epo/agriculture
The World Wide Fund for Nature web site—this site is particularly useful for information on agricultural sustainability and conservation. It also provides information on the European Common Agricultural Policy and is a link into issues surrounding agricultural intensification in the Mediterranean along with particular issues such as the CAP olive oil regime.

EXERCISES

1 Using the data in Table 6.4, arrange the following list of animals order of increasing production efficiency:

deer, dragonfly, mouse, pigeon, snail, stickleback

2 Select the correct answer to the following statement. In terms of net primary productivity the most productive ecosystem on Earth is

(a) the tropical rainforest.

(b) the open ocean.

(c) intensively cultivated areas.

(d) algal beds and reefs.

(e) swamps and marshes.

3 Insert the appropriate words to complete the following paragraph (use the list of words below; some words may be used more than once and others not at all)

The route by which ____ moves through an ecological community can be described by a ____. In the first stage, radiant energy is converted into ____ energy by ____. These are then consumed by ____, animals that feed on ____ material. These, in turn, become food for ____ that incorporate the energy into their own tissues. The topmost ____ level is occupied by____, carnivores that prey upon other____.

animals	food chain	radiant	trophic
carnivores	herbivores	secondary producers	
chemical	plant	secondary consumers	
energy	primary producers	tertiary consumers	

4 Consider the table below and explain what it means in terms of decomposition in the various ecosystems listed.

Vegetation type	Litter layer mean biomass (tonnes/ha)
Tundra shrub	83.5
Taiga pine forest	44.5
Temperate oak forest	15.0
Subtropical forest	2.0
Tropical rain forest	0.0

5 Pair the following measures of efficiency a–c, with the matching half of the equations (i)–(iii).

(a) Assimilation efficiency

(b) Growth efficiency

(c) Production efficiency

(i) $\dfrac{\text{Energy fixed in tissues (P)}}{\text{Energy assimilated (A)}}$

(ii) $\dfrac{\text{Energy fixed in tissues (P)}}{\text{Energy consumed (C)}}$

(iii) $\dfrac{\text{Energy assimilated (A)}}{\text{Energy consumed (C)}}$

Now answer the following question: A weasel receives 24.8×10^3 kJ and fixes 546 kJ in its tissue. What is its production efficiency?

6 Match the following extremophiles with their favoured extreme environment.

Extremophile group

(a) Acidophile

(b) Halophile

(c) Lithophiles

(d) Radiophile

(e) Thermophile

Extreme environment

(i) acidic conditions

(ii) radiation

(iii) rock

(iv) salty conditions

(v) temperature—hot or cold

7 Select the correct answer to the following statement. Energy is released for use in metabolism by the process of

(a) aspiration,

(b) inspiration,

(c) perspiration,

(d) respiration,

(e) transpiration.

8 Construct a simplified food web of a garden lawn ecosystem using the information provided below about the feeding preferences of common garden inhabitants.

Organism	Type of feeder	Diet
Aphid	Herbivore	Live plant material
Blackbird	Omnivore	Worms, slugs, caterpillars, spider, and centipedes
Caterpillar	Herbivore	Leaves
Centipede	Carnivore	Millipedes, aphids, woodlice, and springtails
Earthworm	Detritivore	Detritus
Ladybird	Carnivore	Aphids
Millipede	Detritivore	Detritus
Slug	Herbivore	Leaves, stems, and roots

Spider	Carnivore	Springtails and caterpillars
Springtail	Herbivore/detritivore	Living and dead plant material
Woodlice	Detritivore	Detritus

9 Place the following agricultural systems in order of their increasing management and energy requirements:

hunter-gatherer, infield–outfield, intensive agriculture, nomadic pastoralism, transhumance.

Tutorial/seminar topics

10 Is there such a thing as sustainable agriculture?

11 Is there an ecological justification for becoming vegetarian?

7 BALANCES

The eruption of Mount Ruapehu, New Zealand.

BALANCES

Geologists delight in describing soil as 'rock on its way to the sea'. This is not just evidence of a wry sense of humour but a very accurate picture of one long-term process of the planet. Rocks inevitably abrade through the action of wind, water, and ice. Their particles, though briefly held in soils, eventually find their way to the bottom of seas, perhaps to be resurrected as new sedimentary rocks. Some of their elements may be released into solution and be diverted, through plants and microorganisms, into other parts of the biosphere (Figure 7.1).

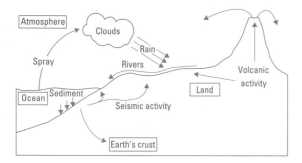

FIG. 7.1 Major geochemical processes. Minerals move between land, sea, and the atmosphere. Whilst some elements may be lost to the Earth's crust, others are returned to the ecosystem by volcanic and seismic activity.

In Chapter 6, we described the central role that energy plays in driving living systems and how it fuels metabolism and the building of new tissues. For this to happen, the raw materials have first to be acquired. Elements such as carbon, hydrogen, oxygen, nitrogen and phosphorus are needed to make the macromolecules from which living tissues are built. For the Plant Kingdom, the only source of these nutrients is minerals and for most animals their supply ultimately comes from plants. Without life, elements move in geochemical cycles of erosion and deposition. With life they move through biogeochemical cycles, diverted through living organisms, on their way to the sea.

The global flux of nutrients is largely determined by the energy available to drive key processes, both biotic and abiotic. Energy captured in photosynthesis also powers the processes that capture nitrogen, phosphorus, and other components, primarily through the activity of microorganisms within the soil or sediments and through plant uptake. Without these nutrients photosynthesis and primary productivity will be limited, and less energy will then be available to fuel nutrient movements through the rest of the community.

Occasionally, an excess of nutrients occurs which leads to changes in the structure of a community and we see a disturbance in what is normally considered to be a 'balanced' ecosystem. Later, we examine cases where the intricate links between the supply and demand for nutrients and the production it fuels have been disrupted, and consider how this damage may be cured or endured.

Sometimes, the nature and scale of environmental degradation is so great that ecosystems require wholesale reconstruction to restore both their abiotic and living components. Then, we attempt to rebuild communities from their constituent parts, and in the process, test our understanding of successional and community processes. Sometimes, nutrients are in such short supply that ecosystems need help re-establishing the complex machinery of their biogeochemical cycles, to reassemble the parts, the species, needed to make the whole thing work.

7.1 The nutrient cycles of life

We can think of the biosphere as being something like a machine (Figure 7.2), one that builds new tissues and new organisms, driven by the energy captured from sunlight. The life that emerges from the raw materials of minerals and energy is not only a product but an integral part of this machine. The interactions of the biota with their abiotic environment and with each other determine its operation. Through the

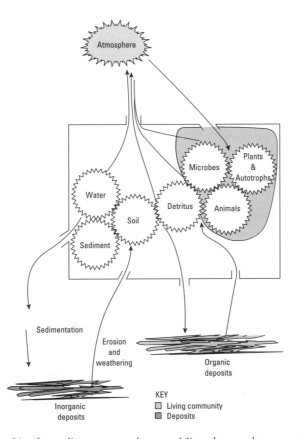

FIG. 7.2 The ecosystem as a machine for cycling matter and energy. Minerals move between the atmosphere and organic or inorganic reserves via the living and non-living systems that drive the system.

activity of living organisms, particularly the microbial communities, materials are collected, processed, and made available to different parts of the biosphere. In the process, living tissues are built and ecological communities constructed.

When energy is plentiful and conditions are equable, the speed at which the machine works is determined by the availability of key components: how much and how quickly each can be supplied. These elements are needed to construct the molecules needed for living tissues. Without them, production grinds to a halt or proceeds very slowly. Conditions that are not equable—not warm and wet—will also slow ecological processes. Polar regions are bathed in weak sunlight, adequate for photosynthesis but the low temperature reduces production and slows decomposition considerably. For example, in the taiga or northern forests (Section 8.2), calcium cycles through the soil and trees in about 43 years, compared with an average of just 10.5 years in tropical rainforests. For calcium the taiga is operating at a quarter of the speed of its tropical counterpart.

Different elements move through the biosphere at different rates. These are known as their **flux rates** and their speed depends upon the physical and chemical properties of each element, and the use to which living organisms put them. Some elements like nitrogen pass through the system relatively rapidly, whereas phosphorous cycles much more slowly. We can classify nutrient cycles also by the spatial scale over which they move. Phosphorus tends to move over short distances, usually within a very localized cycle, primarily because of the insolubility of its compounds. Nitrogen, on the other hand, is far more soluble and this makes it highly mobile: its salts are rapidly taken up by plants or soil bacteria, and it is readily lost with water percolating down through the soil.

Sedimentary deposits are often highly localized, concentrated in particular locations by biological activity. The nutrients locked in these deposits form a reserve that may not be tapped for tens of millions of years, until erosion or human activity releases them again. Some plant and animal remains enter the soil or aquatic sediments to form a labile (changeable) fraction and their nutrients recycle at much faster rates, especially if these deposits are regularly disturbed, say by an ocean current.

The supply of these nutrients has immense economic significance. Most of the nutrient capital of an undisturbed tropical forest is held in the standing crop, its vegetation, and animal life, with only a small fraction in the thin soil. For centuries, the indigenous peoples of these forests have worked with these nutrients cycles, establishing agricultural practices and traditions that limited their land use according to the nutrient budget of the forest—never clearing too much too quickly and regulating the size of their own group by social conventions. As a result, the forest recovers quickly when the people move on.

More recent colonizers have ignored ecological reality at considerable cost. Clear-felling the forest, or burning large areas followed by the use of Western agricultural practices, may produce a bumper harvest for one or two years, but thereafter yields drop. The small nutrient capital of the soil has been depleted and farmers then have to apply inorganic fertilizers to maintain economic returns. The nutrient capital of the forest was lost when the above-ground community was destroyed.

Tropical grasslands (Section 8.2) also show how nutrient supply regulates entire ecosystems. Working in the Serengeti, Sam McNaughton found that the availability of phosphorus, sodium, and magnesium indirectly controlled the density of wildebeest. Areas where the vegetation had higher levels of these elements were able to support greater numbers of large herbivores. Like all living organisms, the wildebeest act as nutrient pools—accumulating nutrients, re-distributing some as excretory products, passing some on to the future as young and surrendering the balance to the environment on death—either directly through decomposition or indirectly via carnivores and scavengers.

Water

Water plays a key role in liberating and moving many of the important nutrients, but is itself driven through a cycle—the **hydrological cycle** (Figure 7.3)—by solar power. Water enters the atmosphere by evaporation and from the transpiration of plants, both of which are driven by solar energy as heat and radiation. Indeed, the movement of water is one of the main components of the engine that disperses heat around the Earth (Box 8.1). High inputs of energy in the tropical regions raise temperatures and vapour pressures, forcing the moist air to higher altitudes or latitudes

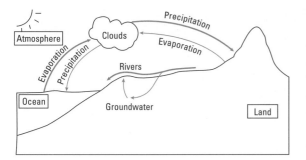

FIG. 7.3 The hydrological cycle. Water evaporates from the sea and is deposited on land where it enters the ground and eventually returns to the oceans via rivers. The thickness of the lines denote the relative amounts of water moving from one compartment to another.

where it cools and its water vapour condenses. The differential heating of land and sea create regular patterns of air movements (Figure 8.9) and where moist air cools its water precipitates as rain or snow.

Bands of precipitation form where cold and warm air masses regularly meet. Patterns of precipitation also depend on distance from the sea and topography

(the form of the land). Continental interiors, distant from any large mass of water, are typically deprived of rainfall. Similarly, land on the leeward side of mountains may receive little precipitation because the incoming air has shed its moisture as it was forced to rise and cool over the mountain range. Such areas are said to be in a 'rain shadow'.

An estimated 97 per cent of the planet's water is held in the oceans with just 0.001 per cent in the atmosphere at any one time. Less than 1 per cent occurs on land as ground- or soil-water or that flowing in rivers and lakes. The rest (around 2 per cent) is frozen in polar and glacier ice. Much of the water we use comes from streams and rivers or from groundwaters—that is, 'fossil water' within the aquifers of water-laden rocks. Humankind has developed a variety of technologies for exploiting aquifers, using natural springs or bore-holes to draw up water from deep rocks. When our abstraction exceeds the natural recharge rate the aquifer starts to be depleted. Our increasing dependence on water that fell as rain many centuries ago means that we can no longer afford to regard them as a renewable resource (Box 7.1).

BOX 7.1   The drying of Doñana

Doñana National Park lies on the Atlantic coast of Spain at the mouth of the Guadalquivir River (Figure 7.4). The 50 000 ha site was designated a national park in 1969 in recognition of its ecological value as one of finest examples of coastal wetlands in Europe. Home to over 361 species of birds, Doñana is internationally famous for the 6 million birds which visit the reserve on their migration between Africa and Northern Europe. With over 750 species of plants providing food and shelter to a wide range of animal species, it is also an important habitat in its own right.

The original rationale for the setting up of the national park was to protect Doñana from the threat of agricultural intensification.

However, this has not stopped development taking place along its perimeter and today Doñana is surrounded. This modern development has brought with it an insatiable demand for water—a demand that now threatens to turn Doñana into a desert. To the north lies the intensive farmland of El Rocio—much of it under glass—providing salad crops and strawberries for the

Northern European market. To the west, the holiday resort of Matalascanas witnesses a seasonal population increase of 200 000.

FIG. 7.4 Doñana National Park and World Heritage Site is an extensive area of wetlands and coastal dunes that have formed on the delta of the Guadalquivir River on the Atlantic coast of Spain.

BOX 7.1   Continued

Agriculture and tourism are both heavily dependent on water and draw it from a series of boreholes on the perimeter of the National Park. In excess of 4 million cubic metres are extracted each year, faster than the natural recharge of the groundwater. In addition, a sequence of 30 dams upstream on the Guadalquivir River further restrict water entering the site. As a consequence, the water table has fallen over the past 30 years drying out Doñana in the process—turning wetlands into coastal dunes. All of this has not been helped by the destruction of the native Phoenecian juniper (*Juniperus phoenicea*) woodland and subsequent commercial afforestation by umbrella pine (*Pinus pinea*) and eucalypts (*Eucalyptus* spp.), which represent another drain on Doñana's water table.

In some places, pine has invaded areas that were formerly dominated by heather (*Calluna vulgaris*) and green heather (*Erica scoparia*), so its spread closely matches the drying of Doñana. In their recent investigation, Juan Carlos Munoz-Reinso and his colleagues used specimens of *P. pinea* to pinpoint the time Doñana went into water deficit. They aged specimens growing in dried up pond beds and found that the problem started in 1973–74. Around this time, conditions became suitable for the pine to grow on sites that would otherwise have been too wet for it. Their surveys have also followed subtle changes in the species composition of Doñana's plant communities (Figure 7.5). For example, species that favour a winter water table level of 1–2 m (such as *Calluna* and *Erica*) have progressively given way to new communities dominated by *Halimium halimifolium*—a species characteristic of winter water tables at depths of 3 m or more.

As Doñana's water table continues to fall (at rates of 0.1–0.5 m per year), the nature of the National Park will continue to change and these may already be irrevocable. The growing number of self-sown trees further dry the soil and so deflect some areas along a successional trajectory towards woodland (Section 5.3). This is further complicated by the potential impact of climate change, such as changes in temperature, rainfall patterns, and increased sea-level.

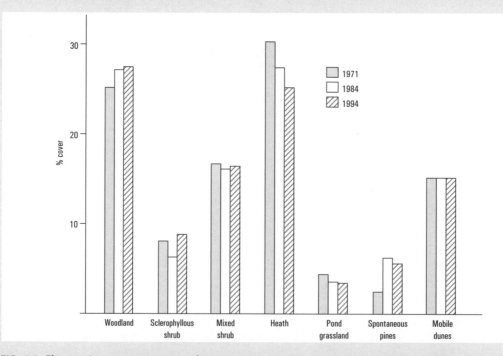

FIG. 7.5 Changes in percentage cover of seven vegetation categories in Doñana during the period 1971–94. The most notable changes have occurred in heath and pond grassland, which have declined as a result of falling water tables. Meanwhile, there has been an increase in pines invading the site as it dries and a smaller increase in juniper woodland and sclerophyllous shrub in response to the drier conditions.

Even where water is plentiful, its overuse can precipitate ecological disasters. This is especially true in arid and semi-arid areas. Here, evaporation exceeds precipitation and under these conditions, applying excessive water through irrigation systems can draw salts to the surface, which then crystallize as a saline crust. The capacity of water to dissolve and displace nutrients leads to salinization, the build up of some elements (especially sodium and magnesium) to concentrations that are toxic to plants and microorganisms, rendering the soil useless for agriculture. This can happen very rapidly, especially when modern technologies are used to irrigate large areas without due regard for their natural ecology (Section 8.1).

The problem of salinization is far from new. Archaeological evidence suggests that a number of ancient civilizations declined because their soils became saline through mismanagement (Section 5.1). Around 6000 years ago, the people of Mesopotamia (present day Iraq) built their civilization on the agricultural lands between the Rivers Tigris and Euphrates. In order to grow their staple crop, wheat, they developed sophisticated irrigation schemes. Unfortunately, the evaporation from these well-watered soils allowed salts to accumulate in their upper layers and within 2500 years the land had soured. The farmers began to cultivate barley, which is more salt tolerant, but the soil eventually became too saline even for this and the fields were abandoned. Without productive lands the people had to move north into Babylon where the cooler climate makes the soil less susceptible to salinization.

History continues to repeat itself and today over 76 million hectares of salinized land exists across all five continents (Table 7.1). If this continues, the potential ecological and economic repercussions could be considerable. In Western Australia, for example, 11 per cent of the wheat belt is suffering from the effects of salinization and this is expected to rise to 30 per cent within the next 50 years. The problem here lies in a farming system that depends on sheep and cereals. One solution is to farm more sustainably by using species adapted to the semi-arid conditions, thereby avoiding the need for irrigation. In this case, native eucalyptus are being promoted as a suitable crop for the drylands. Not only does this make sense ecologically but also fills a gap in the economy as Australia currently imports the most Australian of products . . . eucalyptus oil!

TABLE 7.1 The global extent of salinization

Continent	Area salinized (M ha)	% of world salinization
Africa	14.8	19
Asia	52.7	69
South America	2.1	3
North and Central America	2.3	3
Europe	3.8	5
Australasia	0.9	1

The long-term outlook does not look good for water resources, as a number of international agencies suggest that water will become even scarcer as industry and agriculture compete with a growing population in particular areas. Water shortages and water treatment measures are becoming increasingly critical to the economic well-being of regions with a mediterranean climate and are part of the political bartering between neighbouring countries, from Southern California and Mexico to the Middle East.

The principal nutrients

Carbon

The patterns of nutrient cycling and the proportion fixed in their reservoirs change with each element and with time. The carbon content of the Earth's atmosphere has been far from stable over geological time. In the past, most notably the Silurian era, high atmospheric concentrations of carbon dioxide promoted high primary productivity, which eventually translated into the thick deposits of carbonates and hydrocarbons of the Carboniferous era. As we shall see later (Section 8.3), the rate at which carbon cycles, and its residence time in the atmosphere, regulates the planet's temperature.

Carbon makes life on Earth possible. Without its capacity to form molecular chains and rings, life would not exist. Dry a human being or most other living things and almost half of its weight will be carbon. Marine plankton and the major forests are the principal routes by which carbon is removed from the atmosphere. Eventually, it returns to reserves in rocks and sediments (Figure 8.19). A large proportion of global carbon is locked away as calcium carbonate in the shells and skeletons of marine organisms, to be compressed on the sea floor into chalk and limestone. Elsewhere, the remains of the long dead exist as hydrocarbons, the gases, and oils used to fuel today's industrial societies. Closer to the surface, the more recently dead plants and animals are held as soil organic matter and peat. Overall, perhaps as much as 40 000 times the size of the atmospheric pool is fixed in the oceans and these surface deposits.

Ultimately, all the carbon of living systems is derived from the atmosphere, yet it exists here at very low concentrations—a mere 0.03 per cent for its most abundant gas, carbon dioxide. Carbon fixed in sedimentary rocks has a slow flux rate, taking perhaps 100 million years before it re-enters the biosphere. Unfortunately, the rate at which we have been releasing carbon from these sinks over the last 200 years has led to changes not only to the carbon balance of the atmosphere, but also its energy budget (Section 8.3).

The impact of the planet's photosynthesis on global atmospheric concentrations of carbon dioxide leads to a distinct twice yearly oscillation of around 5 parts per million. This planetary equivalent of inhalation and exhalation is due to the disparity in the land area between the northern and southern hemispheres (Figure 7.6). Terrestrial processes in the north dominate the pattern, so that carbon dioxide levels are low when its photosynthesis is high, but rise during its winter when its respiration exceeds primary production.

Microbial breakdown of the organic matter in the soil is one major source of carbon dioxide for the air. Ploughing and forest clearance expose the lower soil layers to the oxidizing atmosphere and provide soil bacteria with the oxygen needed for aerobic respiration. Overall, agricultural methods based on regular soil disturbance and short-lived crops tend to deplete the organic content of the soil and accelerate the release of carbon dioxide.

Oxygen

With the advent of photosynthesis about 2.5 billion years ago, the increasing abundance of oxygen in the atmosphere selected physiologies that were adapted

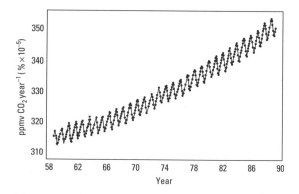

FIG. 7.6 Recent changes in atmospheric carbon dioxide attributed to human activities have been measured at Muana Loa Observatory in Hawai'i. Here, annual oscillations reflect seasonal change in photosynthesis and respiration. Since 1958 there has been a consistent increase in mean carbon dioxide concentration of the atmosphere (see also Figure 8.17).

to its corrosive presence. All higher organisms respire aerobically—that is, they use the capacity of oxygen to give up electrons and to bind with hydrogen to pull apart carbon molecules, which, in a controlled way, can release their stored energy. For this reason, the oxygen content of the atmosphere is closely linked to the biological activity of the planet, both in its production by photosynthesis and its depletion by respiration. As a result, its movement is tightly linked to the carbon cycle. However, because oxygen forms such a large proportion of the atmosphere (21 per cent) we do not detect annual shifts in its concentration, only changes through geological time.

Oxygen is also tightly linked to the other nutrient cycles of the planet. Where anaerobic (oxygen-poor) conditions dominate, some bacteria are able to use sulfate and nitrate to respire. As they do this, they release carbon dioxide, so liberating the oxygen in these compounds back into the atmosphere.

Sulfur

The fire and brimstone of the sulfur cycle makes it a truly elemental cycle—linking earth, air, fire, and water (Figure 7.7). Significant amounts are added to the atmosphere from non-crustal sources, through volcanic eruptions while sediments and the soil are again the main reservoirs. Sulfur forms only a small fraction of living tissues, but plays a crucial role in

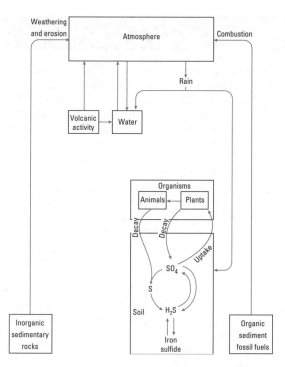

FIG. 7.7 The sulfur cycle. The primary sources are releases from sediments and volcanic eruptions. The burning of fossil fuel has added considerably to localized sulfur pollution and the problem of acid deposition.

regulating the structure of many amino acids and proteins. In the atmosphere, its compounds play several important roles: dimethyl sulfide, produced by marine phytoplankton, promotes cloud formation by acting as a nucleus for the condensation of water droplets.

Sulfur is also released as hydrogen sulfide through the action of sulfate-reducing bacteria in the anaerobic conditions of waterlogged soils. Each year, around 100 million tonnes of this gas are produced from soils, muds, and sediments by species such as *Desulfotomaculo* and *Desulfovibrio*. Anaerobic muds are characteristically stained black by the reduced iron sulfide and give off the rotten egg smell of hydrogen sulfide. Other microorganisms from the upper oxygenated layers oxidize sulfur compounds back to sulfates (e.g. *Thiobacillus*). These chemoautotrophs are using these inorganic compounds as their source of energy instead of organic (carbon) compounds and, therefore, function without the intervention of photosynthesis.

The concentration of sulfur in the air has increased on a local scale with the burning of fossil fuels, particularly coals rich in sulfate minerals. Although largely confined to industrial areas, sulfur dioxide may travel hundreds of miles from its source, contributing to the regional problems of acid rain

---

BOX 7.2   Acid rain

For a long time, the problem of acid rain was thought to be largely a consequence of the sulfur-rich gases produced by burning some fossil fuels (especially poor-quality coals) and the smelting of metallic ores. We now know that this picture is far too simple. Acid rain should be more properly described as acid deposition, as it consists of both wet and dry deposition, and its effects are not attributable to sulfur alone.

The other significant element is nitrogen, especially the nitrogen oxides (NO_x), which are produced by the internal combustion engine and power stations. Nitrogen oxides now account for 30 per cent of the acid deposition in Europe. With the reduction in sulfate emissions from power stations over the last two decades, NO_x have become increasingly important. Acid deposition is a regional pollution problem and needs international agreements to limit atmospheric concentrations. Uncontrolled,

the emissions from power stations and motor vehicles can cause considerable damage to forests, lakes, and streams. Not only does acid deposition add large amounts of sulfur and nitrogen to these systems, upsetting their nutrient balance; the fall in pH also increases the availability of toxic metals such as aluminium.

Rainfall is naturally acidic, as carbon dioxide dissolves into it during its descent through the atmosphere. However, a significant increase in the acidity of rain, above natural levels, was noted in Britain within a 100 years of the start of the Industrial Revolution. Today about two-thirds of the acid deposited over Britain is in the form of dry deposition. The remainder comes from wet deposition, where SO_x and NO_x compounds are resident long enough in the air to combine with moisture to form dilute sulfuric and nitric acids. This typically produces a rainfall around one pH unit more acidic

---

BOX 7.2   Continued

(a 10-fold increase in its concentration of reactive hydrogen ions) and can have a pH as low as 3.1 in the worst situations.

Nitrogen and sulfur inputs can also have a fertilizing effect in some ecosystems. Nitrogen inputs from the air in Europe represent a 500 per cent increase over background. This can lead to shifts in the species composition of some plant communities, particularly in areas where the nutrients are in short supply. Many plants growing on nutrient-poor sandy or chalky soils rely on rainfall as their principle source of nutrients (**ombrotrophs**) and are thus adapted to low nutrient levels. Abundant nitrogen means other species can invade and out-compete the native flora. Nutrient-poor ecosystems such as grasslands and heathland may change near major sources of nitrogen, such as densely trafficked roads, and may become dominated by coarse, rank vegetation.

Acid rain damages ecosystems, not only because of the nutrients they add, but also due to their acidity. Generally, the acidification of the soil mobilizes soil nutrients, promoting an initial surge of growth. However, this acidity displaces key nutrients and cations (such as calcium and magnesium), which may then become limiting to further growth. Nitrogen oxides entering through plant stomata can have more direct effects on plants, creating localized acidity and damaging leaves and new shoots (Plate 7.1). The flush of nutrients eventually finds its way into water courses where it causes algal blooms, comparable to eutrophication (see Section 7.3).

Increased availability of metals is one of the longer-term effects of acidification. Scandinavian lakes have suffered particularly from acidification and excess aluminium has been responsible for many incidences of fish-kill. The predominant wind directions means that this area receives much of the atmospheric pollution of north-western and central Europe. The region is dominated by coniferous forests and thin soils, with little capacity to buffer the incoming acidity. The result has been a large number of lakes becoming fish-less since the 1960s. Sweden, Finland, and Norway have had to resort to adding large amounts of lime to these ecosystems to try to protect or restore them.

(Box 7.2). Sulfate aerosols are known to be important in the energy balance of the atmosphere, producing regional cooling by their effects of cloud formation near their source (Section 8.3).

Phosphorus

Phosphorus is an earthbound element and lacks any significant atmospheric component (Figure 7.8). Geological deposits (mainly in the form of its calcium salt, apatite) account for a global resource of 2000 billion tonnes, but its rate of release by weathering and erosion is very slow (just 100 million tonnes per year—less than one 500 000th of the total reserve).

The sea is the richest source of available phosphate. Indeed, most phosphate-rich rocks tend to be of marine origin. However, the main route back to land for phosphate is through seabirds, feeding on phosphate-rich fish, then roosting and defaecating on land. Their faeces form guano, which, in some locations, has accumulated to such a considerable depth that it is commercially extracted for use as a fertilizer.

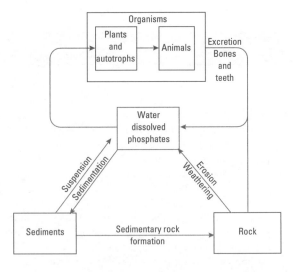

FIG. 7.8 Phosphorus cycle. Phosphorus occurs mainly in the form of phosphates, insoluble compounds that are slowly weathered from rocks or re-suspended from living organisms. This cycle differs from most of the others as it lacks a significant atmospheric phase.

Plants and animals hang on to phosphate and concentrate it to many times the level of the surrounding environment. Phosphorus is an essential component of proteins, nucleic acids, cell membranes, teeth, and bones, but under normal conditions is nearly always in short supply. It plays a central role in a host of cellular processes, not least because it is the key component in the currency of biological energy transfer—adenosine triphosphate (ATP). It enters the food chain primarily through plants and phytoplankton. An atom may remain in an animal for many years and within a tree for centuries. On its release, it can be recycled into the living system or become immobilized in the soil where it might remain for many thousands of the years. Large amounts of iron sulfide in a soil rapidly lead to the immobilization of phosphorus by the formation of insoluble compounds. Insolubility is the fate of most free phosphorus and is the reason why this element is the main limiting nutrient in many ecosystems.

As before, microorganisms and fungi also play an important role in maintaining the circulation of this nutrient. A number of plants exploit their capacity by forming symbiotic associations (Section 4.2) with fungi to secure additional phosphorus. In these associations, known as **mycorrhizae** (*myco*—fungus, *rhiza*—root), the fungus either grows around (ectomycorrhiza) or around and within (arbuscular mycorrhizal fungi) the plant root. With the latter, the fungus penetrates the root forming structures within it (Figure 7.9). The plant supplies sugars and other nutrients to the fungus in exchange for the ready supply of phosphorus scavenged by the fungal network extending beyond the root. The fungi also help by supplying moisture to the roots, thereby conferring drought resistance to the host plant. The importance of these associations has been highlighted in the development of restoration ecology (Section 7.3), since we now know they are essential for the establishment of some plant communities.

Nitrogen

Another, possibly more important, mutualistic association in the soil involves the capture of the other major plant nutrient, nitrogen. The flux of nitrogen once again exerts considerable control on ecosystem productivity and processes (Figure 7.10). Next to carbon and oxygen, nitrogen is the third most abundant

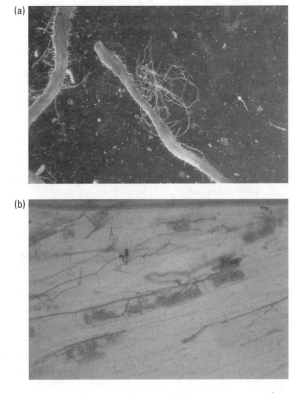

FIG. 7.9 Mycorrhizae: plant and fungus in partnership. Arbuscular mycorrhizal fungi (AMF) growing (a) around and (b) inside the root.

element in biological molecules, a major component of proteins and nucleic acids. But it is so difficult to coax reactive nitrogen out of the atmosphere that some plants have again resorted to symbiosis, this time with some very ancient genera of bacteria and actinomycetes, collectively known as the nitrogen fixers.

The bulk of the atmosphere is nitrogen (around 78 per cent) and it is from this reserve that living systems to derive their supply. Atmospheric nitrogen is largely in the molecular form (N_2), which, with its three shared electrons, is difficult to split into atomic nitrogen. Thunderstorms, and the massive discharges of energy they produce, do this, though only relatively small amounts of nitrate enter the soil this way. Most organisms cannot afford the energetic cost of splitting the nitrogen molecule and only a few microorganisms have the enzymic equipment necessary.

Bacteria and cyanobacteria (blue-green algae) are responsible for the bulk of biological nitrogen

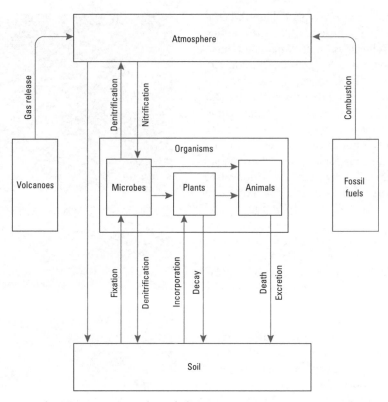

FIG. 7.10 The nitrogen cycle. Nitrogen passes through living organisms on its way to the atmosphere from the soil. Additional sources are oxides of nitrogen produced by volcanoes and the burning of fossil fuels.

fixation from the atmosphere. They occur in both terrestrial and aquatic environments and can be either free-living, using organic detritus as their energy source, or have a symbiotic relationship with certain plants. All possess the enzyme nitrogenase, which is responsible for the processes by which molecular nitrogen (N_2) is split and combined with hydrogen to form ammonium.

The most studied nitrogen-fixing bacteria belong to the genus *Rhizobium*. These survive in the soil as saprophytes living on dead organic material. When the opportunity arises they infect the roots of leguminous plants—species belonging to the pea and bean family. Here they are housed inside root nodules (Figure 7.11) and live as symbionts within their host. The plant not only provides shelter (especially an anaerobic environment) but also sugars that provide a carbon source for the *Rhizobium*. The atmospheric nitrogen fixed by the bacteria makes this contract a

worthwhile investment by the plant since their private income of nitrogen gives many legumes a competitive advantage over their neighbours in nitrogen-poor soils. Even so, if these costs can be avoided, both the *Rhizobium* and its host will switch, taking up any abundant nitrate that might become available in the soil.

However, nitrogen fixation using symbionts is not confined to legumes. Root nodules occur in a number of other plants. Alder (*Alnus glutinosa*), for example, has a symbiotic relationship with a nitrogen-fixing actinomycete bacteria, *Frankia*. There are also a variety of free-living bacteria, such as *Thiobacillus*, and cyanobacteria, which fix nitrogen for their own use, giving these organisms the competitive edge when nitrogen is scarce but other key nutrients are abundant.

That nitrogen fixation can occur in a variety of plants encourages us to try, through genetic engineering, to introduce the association into the major crop

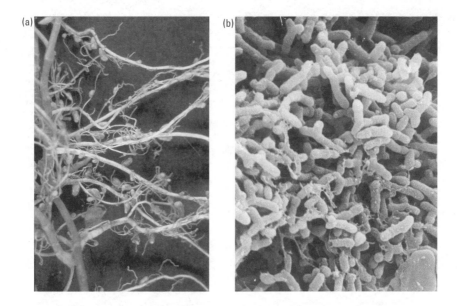

FIG. 7.11 Root nodules. (a) Root nodules of clover (*Trifolium* species), which have formed from root epidermal cells as a result of infection by *Rhizobium* bacteria. (b) Using electron microscopy the y-shaped bacteroids are clearly visible within the nodule.

plants. If we succeed, we could reduce our reliance on nitrogenous fertilizers and the large energy costs associated with their production. Considerable effort is being directed at cereals in particular, which, ordinarily, need considerable amounts of artificial fertilizers. In the process we may learn even more about the mechanism of nitrogen fixation and this remarkable symbiotic relationship might have evolved.

Nitrate is the main form of nitrogen taken up by plants, but this is highly soluble and readily leached from the soil. In moist climates, soil nitrogen is readily depleted and competition for available nitrate governs many communities of plants and animals. This has lead to some extreme strategies for supplementing an individual's nitrogen supply. Carnivorous plants like the Venus fly trap (*Dionaea muscipula*) and pitcher plants (*Sarracenia* and *Nepenthes* species) are found in waterlogged bogs and swamps, where nitrogen concentrations are low (Plate 7.2). Insects, caught by the carnivorous plants, help to make good this supply. Directly or indirectly, all animals ultimately depend upon plants for their nitrogen, so this is something like justice, with these plants at least, getting their own back.

Plants find themselves in competition for nitrogen, not only with each other but also with the soil microbial community. Different genera of bacteria use different nitrogen compounds to derive energy from its bonds or produce new cells. Heterotrophic bacteria, using organic compounds as their source, will mineralize organic nitrogen to ammonium both as a source of carbon compounds and of energy (Figure 7.12). The ammonium they do not need for their own cell proliferation will evaporate unless is used by other bacteria, especially the nitrifiers. *Nitrosomonas* converts ammonium to nitrite (NO_2), which may then be further oxidized to nitrate (NO_3) by *Nitrobacter*. Together these two processes are called **nitrification**. Any nitrate not taken up by plants or microorganisms may undergo **denitrification** through a process of reduction, in which oxygen is progressively removed from nitrate, nitrite, and nitrous oxide (N_2O), eventually yielding molecular nitrogen (N_2), which returns to the atmosphere. Denitrification is carried out by a variety of bacteria (including *Bacillus* and *Pseudomonas*) and is a very important process in oxygen-poor environments such as waterlogged soils where it is the major cause of the shortage of nitrogen.

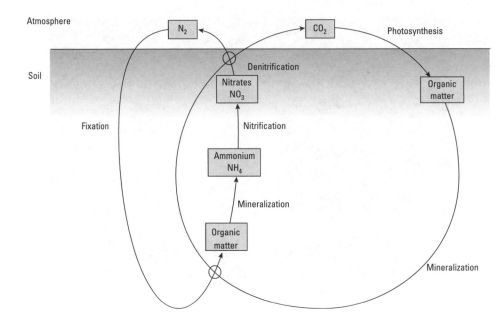

FIG. 7.12 Microbial interactions between the nitrogen and carbon cycles. The carbon cycle fuels the nitrogen cycle in both the fixation and denitrification phases of the cycle.

Large amounts of nitrogen (around 60 million tonnes) also enter the atmosphere from the burning of fossil fuels, making a significant contribution to acid rain (Box 7.2). On the ground, our inorganic fertilizers add a highly concentrated source to the soil in a form that can be readily leached away. These have produced highly localized concentrations of nutrients and, in doing so, upset ecological communities adapted to low-nutrient conditions. A nutrient that becomes abundant in a community of species adapted to low nutrient levels inevitably leads to change in its configuration.

7.2 Eutrophication—too much of a good thing

It is apparent to any farmer or gardener that the fertilizer applied to the soil one year is not there several years hence. For those of us without green fingers, it might never have been there in the first place. Runoff and nutrient leaching mean that the money we spent trying to produce something on our patch of land invariably goes to fertilize something else, somewhere else.

Very often, it is primary production in the waters draining away that snatches up these nutrients. Enriched by fertilizers, or other sources of nutrients, rivers, streams, and lakes change their community structure and patterns of cycling, in a process termed **eutrophication**. This is nutrient overload—an example of how a nutrient in the wrong place at the wrong time and the wrong concentration can become a serious pollutant. Eutrophication can be a natural process too, as freshwaters become enriched with age, as part of the successional process (Section 5.3).

Human activity is by far the greatest cause of such imbalances in freshwaters and, indeed, can cause eutrophication in soils, when fertilizers stay where they are put and a build-up occurs. The main culprits are phosphorus and nitrogen, with the former being the most serious because it is, ordinarily, so unavailable in natural systems.

Eutrophication leads to a dramatic deterioration in water quality that is all too obvious at certain times of the year. The water becomes green and soupy as the phytoplankton density increases, sometimes forming bright green flocs—floating mats of cyanobacteria (blue-green algae). These flourish because of their ability to fix nitrogen from the atmosphere and will blanket out competition from other phytoplanktons unable to do this. After a while water in this state begins to smell foul. This is a sign that the oxygen content of the water has been depleted due to the death and decomposition of the cyanobacteria and that other life in the water is struggling to respire aerobically.

We can easily measure the decline in the oxygen content of the water, but must also measure the amount of oxygen that will be needed to degrade all the organic matter in the water. This is the **biochemical oxygen demand** (BOD) and is an indication of the balance of the system. A high BOD implies a high organic content, requiring large amounts of oxygen for decomposition and a depleted supply for the rest of the biota. When the living community dies their tissues add to this oxygen demand. So the system can quickly spiral down towards a community of the few species able to survive turbid, oxygen-poor waters.

Whilst BOD represents a useful snapshot of current conditions, it says little about past pollution events. Evidence for these though can be found recorded in the community of invertebrates found in the water. Some species will only survive in oxygen-rich waters. These can act as **indicator species**, whose presence or absence provides us with clues as to the state of the water and its recent history. For example, stonefly and mayfly larvae are only found in well-oxygenated waters. A community dominated by species such as oligochaete worms—groups that can tolerate very low oxygen levels—mean the water has a high oxygen demand and will be carrying an unbalanced carbon load. Different methods of counting species and groups are used in different parts of the world to measure water quality (**biotic indices**), but most use the species composition of the freshwater community to judge levels of pollution, particularly organic pollution from sewage discharges.

Restoring eutrophic waters

When people think of pollution and effluents they picture industrial plants billowing fumes and happily forget the number of times they flush more nutrients down the toilet. Likewise, we tend not to connect the food on supermarket shelves with fertilizer runoff or oceans of slurry from livestock houses on intensive farms. Even organic and free-range livestock can contribute to eutrophication when areas are overstocked and their wastes enter the drainage system.

Clearly, the best way of dealing with any form of pollution is to control it at source. Domestic and industrial sewage account for over 80 per cent of the phosphorus input into the environment, primarily from human waste and domestic detergents. Over the last 30 years, technology and legislation has led to the marketing of low-phosphate detergents, which today are commonplace. Even so, eutrophication remains a major problem, from domestic, industrial and agricultural sources of pollution.

A classic example of eutrophication occurred in Lake Washington near Seattle. As the city expanded in the 1940s and 1950s its sewage discharges into the lake led to a deterioration in water quality, including a 75 per cent reduction in clarity and massive explosions of the filamentous cyanobacteria *Oscillatoria rubescens*. Public concern led to a campaign to reduce effluent inputs into what had become known as 'Lake Stinko'. The phosphorus levels in the water were shown to be the key to the problem and a programme to reduce these was implemented. It took 7 years to reverse these trends.

Large-scale experiments have since been carried out in an attempt to understand the causes, effects, and potential cures for eutrophication. In 1968, the Experimental Lakes Area was set up in Ontario in which a series of small experimental lakes were deliberately polluted and subsequently monitored. In one experiment, a lake (ELA 226) was divided in half using a plastic membrane and fertilized, one-half receiving carbon and nitrogen and the other carbon, nitrogen, and phosphorous. The half that received the phosphorous developed a dramatic algal bloom—a result that helped confirm the role of phosphorus in eutrophication. Later experiments proved that cyanobacteria, no longer checked by a shortage of phosphorus, exacerbate eutrophication by 'fixing' atmospheric nitrogen and adding to the nutrient overload in the lake.

Recently, experience of other eutrophication events suggest that the consequent changes to the plant and animal community requires more than simply halting the nutrient inputs. The Norfolk Broads of East Anglia are low-lying wetlands, the product of peat digging and reed growing in the Middle Ages. Over the centuries these have developed into a community rich in specialist species and the associated fenlands have a number of rare and distinctive plants and animals, including the water soldier plant (*Stratiotes aloides*) and the fen raft spider (*Dolomedes plantarius*). However, the area borders some of the richest and most intensively farmed soils in Britain and over the last 60 years these soils have received immense inputs of fertilizer. In the 1970s, the Broads began to show signs of eutrophication. Early attempts to address the problem by controlling discharges failed. One reason for this was the large capital of phosphorus that had accumulated in the sediments. Under anaerobic conditions this was mobilized and released back into the water, again allowing cyanobacteria and other phytoplankton to flourish. This produced a cloudy shallow water in which rooted macrophytes could not grow. Animals adapted to clear open waters disappeared, to be replaced by those able to live in the turbid conditions. One strategy has been to pump out the enriched sediments, using a machine called the 'mudsucker', but even this failed to return clear water to the Broads (Figure 7.13).

Now the Broads Authority uses a management strategy designed to combat eutrophication at several levels. First, at source, by designating Nitrogen Sensitive Areas in which incentives are used to encourage farmers to adopt more traditional, less intensive agricultural practices. Second, domestic sewage is treated to reduce phosphate before discharge. Third, the Broads are being dredged by mud pumps such as the mudsucker. On nature reserves, ecologists use different methods according to the severity of the eutrophication. In mild cases, reeds are harvested and removed, to 'crop off' excess nutrients. Electric fishing is also used to remove those species that stir up the sediments (and the nutrients represented by their biomass) and clear water species (such as the predatory pike—*Esox lucius*) reintroduced. In the most grossly polluted areas the mudsucker is used in conjunction with phosphate precipitation techniques by which iron and calcium sulfates are introduced to produce highly insoluble phosphate salts that sink into the sediments, making the phosphate unavailable and eliminating one of the major sources of eutrophication.

FIG. 7.13. The Broads Authority mudsucker in action. This machine is used to remove nutrient-rich sediments from eutrophic bodies of water as part of the wetland restoration programme.

Other strategies used on lakes in Holland involving the manipulation of the animal community have met with some success. The idea of **biomanipulation** is to promote a phytoplankton and rooted plant community typical of a clearwater system by manipulating its species composition. In this case, it meant introducing certain fish and removing others. Again, bottom feeding species, such as bream (*Abrama bramis*) are removed, and other species of fish and mollusc introduced to promote clear water conditions. With a thriving zooplankton community (and absence of cyanobacteria), fish that feed on the zooplankton, and the predatory fish that feed on them, the clear waters that the rooted plants need can be promoted. When these macrophytes grow the nutrient supply to the phytoplankton becomes considerably reduced, and the whole community is less readily pushed into a eutrophic state.

Seas too can become eutrophic, especially when they are enclosed and are slow to recharge with new water from adjoining oceans. The Mediterranean is connected to the Atlantic Ocean by the Straits of Gibraltar and has a long recharge period—in this case 80 years. Despite its relative enclosure the Mediterranean is classed as oligotrophic; that is, nutrient-poor, and this is indicated by its relatively low productivity, averaging less than 150 g of carbon per square metre each year. This is because relatively few rivers discharge their nutrients, sediments, and pollution into it. However, in the Mediterranean eutrophication tends to be localized, especially in the Adriatic Sea between Italy and Greece. About one-third of the freshwater input into the Mediterranean arrives in the Adriatic, mainly from the River Po draining the Alps and Northern Italy and from the rivers draining the Balkans in the East. These are responsible for the higher than average levels of nutrients and pollutants found in the Adriatic and at the local level, there can be considerable eutrophication, such as in the Venice lagoon (Box 7.3).

BOX 7.3   Venice in peril

Eutrophication in the Mediterranean tends to be localized especially in coastal waters, close to urban areas with discharges direct to the sea. One area where eutrophication is most obvious is around in the Venice lagoon. Extending over 55 000 ha, the lagoon has an average depth of only 1 m. It is fed on the landward side by a series of rivers and is linked to the Adriatic Sea by three inlets (Plate 7.3). As a consequence, the water within the lagoon is warmer, two-thirds less salty, and far more eutrophic than the adjoining sea.

Like any other city, Venice generates liquid waste of various descriptions, most of which ends up in the lagoon. Added to this are inputs from surrounding agricultural, industrial, and urban sources, which together account for between 80 and 90 per cent of the nutrient load entering the lagoon (Table 7.2). All these inputs have turned it into a large, shallow container filled with nutrients and bathed in a weak solution of saline. As a result, the Venice lagoon can readily out-stink the likes of Lake Washington in its former state (Section 7.3), especially during hot, calm periods in the summer months. The smell comes from the breakdown of organic matter. In hot weather, this breakdown starts out as an aerobic process but quickly uses up the available oxygen and becomes anaerobic. Sulfur bacteria then take over

TABLE 7.2 Table of the annual load of nutrient inputs into the Venice lagoon

Source	Nutrient inputs into the Venice lagoon (tonnes per year)		
	N	P	Organic material
From Venice	700	200	2 400
From mainland	7300	1200	12 000
Total	8000	1400	14 400

producing hydrogen sulfide gas as a breakdown product (Section 7.1).

The excess nutrients cause algal blooms of sea lettuce (*Ulva rigida*), which, over the years, have displaced the lagoon's natural vegetation. By 1992, dwarf eelgrass (*Zostera noltii*)—a species characteristic of good-quality brackish water—had been swamped by the *Ulva* and could only be found hanging on as small clumps near the edges of the inlets. Meanwhile, algal macrophytes, extremely responsive to high concentrations of nitrate, became the dominant primary producer in the lagoon, adding even more organic matter.

BOX 7.3   Continued

The authorities established the *Consorzio Venezia Nuovo* (CVN) to lead an integrated management programme aimed at reversing the environmental deterioration and restoring the ecosystem of the lagoon. Aldo Bettinetti and his colleagues identified the key problems along with some possible solutions. They concluded that the progressive urbanization of the surrounding mainland was a key contributor to the lagoon's problems (Table 7.1). In the past, a natural buffer zone of woodland and marshland helped protect the lagoon from sediments and pollution, but this was reclaimed for urban and industrial uses in the middle of the twentieth century and now supports a population in excess of 1.4 million and includes the large industrial complex of Porto Marghera.

Bettinetti's team mapped out the cause and effect of eutrophication within the lagoon (Figure 7.14) and with the help of geographical information systems (GIS) located vast spreads of *Ulva* and the beleaguered colonies of *Zostera*. Their findings reveal that 90% of the lagoon is prime *Zostera* habitat and that it can easily outcompete *Ulva* in low-nutrient conditions. It is hoped that the removal of *Ulva* will let the eelgrass re-establish, so keeping the *Ulva* at bay and they recommended regularly harvesting *Ulva* to crop off excess nitrate and phosphate—a similar strategy to that adopted in the restoration of the Norfolk Broads and other eutrophic wetlands (Section 7.2). To be effective the *Ulva* must be harvested at the peak of its biomass, whilst aerating the water to maintain the oxygen concentration above 5 mg/l.

Medium-term measures, on the other hand, seek to reduce erosion so that the shallow nature of the lagoon can be maintained as suitable sites for *Zostera*. In the most polluted areas excess nutrients are removed by removing the top layer of sediments. There are also plans to reintroduce populations of filter feeders to help remove nutrients and reduce the turbidity of the water. Similarly, *Zostera* will be introduced into its former strongholds. Ultimately, long-term measures will reduce the nutrient inputs from the city and the surrounding developments at source by providing better effluent treatment facilities.

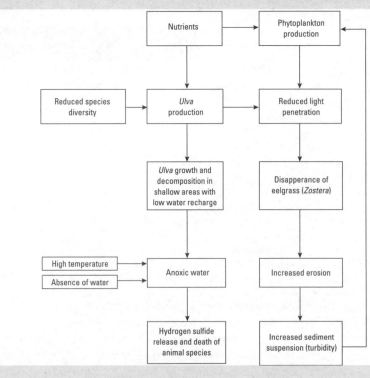

FIG. 7.14 A flowchart of the causes and effects of eutrophication in the Venice lagoon.

More eutrophic still is the Black Sea, connected to the Mediterranean via the Bosporus. Once home to a diverse marine community, years of pollution and overexploitation have rendered it grossly eutrophic with anoxic (oxygen-depleted) conditions over much of its sea floor. As well as wastewater and fertilizer runoff, the Sea also receives a variety of pollutants ranging from oil spills to radioactive waste. Notably, its primary productivity is very high (300 g C/m^2/ year). From space the Black Sea is visibly darker than the more oligotrophic waters of the eastern Mediterranean (Plate 7.4). For decades it was thought that the damage was irreversible but there has been a decrease in the volume and incidence of phytoplankton blooms suggesting that the Sea is slowly stabilizing and might yet recover.

There can be few pollution incidents as dramatic or distressing as the sight of a broken tanker and the oiled seabirds in its wake (Plate 7.5). World trade in oil results in our transporting it many thousands of miles and sometimes we spill it—either accidentally or on purpose (Box 7.4). More mundane and far more common is the continual pollution associated with the loading and unloading process, or the surreptitious washing-out of tanks at sea: an estimated total of 650 000 tonnes of oil is discharged into the Mediterranean (Table 7.4). Around 17 per cent of global marine oil pollution occurs in the Mediterranean, though the sea represents just 0.7 per cent of the surface waters of the planet. Even more deplorable is the deliberate targeting of marine and terrestrial oil installations in times of war (Plate 8.3).

With each oil spill we have learnt more about their causes and consequences. Depressingly, each incident provides us with yet another opportunity to test our techniques. As a result, our methods have changed considerably over the last 40 years. Today, we seek to contain spills as far as possible and rely on natural degradative processes to what enters the environment. Human intervention in such cases is best confined to mopping up the oil that comes ashore in the most severely contaminated areas.

BOX 7.4 Oil pollution—a catalogue of disasters

Faced with a major spill, the obvious response is to contain the oil and to clean it up with whatever means available. Back in 1967, when the *Torrey Canyon* broke its back on the Cornish coast, a variety of techniques were used, including bombing the vessel with napalm to set fire to the oil (Table 7.3). The remainder of the slick was treated with more kitchen-sink technology—in the form of detergents and dispersants, to break it up. Over 10 000 tonnes of these chemicals were used, but they turned out to be more toxic to the wildlife than the oil they were meant to combat. Some parts of the coastal community took up to 10 years to recover from their effects.

TABLE 7.3 Major marine oil spills during the past four decades

Vessel	Date	Location	Extent polluted (km)	Size of spill (tonnes)	Bird deaths
Torrey Canyon	March 1967	Cornish coast, England	225	117 000	30 000
Amoco Cadiz	March 1978	Brittany coast, France	360	233 000	4600
Exxon Valdez	March 1989	Prince William Sound, Alaska	700	35 000	100 000–300 000
Braer	January 1993	Shetland Isles, Scotland	235	86 000	6313
Sea Empress	February 1996	Milford Haven, Wales	100	75 000	22 132
Erika	December 1999	Brittany coast, France	400	11 100	150 000–300 000
Prestige	November 2002	Spanish and French coast	>1000	77 000	100 000–200 000

BOX 7.4   Continued

TABLE 7.4 Routine oil pollution within the Mediterranean

Source of pollution	Quantities (tonnes)
Tanker operations (especially deballasting)	450 000
Navigation accidents	65 000
Oily bilge waters, sludge, and used lubricating oils from ships	60 000
Ship repairs (cleaning of tanks and piping)	35 000
Pipelines	20 000
Petroleum handling installations in ports and special terminals	15 000
Offshore drilling	5 000
Total	650 000

A slick on the water surface limits oxygen diffusion into the water, leading to anoxic conditions beneath. Oxygen levels also fall as the microbial community begins to decompose the oil. As a macropollutant, oil physically obstructs living organisms, smothering, and weighing them down. More direct poisoning follows if an animal ingests it. Oil is a complex cocktail of hydrocarbons and, generally, those compounds that are the most soluble in water tend to be the least toxic. Detergents too are toxic and like many compounds in oil interfere with cell membrane function. Because membranes are partly composed of lipids (fat molecules) any compound capable of breaking up petroleum hydrocarbons will also cut through cell membranes. Today, detergents and dispersants are used only as a last resort.

The next challenge came with the *Amoco Cadiz* spill in 1978. This was by far the largest accidental spill, releasing a quarter of a million tonnes into the English Channel. Reluctant to use detergents, physical mopping up of the oil was used along with a reliance on natural degradative processes. The worst of the pollution was weighed down with sawdust and chalk and allowed to settle on the seabed. Meanwhile, oil was physically removed from the open sea and shore. Detergents were restricted to economically important areas, such as harbours. The strategy of relying on the resident microorganisms worked, with 80% of the main hydrocarbon groups were degraded within the first 7 months of the spill.

A similar integrated clean-up operation was used after the *Exxon Valdez* ran aground in Alaska. Pollution control experts tailored their response to the nature of each beach, some of which were left to recover by themselves, except for the addition of fertilizers to encourage microbial

growth. Other areas were washed down with high-pressure water jets and some were mopped with a new absorbent cloth, in an operation called 'rock-polishing'.

Sometimes nature finishes the job almost completely. This happened when the *Braer* struck the rocks off the Shetland Isles in winds reaching Storm Force 11–12, and gusting to Hurricane Force 13–14. The oil was quickly whipped into an oil–water emulsion (known as a mousse), which assisted in its dispersal. Although the severe conditions meant that little damage occurred to the coastal communities, the wind did carry the oil further inland.

Fate, however, was not so kind with the spills from the *Sea Empress* and *Erika*, which polluted the Welsh and Brittany coasts, respectively. The *Erika* became infamous for having produced the greatest number of bird deaths for the least amount of oil. Recently, the spill resulting from the sinking of the *Prestige* entered the record books for having contaminated the longest stretch of coastline ever (over 1000 km), from the Galician and Cantabrican coasts of Spain through to the north-western coast of France. Despite the advice of experts urging the grounding of the ship and the pumping out of its tanks, the ship was forced offshore and allowed to sink.

As with all oil spills, damage to a region's ecology inevitably affects its economy and the *Prestige* spill is no exception. Fishing has been prohibited along 798 km of the Galician coast affecting the livelihood of 16 140 fishermen and their dependents and 680 beaches have been badly affected. The loss of the *Prestige* was particularly controversial, coming at a time when the European Union had outlawed single-hulled tankers . . . but with a cut-off date of 2015 to allow fleets to be replaced!

A large range of microoganisms can be relied upon to attack these hydrocarbons. Hundreds species of bacteria and fungi from 70 genera will readily degrade oil. These evolved to break down hydrocarbons that naturally occur as breakdown products of biomass. Local populations grow faster if they have been exposed to oil in the past, so an area subject to frequent spills can have an active community of these microorganisms. This, along with the higher water temperatures, helped the Persian Gulf recover from the deliberate spill in 1991. To these bacteria, oil is just another carbon source and, once given nutrients (usually nitrogen or phosphorous), they readily utilize it.

Eutrophication in terrestrial ecosystems

Eutrophication affects terrestrial ecosystems too, with agriculture being the major cause of the enrichment. Once again, ecologists attempt to restore these ecosystems by removing the excess nutrients. Eutrophication is particularly problematic when it affects communities of naturally nutrient-poor soils, such as heathland and chalk grassland. In such situations, competition for nutrients is intense and the plant community will comprise species adapted to these conditions. These communities change when nutrients become abundant, as some species are able to increase their numbers or growth faster than their neighbours. Typically, nutrient-rich habitats are dominated by a small number of fast-growing species with many of the rare species, adapted to the nutrient-poor conditions being lost.

The move towards intensive agricultural practices has brought major changes to many temperate communities (Section 6.5). The problem is not just restricted to farmland. Amenity areas such as parks, gardens, and public open spaces can also have high levels of nutrients following many years of fertilizer applications, along with the inputs from frequent visits by domestic pets. Indeed, research has shown that the highly manicured golf course greens are so rich in phosphate that they could be bagged up and legally sold as fertilizer!

The residual fertility that results from eutrophication is a challenge to ecologists involved in habitat restoration. An oversupply of nutrients leads to a build-up of biomass. Habitats traditionally managed by cutting, grazing, or burning can suffer if this management is suspended and the standing crop is allowed to recycle back into the system. Cropping-off excess nutrients by removing this biomass and its nutrients can help, but it may not be enough. Another technique is to bury the rich topsoil or remove it completely. Sometimes, grazers that specifically target nutrient-rich plants can be used to create space for the less vigorous plant species. For example, highland cattle are being used on lowland heaths in England to browse species such as gorse, a legume that fixes nitrogen in its roots and thus enriches the soil. Cattle also represent a reserve of major nutrients that is easily removed from the site.

7.3   Restoration—'the acid test of ecology'

Restoration ecology helps us redress some of the damage we so frequently inflict on our environment. This is ecology as technology—a science applied to solving real-world problems. As a science, restoration ecology also answers questions about ecosystem function and uses these insights in the techniques of restoration. It is something of an art too, since restoration requires the creative skills needed to design entire landscapes and their ecosystems.

Much of modern science is based on a reductionist approach, understanding the system by taking it apart and in some cases, rebuilding it again. Whilst there have been many elegant experiments that have examined the nuts and bolts of the environment, we cannot take whole ecosystems apart—natural ecosystems are far too precious to be used in destructive experimentation. However, there are numerous habitats that have suffered degradation and which offer ecologists a convenient opportunity to study the ecological processes of a self-sustaining community as they try to restore it. The scientific knowledge gained from such projects is considerable and ranges from nutrient cycling to species interactions and population dynamics. Such opportunities present ecologists with a chance to test their theories in the real world. All this has led Tony Bradshaw to refer to restoration as being 'the acid test of ecology': a means of testing theories and hypotheses, thus adding to our understanding of ecological systems.

The three Rs of restoration

Taken at its most literal, restoration means returning something to the way it was before a change took place. Indeed, the aim of many projects is to reinstate lost habitats and the plant and animal communities they support. In practice, restoration ecology often encompasses several aims:

- stabilization of land surface;
- pollution control;
- visual enhancement;
- creation of amenity value;
- increase of biological diversity;
- improvement of ecosystem function;
- establishment of productivity.

Much of our knowledge and techniques have come from **reclamation** projects on sites degraded by extractive or industrial processes. Mining or quarrying often results in large amounts of waste material (spoil) that can be both unsightly and a significant source of pollution. The aim here is use these wastes to create something more aesthetically pleasing and ecologically sustainable (Plate 7.6).

A number of options are available. Restoration might involve returning a site to its pre-industrial state, perhaps in some cases even to a natural or semi-natural habitat. Restoration to these states can be costly and complex and it may be that we choose partial restoration or **rehabilitation** instead. These might, with time, mature to a full restoration. Other possibilities include making the land fit for agriculture or housing. These do not require a natural community and the end points are more properly described as **replacement** habitats, a compromise somewhere between the existing degraded state and a restored site. Figure 7.15 shows the range of alternatives. In some situations, it may be best to let nature take its own

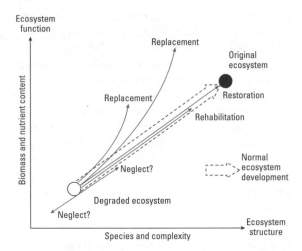

FIG. 7.15 The options available in ecological restoration. As degraded sites are reclaimed they show an improvement in both the structure and function of the ecosystems. Complete restoration might not always be the aim of a project. Partial restoration in the form of rehabilitation or alternative replacement habitats may be appropriate endpoints for a project.

course and for the site to recolonize naturally. However, this is not an option where a substrate is unstable or a significant source of pollution.

The time factor

Restoration techniques can change entire landscapes within a relatively short time, yet these changes can themselves be long-lasting. Our decisions at the planning stage are, therefore, critical. We need the foresight to visualize the consequences of our decisions over ecological time. Often, it is impractical or impossible to return the habitat to its pre-impact state and we need to consider the options available to us (Figure 7.16).

Our choice of endpoint is governed by a number of factors, including the history of the site, the existing conditions, and the resources at our disposal. If the site is stable it may be possible, and indeed preferable, to let a community develop on its own. The most complicated option seeks to anticipate future successional changes, predicting the way an ecosystem might develop. We then have to consider the timescale involved in the restoration process,

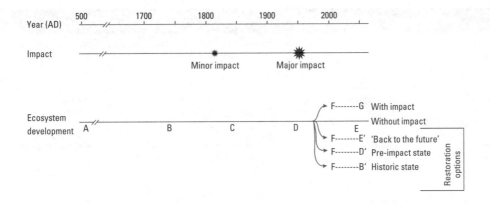

FIG. 7.16 The restoration continuum. The diagram shows ecosystem transitions both with and without impacts and restoration work. Without the major impact (around 1950 on the timeline), the ecosystem would have developed from stage D to E. The impact deflects it to F and if left unchecked this would develop into G (which might be undesirable, such as a badly eroded site). Restoration offers a range of options to aim for. There are three possible options are available to restoration. The first is to return to the immediate pre-impact stage (D′) or to an earlier successional (historical) stage (B′). A third option follows the transition that would have occurred if the impact had never happened (E′)

including the time required for establishment and seek to place our restoration plans on that trajectory. Here, we restore a site to some previous state with what might be called the 'back to the future' option.

Such multiple endpoints (Figure 7.16) present us with an insight into successional and community processes. Indeed, they serve to show that communities are not fixed but flexible and operate under the influence of time and chance. Endpoints too change over time.

Restoration methods

In ecological terms, restoration is a form of accelerated succession with the ecologist trying to establish the new ecosystem as completely and as quickly as possible. Many restoration projects seek to reclaim spoils, such as the waste materials from mining and other excavation activities. These tend to be devoid of organic matter and nutrients and, therefore, demand something close to a primary succession (Section 5.3) in their restoration. In other cases, sufficient organic matter and even some topsoil may

remain, along with a reserve of nutrients to allow a secondary succession to occur. Such a site, if near to a source of colonizing species, may be left to restore itself naturally, providing there is no risk of the spoil moving or becoming a nuisance.

Spoils vary considerably in their chemical and physical properties (Table 7.5). Consequently, they differ in their nutrient requirements and site preparation. Often sites need to be landscaped and the spoil broken and mixed to form a substrate with a suitable range of particle sizes. In some cases, ecologists may be working with a relatively inert substrate such as the sand from china clay operations, crushed concrete, and brick rubble. Others, such as colliery spoil or smelter wastes, may be excessively acidic or alkaline and rich in a range of toxic metals. Most have few, if any, plant nutrients. Initially, artificial fertilizers can provide these, although many spoils are free-draining, so that without a proper soil structure these nutrients will be simply washed away. This is one good reason for raising the organic content of the waste, to create a soil structure that will hold the material together. It also assists in binding the nutrients we add as well as feeding the microbial

TABLE 7.5 Problems associated with spoils and their immediate and long-term treatment

Category	Problem	Immediate treatment	Long-term treatment
Physical structure	Too compact	Rip or scarify	Vegetate
	Too open	Compact or cover with fine material	Vegetate
Stability	Unstable	Stabilize/mulch	Regrade or vegetation
Moisture	Too wet	Drain	Drain
	Too dry	Organic mulch	Vegetate
Nutrition			
Macronutrients	Nitrogen	Fertilizer	Legumes
	Others	Fertilizer and lime	Fertilizer and lime
Micronutrients		Fertilizer	
Toxicity			
pH	Too high	Pyritic waste or organic matter	Weathering
	Too low	Lime or leaching	Lime or weathering
Heavy metals		Organic mulch or metal-tolerant cultivars	Inert covering or metal-tolerant
Salinity		Weathering or irrigation	Tolerant species
Plants and animals			
Wild plants	Absent or slow	Collect seed and sow or spread soil containing propagules or plants	Ensure appropriate conditions for colonization
Cultivated plants	Absent	Sow normally or hydroseed	Appropriate aftercare
Animals	Slow colonization	Introduce	Ensure appropriate habitat management

community—essential if a soil is to be capable of decomposition and nutrient cycling.

Most of the time, ecologists will help the process along by creating a suitable substrate and providing nutrients and colonizing species. In its simplest form, this may mean adding soil, fertilizers and seed. Elsewhere, more drastic measures are used. Limestone quarries in the Peak District of northern England have been restored using 'conservation blasting'. First, quarry faces were blown to create natural-looking rock falls, then sewage slurry containing the seeds of lime-loving plants (calcicoles) was sprayed on to the sides of the artificial cliff. The slurry added nutrients and provided a more favourable substrate as well as helping to disperse the seed.

We can also use plants to improve nutrient capture, especially legumes, by exploiting their association with nitrogen-fixing bacteria (Section 7.1). Similarly, species having mycorrhizal associations can be planted to take advantage of their capacity to scavenge phosphates in nutrient-poor environments. Both associations may require the soil to be inoculated with the microbial or fungal partner, perhaps by introducing topsoil where these are known to be abundant or through the use of commercially available inoculum.

Dealing with the toxic components of a spoil requires a different set of techniques. Many colliery spoils are often rich in iron pyrites (iron sulfide). When in contact with the air, the sulfide oxidizes to produce sulfuric acid, lowering the pH of the soil to levels that are toxic to many plants. It is possible to remove large pyritic particles from wastes on some sites, though usually it is treated on site by the addition of lime, which 'mops up' the acidity as it develops. In some cases, it has been possible to combine two wastes, acidic and alkaline wastes (such as colliery waste and basic slag), thereby solving the problem. Another alternative is to mix the waste with non-toxic material, or to bury it beneath a layer of imported soil. The danger here is that this soil itself becomes soured by acids or metals migrating upwards. The soil cover needs to be sufficiently deep to stop this and prevent deep rooting plants coming into contact with the waste.

Using resistant species—plants that can tolerate high acidity or high levels of metals—can help offset some of these problems. A range of species are now available commercially, and may also be collected from around ancient mineral workings or natural rock outcrops rich in copper, cobalt, lead, nickel and so on. These races or ecotypes (Box 2.4) have evolved mechanisms that limit toxic metal uptake. Locally adapted ecotypes can be tolerant to several metals; for example, 'Merlin', a cultivar of the grass red fescue (*Festuca rubra*), is resistant to both lead and zinc and has been used in mining restoration projects throughout the world. Other metal-tolerant grasses include cultivars of common bent grass (*Agrostis capillaris*) 'Goginan'—tolerant of lead and zinc—and 'Parys' a copper-tolerant cultivar. These *Agrostis* cultivars are particularly useful since they are able to grow in acidic conditions, making them ideal for the restoration programmes involving acidic mine spoils. Not only grasses but trees and other flowering plants also can be used in some locations. One of the most dramatic examples of a metallophyte species, which not only tolerates but also actually accumulates a metal, is the 'nickel tree' *Sebertia acuminata* (Plate 7.7). The tree grows on the nickel-laden serpentine soils of New Caledonia and oozes a blue-green nickel-rich latex from its bark. These, and other species, have been used by prospectors as indicators of mineral-rich rocks, an activity known as geobotanical prospecting.

Species selection can make the difference between success or failure in restoration programmes. This requires an understanding of the **autecology** of

species—how individual species respond to both the biotic and abiotic environments. Knowing the characteristics of a plant enables ecologists to anticipate how it might perform in its new habitat. Generally, the ideal plant for a restoration is one that is robust enough to cope with poor conditions and requires little or no management.

As the restoration ecology has developed, attempts have been made to assess and quantify species' performance in restoration projects, thereby taking some of the guesswork out of species selection. Richard Pywell and his colleagues analysed the performance of 58 plant species commonly used in British grassland restoration projects in a series of 4-year field trials. They measured plant performance according to a series of growth traits (life-form, germination characteristics, etc.) and compared these against three well-known classification systems. One was the Grime C-S-R life-strategy classification (Section 4.3). They also used the National Vegetation Classification Scheme (NVC) (Section 5.3) to provide an insight into the species' role within major plant communities. Finally, species were classified according to the Ellenberg system—a pan-European classification of plants according to their soil and climatic requirements.

The aim was to find traits that could act as indicators of good performance in restoration situations, such as an ability to colonize gaps, a strong competitive ability, the ability to regenerate vegetatively and a tendency towards a generalist rather than a specialist life history strategy. Not surprisingly, they found that the most successful traits were those that characterize key stages in the successional process. For example, species that performed well in terms of their initial establishment, such as ox-eye daisy (*Leucanthemum vulgare*), had a good colonizing ability, which included a high germination rate and a tendency towards rurality (r-selected 'weediness'). Later in the restoration process, characteristics more associated with mid-successional communities became more important in determining a species' success. This included the ability to compete against other species, vegetative regeneration, and the ability to persist within the seed bank.

This study also indicated that high residual fertility represented a major challenge to successfully restoring grassland on former agricultural land. The nutrient-rich soil conditions tended to favour generalists such as the ox-eye daisy and yarrow (*Achillea millefolium*), rather than more specialist species such as yellow rattle (*Rhinanthus minor*). For any plant to succeed it had to be able to establish and persist against competitors as the vegetation cover was quick to develop. Using the data, Pywell and his colleagues were able to recommend a 'phased introduction' of species to increase the chances of success.

Perhaps the most extreme form of restoration is **translocation**. Here, an entire habitat is moved from one site to another. This is neither easy nor cheap and is only attempted when a valued habitat is under severe threat. Some habitats are easier to translocate than others. Grassland, for example, can be lifted as turves with extra topsoil and re-laid elsewhere. The same goes for heathland (Plate 7.8). Woodlands too have been moved, although this is more problematic, given the scale of the operation and the species involved.

As well as the expense, restoration ecologists have to consider whether the community is likely to survive in its new location (the local geology, topography, drainage, etc.) and have to be skillful in matching up donor and receptor sites. Clearly, there is little point in transplanting a nutrient-poor habitat where it is likely to receive nutrient-rich run-off from neighbouring farmland. Equally, we need to know what impact the transplanted material will have on the existing communities. There are also ethical questions about moving habitats around and the extent to which a translocated habitat 'belongs' to its new site. In some cases partial translocation is possible, with material (soils, plants, and animals) being taken from one site to assist the development of another. Whatever the scale of the restoration undertaken, an established community will need proper aftercare, including site monitoring and a review of its management as the community develops.

Restoration ecology is playing an increasing role in civil engineering where it offers alternatives to concrete barriers and embankments. Figure 7.17 compares the options for 'hard' and 'soft' engineering, in terms of cost and performance. More often than not, inert structures cost more to construct and have a limited lifespan. **Ecoengineering**, on the other hand, makes use of sustainable, self-renewing materials, such as trees, shrubs, and other vegetation. Plant roots are mechanically very strong and provide increased anchorage and stability as they grow through the soil. Similarly, a mat of vegetation intercepts rain and shelters the ground surface, thus protecting the structure from erosion. These habitats are not only cheaper to construct; but can also be maintained indefinitely provided they are well managed.

People seem to prefer trees to a concrete wall. They have a range of other advantages. The acoustic and mechanical properties of vegetation, along with its

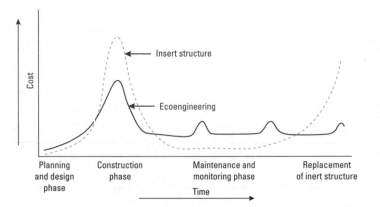

FIG. 7.17 Cost analysis of 'hard' versus 'soft' engineering. A comparison of relative costs and timescales involved in planning, construction, and mature operation of ecoengineered and inert structures

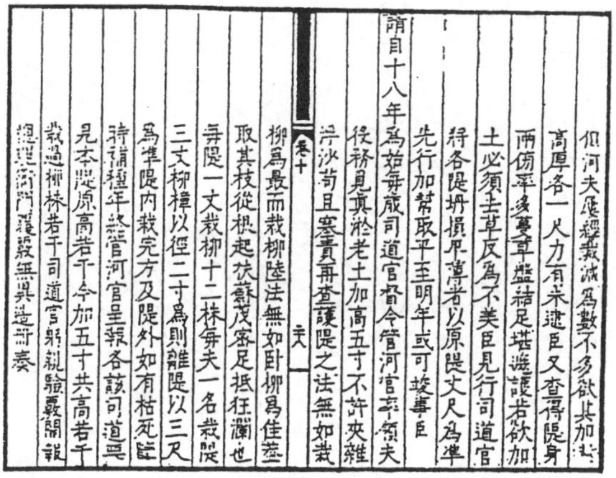

FIG. 7.18 Ancient Chinese manuscript—possibly one of the first documented examples of ecoengineering.

It is observed that both sides of the earth dam embankment surfaces are covered with creeping sage grasses, which provide sufficient surface protection. If it is intended to raise and thicken the embankment, the grasses have to be removed and the surface protection condition could become worse . . . It is revealed that the best method for protecting the embankment is planting willows. Among the six planting methods of willow, the lateral planting is best. Since this method permits the willow's branches to grow much closer from the root system, thus it allows the willow to have much blooming branches to resist the impact of impounding water. Every 1 zhang (320 cm) of embankment should be planted with 12 willows . . . The willow should have a minimum girth of 2 cun (3.2 cm) and stick out from the embankment for 3 chi (96 cm). The planting should start from the inner to the outer portion of the embankment. Any dead willows found should be replaced immediately . . .

From a report by Pan, Minister for River Flood Control, Ming Dynasty (1591).

capacity to trap dust, absorb rainfall, and bind soil, make it preferable to a simple, inert barrier. Despite such innovation ecoengineering is far from new. Back in the Middle Ages the Chinese used willow in their flood protection schemes (Figure 7.18). Then, as ever, people used materials that were readily at hand, such as the willow trees growing along the riverbank. Restoration ecology has rediscovered other, far older land management techniques, many of which disappeared with the advent of the Industrial Revolution.

The overall aim of restoration ecology is to produce ecologically viable ecosystems. However, decisions about endpoints have taken in a political and social context, including the need for local amenities and the wishes of local people. Indeed, some of these non-ecological objectives can present us with difficult choices.

Past successes have led to sweeping claims for restoration ecology and there is always the danger that our ability to restore ecosystems could be used to justify their degradation. The truth is that our ability to restore lost and damaged habitats is limited, not least because of our lack of knowledge and the restricted timescales to which we work. The best we can hope for is an approximation of a natural ecosystem. However, restoration enables us to use ecological processes in the restoration of damaged and degraded ecosystems and in the process we might well learn something of how they are constructed and operate.

SUMMARY

The physical and chemical processes by which the crustal rocks are degraded and rebuilt are supplemented by the activity of living organisms, creating the major biogeochemical cycles. Organisms use the chemical properties of the different elements in their metabolism and tissues. These properties also determine the range and speed by which these elements cycle through the biosphere.

Beyond carbon fixation by higher plants, bacteria, cyanobacteria, and fungi play a crucial role in scavenging and fixing key nutrients and their subsequent chemical transformations. In the warmer latitudes, shortages of some of these nutrients, especially phosphorus and nitrogen, limit ecological activity. Most ecological communities are adapted to these shortages and a variety of plant species have formed mutualistic associations with soil bacteria or fungi to improve their supply of nitrogen and phosphorus.

Considerable changes can occur in communities when key nutrients become abundant. Eutrophication—nutrient enrichment—can occur in both aquatic and terrestrial habitats and leads to major changes in their species composition and may require treatment to reduce or eliminate the pollution. In contrast, the pollution caused by oil spills is often best treated by containment, leaving nature to restore itself.

At the other extreme, on severely degraded sites (such as industrial wastes) recovery happens slowly or not at all, requiring ecological restoration and a range of techniques to accelerate succession. These sites are often short of nutrients and we seek to re-establish the plant and microbial communities that will restore their ecological processes. Other options include partial restoration in the form of rehabilitation or the recreation of an entirely new replacement habitat.

FURTHER READING

Bradshaw, A. D. and Chadwick, M. J. 1980. *The Restoration of Land*. Blackwell, Oxford. (A classic text which considers a wide range of ecological issues concerning restoration.)

Freedman, B. 1995. *Environmental Ecology*, 2nd edn. Academic Press, San Diego. (A useful book which deals with general aspects of ecology and focuses particularly on pollution and related issues.)

Jordan, W. R., Gilpin, M. E., and Aber, J. D. (eds). 1987. *Restoration Ecology*. Cambridge University Press, Cambridge. (This book presents a diverse overview of the subject—it is particularly interesting as it also includes some of the socio-economic, aesthetic, and ethical issues concerning ecological restoration.)

Perrow, M. R. and Davy, A. J. (eds). 2002. *Handbook of Ecological Restoration. Volume 1, Principles of Restoration*. Cambridge University Press, Cambridge. (A comprehensive account of the current state of restoration ecology in terms of the ecological concepts involved.)

Perrow, M. R. and Davy, A. J. (eds). 2002. *Handbook of Ecological Restoration. Volume 2, Restoration in Practice*. Cambridge University Press, Cambridge. (A companion to the first volume which examines case studies in a wide range of ecosystems world-wide.)

WEB PAGES

www.oils. gpa.unep.org/facts/facts.htm
A section of the United Nations Environment Program web site that provides information about oil pollution and techniques used in cleaning oil spills.

www.ser.org
Society for Ecological Restoration web site—includes a useful online primer on the subject along with links to other resources.

EXERCISES

1 Select the most correct descriptor of salinization:

(a) occurs when crops are grown too near to the sea,

(b) occurs when drought causes crops to die,

(c) is due to a build up of salts in the upper layer of the soil,

(d) is due to poor-quality irrigation water,

(e) is solely due to a build up of sodium chloride.

2 Insert the appropriate words to complete the following paragraph (use the list of words below; some words may be used more than once and others not at all)

Sites requiring ____ include ecosystems that have been damaged as a result of both ____ and artificial degradation. In some cases, full restoration to the ____ ecosystem may not be possible and ____ seeks to restore the site to an acceptable state. The facilitation of a self-sustaining ecological function is achieved by a process of ____ ____ provides an ____ ecosystem and this is useful in situations where the pre-impact state was of ____ ecological value (e.g. intensive agriculture). In all restoration projects the overall aim is to restore ecological function to the site, thereby producing a ____ ecosystem.

Accelerated	natural	replacement
alternative	original	restoration
arrested	rehabilitation	succession
limited	remediation	sustainable

3 Examine the map below, then consider the following scenario. The imaginary fishing village Safe Haven is situated close to areas famed for their wildlife and landscape value.
A supertanker has run aground on rocks 6 km offshore and most of its 100 000 tonnes of oil are being driven towards the coast by strong winds. As coordinator of the emergency control centre what are your priorities?

4 Select the most correct answer to the following statement. Nitrogen is an important nutrient but is frequently a limiting factor because

(a) it is mainly found in the air,

(b) its salts are highly soluble and are readily leached out of the soil,

(c) it only enters the soil via electrical discharges associated with lightning,

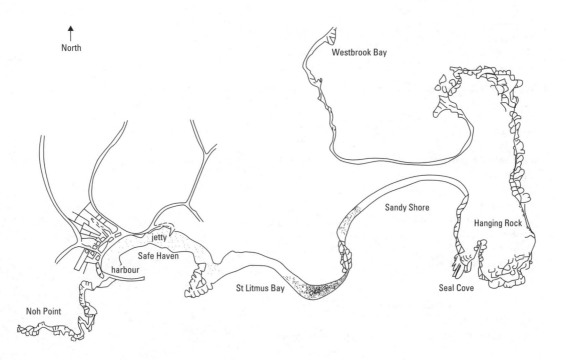

(d) its salts are highly insoluble and therefore unavailable to plants,

(e) mycorrhizae are required to fix nitrogen in root nodules.

5 The following experiment was carried out on nutrient-poor spoils from china clay mines in Cornwall. The graph shows differences in the growth rate of sycamore (*Acer pseudoplatanus*) growing in the vicinity of alder (*Alnus glutinosa, A. incana*) trees. Examine the graph carefully. What is happening here? Has it any use in other restoration projects?

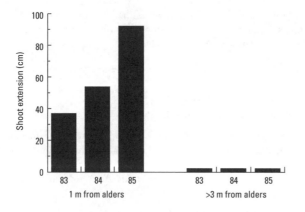

6 List and briefly describe three techniques used in the restoration of eutrophic bodies of water.

7 Select the most correct answer to the following statement. The bacteria *Nitrosomonas* converts ammonium to nitrogen by a process known as

(a) denitrification,

(b) eutrophication,

(c) flocculation,

(d) nitrification,

(e) rhizobification.

8 Match the following conditions of spoil with the appropriate methods of treatment:

Problem

(a) High pH

(b) Salinity

(c) Low level of macronutrients

(d) Too dry

(e) Unstable

(f) Low pH

(g) High level of heavy metals

(h) Too compact

(i) Too wet

(j) Low nitrogen levels

Treatment

(i)   mulch with organic matter

(ii)   drain

(iii)   add pyritic waste, organic matter, and allow to weather

(iv)   irrigate, allow to weather, use halophytic species to revegetate
(v)   rip or scarify prior to revegetation
(vi)   add fertilizer and lime
(vii)   stabilize, mulch, and possibily regrade prior to revegetation
(viii) lime and/or allow to weather
(ix)   add nitrogen fertilizers/legumes
(x)   add organic mulch and vegetate with metallophytes

Tutorial/seminar topics

9 Restoration is said to be the acid test of ecology—explain how the restoration might increase our knowledge of ecological principles.

8 SCALES

SCALES

Travel, it is said, broadens the mind. Choose the right places to visit and the Earth will quickly reveal the immense diversity of its life. Inspired by their visits to the tropics, both Charles Darwin and Alfred Russel Wallace sought to understand the reasons for such variety. Both were familiar with the ecology of lowland England but their respective expeditions to the forests of South America and South East Asia spurred them to understand why there were so many species, and how new species might arise.

Why should the tropical forests have such a wealth of species, compared to the forests of the North? Why does the number of species increase from the poles to the tropics in nearly all ecosystems? The heat and the damp of the tropics promote rapid growth, but, as we shall see in the next chapter, a benign climate alone cannot explain a higher diversity. The major terrestrial ecosystems form broad bands that trace the climatic zones of the planet, but this is not a perfect match and the pattern is interrupted by distinct communities reflecting local details of landform, geology or climate. Nor are they in fixed positions; ecological communities advance and retreat with fluctuations in

the global climate, across land-masses that slide around the planet in geological time. This mobile, three-dimensional puzzle still challenges modern ecologists to explain the present distribution of species and communities, and the evolution and dispersal of species through space and time.

In ecology, space and time are linked; ecological processes that operate over large areas also operate over long time scales. For example, the ecology of the northern latitudes only begins to make sense when we understand the effects of glaciations over the last two and half million years. Yet, most ecological study has been rooted firmly below the ecosystem level, confined to local communities and relatively short time periods. Until recently, these scales were taken to be the limits of our scientific resolution, so that only within this range could we answer questions with any degree of confidence.

Now, ecologists are being asked about long-term change, about the history and the future of life on Earth. Today, ecology uses data from ice cores, the fossil record and satellites. The spatial and temporal horizons of the science are being stretched. In this chapter, we consider the effect of scale on ecological processes and introduce some of the principles being developed to understand large-scale ecology.

We start by examining the relatively new subject of landscape ecology and the connection between spatial and temporal scales in ecological change. We then go on to describe the major terrestrial ecosystems and show how their distribution reflects the prevailing global pattern of climate. Finally, we explore the interdependence of the biosphere and the atmosphere. Our understanding of this interaction is crucial if we are to make sensible predictions and adopt policies that anticipate climate change over the next 50 years.

8.1　Landscape ecology

Size and timing is everything in ecology. Ecologists have to be sensitive to the implications of scale, as biotic and abiotic factors change from species to species. An insect that can walk on water has adapted to a physical property (surface tension) that has no significance for the mobility of a much larger animal. Individuals may span short time scales—the mayfly has just 12 hours to mate and lay its eggs—and small distances—the damselfly selects particular leaves and lays its eggs millimetres beneath the water surface. Yet other animals, including some insects, fly halfway across the world to mate and lay their eggs.

For the fullest picture of the selective pressures operating on a species, we have to see the world from the species' perspective and the scales to which it responds. For most of its life, the world of a fig-wasp is a fig. However, the size of an organism is not always a good indicator of the distances over which it ranges. For example, monarch butterflies travel hundreds of kilometres across North America to find the roosting sites from which they emerged. Even the wingless flea can travel across the globe attached to the feathers of its host. The chances of a bird surviving a migration from one hemisphere to another may depend on the insects that fed on it in the nest or the parasites it now carries south.

The movement of animals, the seeds, pollen, and other plant tissues are part of the transfer of energy, nutrients and DNA occurring within and between ecosystems. Ecologists may like to think they are dealing with discrete and closed systems, but ecosystems, species and individuals are leaky and open systems. Genes, nutrients and species move across boundaries and these inputs and losses have to be part of our accounting of the balances within each ecosystem.

The greatest frequency of exchange is likely to be between adjacent ecosystems within a naturally defined landscape. **Landscape ecology** is a fusion of geography and ecology, a study of the spatial arrangement of ecosystems and the large-scale processes that unite them. For example, the slope of a valley governs its rate of drainage, the depth and movement of its soil, the supply of nutrients and the stability of its surface. Its altitude and forest cover influence how much rainfall it receives and the chemistry of its waters depend primarily on the soils through which they drain. These soils, in turn, are likely to reflect their bedrock and vegetation cover and also those of adjacent communities higher up the valley. This exchange of materials between neighbouring ecosystems means the character of an ecosystem is shaped by the landscape in which it sits (Figure 8.1).

Landscapes range over kilometres and are most easily delineated by obvious discontinuities—such as a catchment or a significant change in geology or land-use. Within a catchment, water is an important agent for transporting material and such natural demarcations describe some important boundaries. Sometimes, however, these are poorly defined. In the prairies or the Canadian tundra, for example, there can be few obvious discontinuities over large areas, and even the division into catchments becomes almost meaningless. Where there are no sharp boundaries, abiotic change occurs along gradients and communities grade into each other. In contrast, some communities are effectively isolated by their boundaries: the highly localized conditions in the Californian or Chilean coastal valleys mean that adjacent and very similar ecosystems often have distinct plant species, an indication that little genetic exchange occurs between them (Section 5.1).

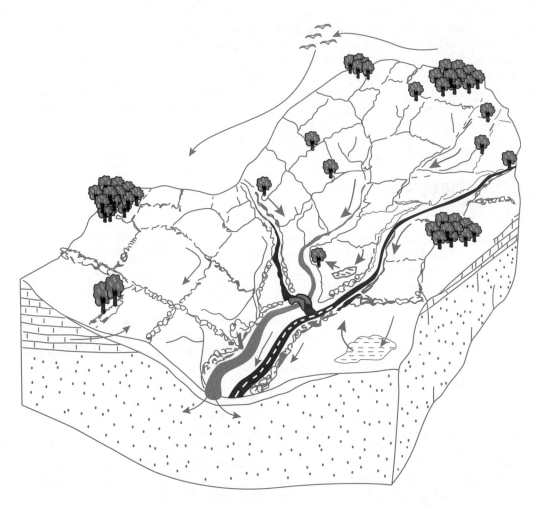

FIG. 8.1 Large-scale processes in landscape ecology, in this case for a temperate agricultural area. Genes, energy and nutrients move using wind, water, or the mobility of the biotic components. Networks of corridors—rivers, hedgerows, or roadside verges—provide habitats down which species can disperse across the landscape. All the living organisms will be affected by large-scale effects, such as the climate within the valley, or the movement of soils and water down the valley.

Because natural boundaries represent different sorts of barriers to different species, some ecologists define 'landscapes' according to the scales relevant to the organism they are studying. In this case, the most appropriate spatial scale ranges somewhere between an individual's home range and the species' regional distribution. More often, a landscape is seen as a mosaic of habitats that extend over a large area, each home to a community of species adapted to its particular abiotic conditions. The component ecosystems within a landscape share the same past and the same climate and are subject to similar geomorphological processes or management practices. A series of distinct landscape types may be collected together in a region.

The different ecosystems within a landscape are most easily distinguished by their abiotic characteristics and then by their plant communities. Thus, within a landscape we might distinguish rivers from lakes and ponds, and then lakes by their degree of eutrophication or the plants emerging from their shallows (**macrophytes**). In such well-defined ecosystems, much of the flow of energy and materials will occur within an ecosystem and rather less between different ecosystems. We might also expect more communication between equivalent habitats, if only because particular species favour particular ecosystems. For example, a species of wildfowl may only roost and feed on oligotrophic lakes because these have large areas of macrophyte cover around their shores, avoiding the open-water conditions found on eutrophic lakes (Section 7.2).

Different habitat patches have different degrees of permanence; over hundreds or thousands of years a pond or lake may become dry land through the natural processes of sedimentation and succession (Box 5.5). Such changes mean that the age of a patch is often a key determinant of its species diversity and landscape ecologists distinguish different patch types, according to their history (Table 8.1).

The archetypal landscape of lowland England is a patchwork of fields sewn together with hedgerows and streams, dotted with islands of woodland and water (Figure 8.2). The vineyards of the Mediterranean are bordered by bramble and scrub, interrupted by rocky outcrops, and with woodland confined to steep slopes and upland areas (Figure 8.3). In both cases, we can describe a series of landscape elements demarcated by their dominant vegetation, and representing different resources for their animal communities.

Landscape ecologists recognize three basic spatial elements—the patch (say, an area of woodland), the matrix (the background ecosystem which dominates the landscape—the fields or vineyards), and the corridors that traverse the matrix, connecting the patches (such as hedgerows or stream-banks) (Figure 8.1). Between these elements are more or less distinct boundaries or **ecotones**. Some plant and animal groups exploit the sharp transitions found at ecotones; many plants, for example, favour woodland margins, enjoying the open canopy and the reduced competition for light and nutrients.

Various ways of quantifying their spatial components have been developed to compare landscapes. The length of patch boundaries, the area of the patches, the

FIG. 8.2 The patchwork appearance of traditional English agriculture.

TABLE 8.1 Forman's patch types	
Disturbance patch	Recently created by some form of disturbance, e.g. forest fire
Regenerated patch	Patch recovering from disturbance, e.g. forest clearance in the process of re-growth
Environmental patch	Some sharp discontinuity, e.g. change in soil type that causes a distinct change in the plant community
Remnant patch	Area remaining of the original habitat isolated by an encroaching matrix, e.g. woodland patch amongst agricultural fields
Introduced patch	A habitat created by human activity, e.g. a woodland clearance

FIG. 8.3 Typical Mediterranean landscape (Languedoc, France).

(a)

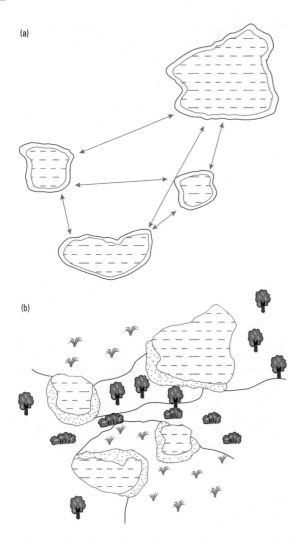

(b)

FIG. 8.4 Simple methods of quantifying the ecological properties of a landscape. (a) **Landscape physiognomy** is the physical layout of a particular element in the landscape. In this example, the habitats of interest are ponds and two measures are indicated here—the length of the boundary of each patch and the distance between each one. (b) **Landscape composition** is a measure of the proportion of the landscape represented by one habitat type. Both of these measures have their counterparts in measuring the structure of single communities, albeit at much smaller scales.

distance between them and the extent to which they are connected by corridors can all be measured using detailed maps, aerial photography or satellite telemetry (Figure 8.4). A highly heterogeneous landscape will be fractured into a mosaic of patches, but highly

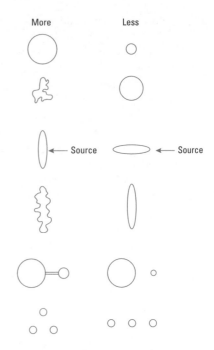

FIG. 8.5 The shape and size of habitat patches determines their degree of connectivity. The exchange of species or materials between habitat patches is increased by patch size, by having a more convoluted shape, by providing a larger target, by having a larger patch boundary, by having a connecting corridor or by having more neighbours.

interconnected if there is a network of corridors and long boundaries. The large fields used in intensive wine production form a very uniform, very homogeneous landscape with short (straight-line) boundaries relative to their area (Plate 8.1).

Connectedness is, in part, a feature of the spatial arrangement of patches as well as the distance between them. We would expect large patches to be more readily colonised than small patches, but this will also depend on patch shape. Small patches with a long boundary have a greater exchange with their surroundings. This traffic is greater still if there is a direct connection or corridor that the colonists can use (Figure 8.5).

Species differ in their capacity to cross the intervening matrix (or make use of any connecting corridors). Often the nature of the matrix serves as a filter, allowing some species to cross but not others; corridors too may be selective (Box 8.1). Birds fly readily between ponds but fish do not make the same journey quite so

BOX 8.1   Fragmented city

Different species see landscapes in different ways. For the birds and rodents of San Diego, the remnant coastal sage brush and chaparral communities offer different opportunities and challenges. Douglas Bolger and colleagues have studied the responses of these groups to the fragmentation of the natural habitat and also the use that species within each group make of corridors that connect patches of the natural Mediterranean-type scrub vegetation.

Within the urban and suburban areas of San Diego these remnant patches vary in size from less than 2 hectares to several hundred hectares and have different degrees of connectivity. Some are isolated and some are linked by strips of the original vegetation. Others are connected by roadsides which have been planted with a few of the native shrubs. Bolger and his co-workers recorded rodent and bird species for 2 years (1992 and 1993) in each of three habitat types—patches, remnant strips, and roadsides. They evaluated how a corridor was being used by each species—either as a thoroughfare to pass from one patch to another, or as a habitat with resources to be exploited. The birds could be divided into two groups—those known to be sensitive to the fragmentation of their native chaparral (and therefore likely to disappear as patch size reduced) and those tolerant of such change.

The vegetation type of the different corridors did not seem to matter for the diversity of either rodents or the birds, either for the remnant strips or the reduced shrubs of the roadsides. However the extent of shrub cover was important for both groups, and most especially for the fragmentation-sensitive birds—these avoided corridors with shrub cover of less than 40% and some species were never found on roadsides at all. These same species avoided San Diego and would not overfly urban or suburban areas. Besides the lack of native shrubs, perhaps the disturbance from the road itself (the noise, the street-lighting, and so on) prevent them using roadside habitats. Only the more tolerant species are found using the corridors, birds that do not seem to demand a dense shrub close to the natural community.

In contrast, all species of rodents recorded in the patches were also found in both corridors types and most were equally common in the corridors and the patches. Most were using these less as thoroughfares more as habitats in their own right, maintaining a reproductive population there. Interestingly, one mouse, *Peromyscus eremicus* was more abundant on the more open roadside corridors, perhaps because they benefit from the greater prevalence of herbaceous plants or because here they escaped their main predators.

In effect, this same landscape has very different properties for these two animal groups. For rodents the corridors are both thoroughfares and for many species, habitats to exploit. If, as seems likely, some species sustain a population in these corridors, they should perhaps be regarded as patches. For many of the birds, however, these corridors do not increase the connectivity of the San Diego landscape and offer no means to crossing the intervening matrix of urban development.

easily. Isolated patches might be considered as habitat 'islands' for some groups and the complement of species found there will depend on its size and the distance to the next patch (Section 9.1).

We can readily measure the degree of connectedness for different agents moving between patches. The simplest method is to count the frequency with which a coloniser reaches a patch, say as the number of visits made to a lake by birds from a distant roost. Collecting such data often identifies unforeseen ecological effects; for example, an increase in woodland fragmentation appears to increase nest predation amongst the resident birds. Another is that insect parasites found in the nest tend to be higher when the birds are forced to roost at the edge of woodlands or in corridor habitats. Both effects result from the increased traffic in pests and parasites moving between patches.

Beyond their value in comparing landscapes, these measures thus have considerable ecological significance. The size of any one population is likely to be closely tied to the size of its patch and its carrying capacity. We have already seen how a metapopulation, divided between a number of patches, may be able to sustain itself even though one or more local populations periodically go extinct (Section 3.7). A highly connected landscape, or a background matrix that is readily traversed, allows most patches to be re-colonized quickly. In this way, the long-term prospects for a species is determined by the heterogeneity and connectivity of a landscape. Indeed, for an endangered species, it can be crucial, either when trying to find a mate or when disease is spreading through the metapopulation (Section 3.7, Box 3.4). Connectivity can cut both ways—isolated patches may provide refuges from an infectious disease or condemn their populations to a paucity of genetic variation.

The extent to which a landscape is partitioned into distinct patches is described as its 'graininess'. The 'grain' of a landscape varies with scale and different species will resolve different grain sizes according to their specific habitat demands. A small patch of woodland may consist of areas of closed canopy or open glades, each with a particular plant and invertebrate community. Some of the insects are confined to the glade for the whole of their life cycle. In contrast, a bird over-flying the landscape may distinguish only woodland and open field. Moving through the glade a butterfly lays its eggs on leaves selected according to the plant species and their nutrient status. At a much larger scale, the bird is seeking out nesting sites according the tree position and the territories of other birds. The two animals are responding to very different grain sizes.

At regional scales, enduring changes in abiotic conditions begin to distinguish different landscapes, where processes begin to operate at different rates. Tomas Santos and his co-workers have shown that woodland habitats regenerate from fragments more slowly in the drier areas of Spain. Remnant patches of holm oak (*Quercus ilex*) forest in the centre of the country support far fewer bird species for their size than woodlands amongst the farmland of the temperate north. The central and southern woodlands are subject to the prolonged summer drought of the Mediterranean region (Section 5.1), and this slows their regeneration. Where rainfall is more consistent and abundant, forest development is faster, so that in northern Spain, holm oak patches are more fully developed, with a greater variety of trees and tree heights. These, in turn, provide more niches for the birds (Figure 5.7) and, compared to equivalent sized patches in the south, they have more bird species per unit area. This demonstrates how some features of a landscape's ecology, in this

case, bird species richness, is determined by higher-level and larger-scale factors—most especially the regional and continental climate and its effect on vegetation dynamics.

Measuring the connectivity of landscapes in terms of nutrient or material flows also enables us to describe regional ecological processes. Such studies can indicate the routes by which a pollutant would move between patches and between species. Ironically, it has often been the other way around—we have learnt much about these processes from following the passage of a pollutant through a landscape. The most spectacular example was the explosion of the Chernobyl nuclear reactor in 1986, which contaminated a swathe of landscapes from Central Europe through Scandinavia and into Northern Britain. With the major research effort that followed the explosion, ecologists were surprised by the ease with which radiocaesium moved though the soils and quickly learnt that their models of elemental flows through ecosystems were far from complete. One consequence of Chernobyl was a better understanding of the ecological mobility of caesium.

Another dramatic case is the pollution and over-exploitation of the Aral Sea (Plate 8.2) in Kazakhstan and Uzbekistan, once the fourth largest lake in the world. For the last 70 years its feeder rivers, the Amu Darya and the Syr Darya, have been diverted to irrigate cotton fields. Since 1960, the Aral has shrunk by nearly half, by around 30 000 km². As its freshwater supply dwindled so its waters have become increasingly saline, leading to the loss of its unique (endemic) fish and invertebrate species. The fishery has collapsed and the large steamers that once used to work the lake are today stranded in desert sands, far from any water.

The impact of the shrinking lake today extends many hundreds of kilometres from its original shoreline. Changes in the local climate also followed. Now dry winds blow salt deposits onto the surrounding soil, scorching the vegetation and making the land almost useless. Increasing atmospheric pollution and the poor quality of the Aral waters have created major health problems for the local people. Infant mortality rates are among the highest in the world and life expectancy among the shortest. International efforts are being made to restore water levels, but there is no prospect of restoring the ecosystem.

One might reasonably argue that no detailed understanding of landscape ecology was required to predict many of the changes that followed the shrinking of the Aral Sea, even at these larger scales. Plans to divert or dam rivers in various parts of the world, from the Euphrates in Turkey to the Mekong in South East Asia have also raised fears about their climatic, ecological and economic implications. Nations continue to fund large-scale engineering of landscapes despite the warnings of ecologists, hydrologists and geomorphologists who have seen previous schemes cause havoc. The disastrous flooding of many central European cities in the summer of 2002 was linked to the damming of the Danube and Rhine watersheds. The increased frequency of floods in central China over recent years has been attributed to the massive engineering of its rivers in the last quarter of a century, which continues to this day. Rather like the past engineering of the Mississippi Basin, such large-scale modification of the landscape to solve small-scale problems may only store up a larger-scale problem when the once-in-a-century floods arrive.

Working down from the level of the landscape we are describing a hierarchy of scales, with the degree of patchiness changing according to the process being studied. Generally, movements of energy, nutrients, and information are greater at the smaller scales and thus show greater connectedness. To make reasoned decisions and quantified predictions requires some way of organizing the various types of environmental information over a range of spatial and temporal scales, to incorporate the important interactions at the regional and landscape levels with lower level processes. One approach looks at the relationship between processes operating at different scales and how they interact to impart stability to a landscape. These ideas have been formulated in more rigorous scientific terms as hierarchy theory.

Hierarchy theory

While a landscape may appear unchanging, its component ecosystems are often much more variable. Even within an ecosystem there will be several habitats—the shoreline, the shallows and deep water of a lake—each with a different inventory of species, each changing with space and time. At these smaller scales, measurements taken between habitat patches or in a single patch over short intervals are typically highly variable. Averaged over the whole landscape however, such measures change little, even over long periods. Thus, the variation we observe in an ecosystem depends upon the scale we use to measure a parameter.

Hierarchy theory suggests this is a general property of many complex systems; it sees a system organized into various functional levels, differentiated by the rate at which some process operates. The smaller and lower levels are 'nested' inside the larger-scale and higher-level processes. Moving to the next level up, the rate of a process slows down so that larger-scale, higher-level processes operate more slowly compared to the lower levels. Consequently, the variation we observe at the smaller scales has little significance at the higher levels. Most importantly, this 'buffering' between levels, the theory suggests, serves to regulate the system. To put it at its simplest, large-scale systems such as landscapes appear unchanging because they are big and complex.

Hierarchy theory defines levels not by species but by functional groups, by what they do and the speed at which they do it. Consider the example given in Figure 8.6, where the process is carbon fixation in

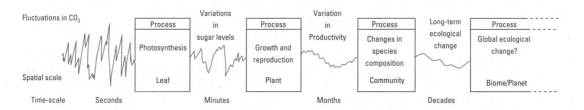

FIG. 8.6 A simple example of the effects of a hierarchy of levels governing an ecological process, in this case carbon fixation by photosynthesis.

plant tissues. Within an individual leaf, fixation fluctuates rapidly as carbon dioxide concentrations near the stomata change from one moment to another. At higher levels in the hierarchy, say the leaves of the whole tree, average rates of fixation, per gram of photosynthetic tissue, vary much less. In the canopy of the forest, over a larger spatial scale, rates are increasingly uniform. As we move up through the hierarchy so the variation decreases.

At the regional level it may take years, perhaps decades, to detect a significant shift in carbon fixation rates, even following some large-scale change. The general level of fixation for the whole forest ultimately depends upon rates occurring in individual leaves, so that even if the forest changes its species composition, variation appears at the higher levels only very slowly, if at all. At the level of the forest or the region, the process of carbon fixation is highly stable.

Constraints work in the opposite direction too. Photosynthesis in individual leaves depends upon local humidity which itself depends upon the density of the tree cover. Forests change the climate of their immediate area, inducing rainfall and raising humidity: about half of the rainfall in the Amazon basin originates from the evapotranspiration of the forest itself. So the speed with which water cycles through the system depends, in part, upon rates of photosynthesis and the opening of stomata to fix carbon. Thus, in this hierarchy, a high level attribute, forest cover, governs a process, carbon fixation, operating at the lower levels (the leaves). Notice here that this model is only good for carbon fixation and we would need to construct an entirely different hierarchy to model some other process.

Other limits on carbon fixation will be imposed by the interactions between species, which have been omitted from our simple hierarchy. For example, very damp conditions on the forest floor will favour a particular assemblage of animals, microorganisms, and fungi. Since the composition of the decomposer community will determine decomposition rates and the flow of nutrients between the soil and the canopy, this too will impact rates of carbon fixation.

The process, as well as the composition of the community, will be constrained by these interactions operating up and down the hierarchy, and the feedback between different functional levels. As a result, hierarchy theory posits that an ecosystem organizes itself, forming itself into a stable configuration driven and limited by these interactions. This may be a general property of any complex system which has these sorts of internal checks and balances, and which will inevitably arrive at a state where it undergoes little further change.

Because of the complexity of interactions within a hierarchy, ecologists have to determine the key large- and small-scale processes that organize ecosystems. An example is forest regeneration following fire or a tree fall. At the local scale, regeneration is constrained by the rates of nutrient release and the scope for colonizing trees to form mutualistic associations with soil fungi (or mycorrhizae—Section 4.2). Over a landscape, regeneration depends also on the distribution and abundance of potential colonizers, the scale of disturbance, the size of regenerating patches and the ease with which colonizers can find them.

We have learnt much about the role of fire outbreaks and the effect of their scale in natural ecosystems in recent years. Like most of the extensive forested areas in North America, the Yellowstone National Park had been subject to fire-suppression patrols. Prior to 1970, a vigilant forestry service quickly extinguished the smallest fire before it could spread. Much of this forest has extensive tracts of lodgepole pine (*Pinus contorta*), a species that becomes more flammable with age. With no small fires to open gaps and remove some of the older trees, the population in the park was dominated by large and flammable trees. At the same time, fuel, in the form of leaf litter, accumulated on the forest floor.

It needed only the extended drought of 1988 and an uncontrolled logging fire to set off the inevitable. Around 70 per cent of the park burnt. The scale of the fires and the dramatic pictures of its destruction led to it being reported as a national disaster. Yet, the burn was not uniform. Massive though it was, different patches burnt to different degrees. Temperatures high enough to kill seeds were achieved in only relatively small pockets. Trees were killed in about 20 per cent of the park whilst less than 1 per cent of its elk population was lost to the flames. Since then there has been a dramatic change in the park with a rapid regeneration of the trees, the ground flora, and the animals. The nutrients and space released by the conflagration allowed opportunist plants and seedlings to flourish. The removal of old, unproductive wood was followed by a flowering of the forest floor and an increase in the resources for its herbivorous animals.

The simple lesson is that frequent but small-scale burns are part of the normal economy of coniferous forests, much as they are in mediterranean-type ecosystems. Small burns do not produce dramatic change, but they do create the gaps that prevent some species from becoming dominant (Section 5.3), and, with a high frequency, ensure a high turnover of species.

Change is a part of the dynamics of all ecosystems, and communities develop that can accommodate the most frequent scales of disturbance. If the range of an individual species is exceeded, more dramatic change may follow at the community level. Then species disappear, the interactions within the community shift, and its organization switches to a different configuration. A large-scale disturbance may lead to the loss or replacement of many species, perhaps reducing a community to a plagioclimax (Sections 5.3 and 9.2).

Today, the northern coniferous forests of North America show few signs of the fires in Yellowstone in 1988 and the park is now back to a former, younger self. Forest fires are short-term events and hierarchy theory suggests that these large-scale ecosystems are susceptible only to persistent and long-term change.

The boundaries of this biome would move if there were persistent changes in the patterns of rainfall or temperature over its latitudinal range. Climate change is slow, but because it operates at the higher levels in the hierarchy it causes major shifts at the lower levels, and produces change which is not quickly reversed.

Diverting the Amu Darya and Syr Darya was the start of a slow process of change. It took 20 years for the Aral Sea to show signs of ecological decline. Now the impact covers many thousands of square kilometres. In contrast, a short-term change may well be absorbed by the system; the fires in the Kuwaiti oil fields after the Gulf War of 1991 were portrayed as a major ecological disaster at the time (Plate 8.3). Despite the large amount of carbon dioxide released (240 million tonnes) in the few months they burned, their effect has hardly been detectable in the global atmosphere. Any shift in atmospheric composition would need large amounts of carbon to be added over a much longer duration.

Moving from the landscape to the regional and global scale, we have to map the distribution of plants and animals on time scales of millennia. Yet, as we shall see later in this chapter, climates can change surprisingly quickly. When they do, the ecological regions shift with them.

8.2 The biogeography of Earth

Based on the distribution of large terrestrial vertebrates, the Earth can be divided into six distinct realms (Figure 8.7). First identified by Wallace, the boundaries of these zones actually pick out long-standing barriers to animal dispersal that have existed across the globe. These can be closely matched to the drifting of the continents, and such discontinuities were used by Alfred Wegener to support his theory of continental drift when he proposed it at the beginning of the twentieth century.

The first mammals appeared around 200 million years ago and one of their fundamental divisions, between the placentals (Eutheria) and the marsupials

(Metatheria), times the break-up of a massive landmass that once existed in the southern hemisphere. Today the greatest variety of pouched mammals is found in Australia and New Guinea, land masses which together split away from Antarctica 65 million years ago. Yet their oldest group, the possums, are found on the other side of the globe, in the Americas, the only survivors of a diverse marsupial fauna that once occupied South America. It has been the sustained isolation of Australia that has allowed its marsupials to survive without competition from other mammals, at least until relatively recently. The placental mammals did not reach Australasia in any

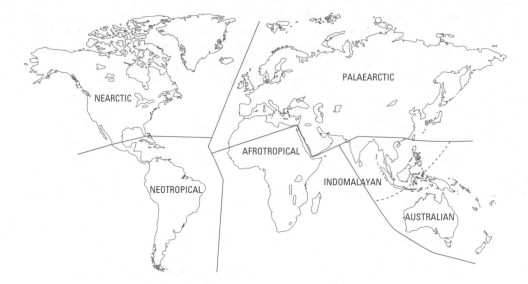

FIG. 8.7 The main zoogeographical realms of the planet, primarily defined by their vertebrate fauna. The mammal fauna is particularly characteristic of an area; before the advent of humans, some orders were entirely excluded from certain realms. The dotted line marks Wallace's line, running through Malaysia and separating the two realms according to their native birds. The line actually follows a deep oceanic trench, marking the junction between two crustal plates. The flora and fauna of the two realms blend across the whole of this archipelago so the demarcation is not well defined, even for the marsupials.

numbers before the most recent human colonists arrived, but marsupials occupy niches they have since had to defend, with varying degrees of success. The northern boundary of the Australasian realm, separating the two mammal groups, Wallace's Line, is one of the sharper boundaries in this classification, and is now known to follow the deep oceanic trench between two crustal plates (Figure 8.7).

Much of the mammalian fauna of the ancient northern continent is common between its realms, simply because there have been no significant barriers to their spread in the recent geological past. Groups that have recently evolved in the tropics show a much more restricted distribution—there are, for example, no native primates (apes, monkeys, and their relatives) in the Nearctic and the only humans to arrive there were *Homo sapiens sapiens*, in recent millennia (Box 5.1, Section 1.1).

In contrast, the distribution of plant communities on the Earth reflects its more recent past and the prevailing climate. Marking out lines of latitude,

major plant assemblages are organized into *biomes*, collections of species that share adaptations to the climate of each region. Within a biome many plants have a characteristic growth form, adapted to the prevailing conditions. The most important factor is the availability of water. Forests dominate where there is sufficient moisture; grassland or low scrub where it is drier.

We can distinguish four major terrestrial biomes—forest, grassland, tundra, and desert—by their regimes of precipitation and temperature (Figure 8.8). For example, deserts are found where evaporation exceeds rainfall. These conditions prevail at 30° either side of the Equator, reflecting the pattern of atmospheric circulation. At this latitude, dry air is descending, having shed its rain in the tropics. Deserts also feature on the western edge of several continents, where cold ocean currents induce coastal fogs that deprive inland areas of precipitation.

Local conditions may confound the simple latitudinal pattern of the biomes (Box 8.2). Proximity to

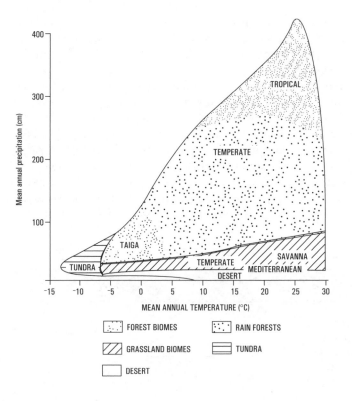

FIG. 8.8 The range of the major biomes defined by temperature and moisture. Forests are only found where precipitation is above about 30 cm per year (in the coldest regions) and 120 cm per year (in the warmest regions). Where water is in short supply grasslands develop.

an ocean or a mountain range can create local climatic conditions that give rise to isolated, often unique plant communities (Plate 8.4).

Here we limit our task by concentrating on the major terrestrial biomes. This is not to deny that there are characteristic aquatic communities—far from it. Most aquatic habitats are defined by their animal community and its adaptations to depth, rates of water flow, salinity and nutrient source. As we see in Chapter 9, many tropical marine communities show the same latitudinal increase in species diversity as that found on land.

We also concentrate on the most productive biomes, and say little here about the hot and cold deserts. The productive ecosystems fall within three basic climatic zones—tropical, temperate, and boreal, principally defined by their average temperate regimes. For each of these zones, sufficient rainfall will produce a characteristic forest, or where it is drier, a grassland; together this produces a simple classification of two major biomes in each of three climatic regions.

Tropical biomes

Tropical rainforests straddle the Equator, roughly within the tropics of Cancer and Capricorn (Figure 8.12). These are amongst the oldest communities on the planet, with some pockets largely undisturbed even during the last glaciations. Their diversity is astounding: in Amazonia, there can be around 300 species of tree in just 2 or 3 km^2. Tropical forests have the highest net primary productivity of all terrestrial ecosystems (Section 6.3, Plate 8.5), yet they grow on some of the most infertile soils.

The soils are poor because they are old and have been leached of their key minerals. A highly competitive decomposer community quickly scavenges dead and decaying organic matter arriving at the forest floor. This makes for extremely rapid decomposition, promoted also by the high temperatures and abundant moisture. Because its organic matter has such a brief existence, the soil has a poor structure and little capacity for chemically binding nutrients.

Consequently, the soil represents a small reservoir of plant nutrients and competition for these is intense.

There is also intense competition for light. Like all forests, the plant community is stratified into layers, with groups of species adapted to different degrees of shade (Figure 8.13). The highest layer is the canopy, perhaps 30–40 m above the ground and formed of the upper branches of mature trees jostling for uninterrupted light. Where sufficient light penetrates the upper layer a sub-canopy of younger trees develops and further down an under-storey of large-leaved species and palms. At the forest floor, among the buttress roots supporting the taller trees, herbaceous plants, particularly ferns, thrive in the semi-dark, dark conditions. Each layer contains seedlings or young trees waiting for sufficient light to join the layer above.

Where the canopy is almost continuous the interior is relatively open because light levels are too low for much undergrowth. A dense 'jungle' of young trees and climbers can only develop where sunlight penetrates well below the canopy, perhaps where a tree has fallen or the forest forms a boundary with a river. Here climbers, creepers and 'hangers-on' (**epiphytes**) all use the main trees to get access to the sunlight, avoiding the costs of building large trunks.

BOX 8.2 The global climate

Consider what the atmosphere does. The main source of heat on the planet is the energy arriving as radiation from the sun. Because the Earth is a sphere, spinning on a tilted axis, with an uneven distribution of sea and land, it warms at different rates in different places. Heat disperses to even out these differences. Conduction is slow though the crustal rocks compared with the convection of water, while the atmospheric circulation (also convection) is fastest (Figures 8.9 and 8.10). Weather is simply the movement of air and water vapour caused by heat imbalances within and between the atmosphere and the Earth's surface.

Heat moves in the oceans as currents: warm water rises as it becomes lighter, and colder, denser waters sink to replace it. This creates persistent circulation patterns between the cold polar and warm tropical regions (Figures 8.10 and 8.11). Heat is transferred to the air mainly through the evaporation of water, producing clouds and eventually precipitation. In the atmosphere, winds blow from high-pressure areas with relatively high temperatures to low-pressures areas. The simple atmospheric circulation between the poles and the equatorial regions is complicated by the forces created by the spin of the earth, the seasons created by its tilt and the configuration of land and sea. The result is a well-defined, if intricate, pattern of wind and ocean circulation distributing heat around the planet.

These patterns are not fixed over the long term, and oceanic currents in particular will move or cease to flow, changing the heat distribution on the planet. To understand long-term climate change, we need to add in the solar flux, (the amount of energy reaching the Earth's surface) and changes in the chemistry of the atmosphere. The solar flux varies over thousands of years because of

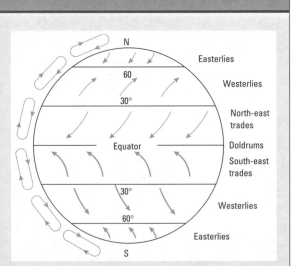

FIG. 8.9 A simplified map of atmospheric circulation. Air warmed at the Earth's surface rises and then cools at altitude. Heating is greatest where the sun is directly overhead. This occurs somewhere between the tropics at various times throughout the year. Air rises and falls at particular latitudes, generating winds that are then deflected by the spin of the Earth. This creates the characteristic wind patterns of the planet, and combined with seasonal factors, zones of precipitation

variations in eccentricity of the Earth's orbit and its relative tilt, as well as variations in the energy output of the Sun. This can account for some climatic change in the recent past. However, other factors are important including geological activity and the composition of the atmosphere.

BOX 8.2   **Continued**

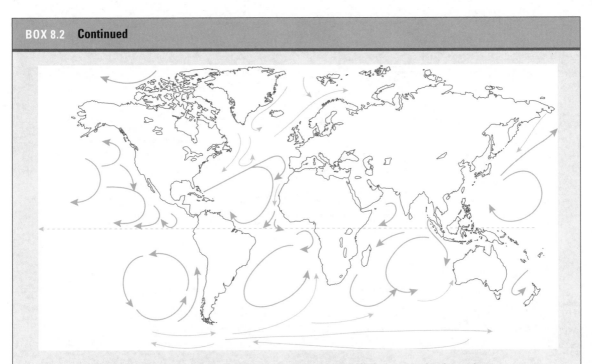

FIG. 8.10 Simplified surface currents of the Earth. Warm currents (dark arrows) develop either side of the Equator and flow clockwise in the northern hemisphere and counterclockwise in the southern. This movement draws in cold polar currents (light arrows).

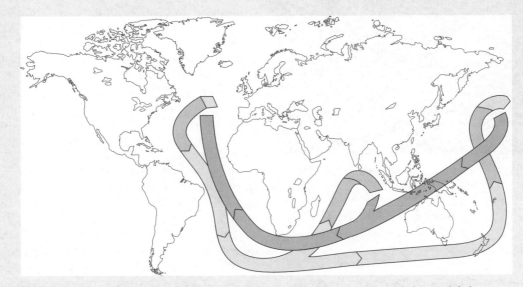

FIG. 8.11 The thermohaline is sometimes described as the 'oceanic conveyor belt', a global circulation of deep and surface water currents of different temperatures and salinities. It is a major distributor of heat energy across the planet. Its mixing of water from different levels is important in determining the degree of stratification and temperature profile of oceanic waters. Cold and saline water moves at depth (light shading), replacing the warm and less saline waters toward the ocean surface (dark shading). The heat transfer of the thermohaline is closely associated with the global climate and changes in its circulation mark different phases in an Ice Age.

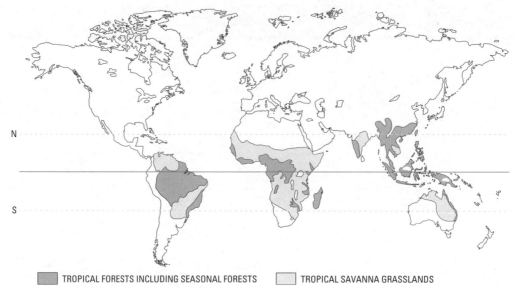

TROPICAL FORESTS INCLUDING SEASONAL FORESTS   TROPICAL SAVANNA GRASSLANDS

FIG. 8.12 The tropical biomes. The hatched lines either side the Equator (the solid line) represent the tropics of Cancer (northern hemisphere) and Capricorn (southern hemisphere).

FIG. 8.13 The stratified structure of tropical rain forests.

Epiphytes attach themselves to the sides of trees, relying on the nutrients in rainfall and the water draining off their host.

Some tropical forests (e.g. in South East Asia) are seasonal and have cycles of leaf-fall prompted by a dry season. Most tropical forests are dominated by broad-leaved trees that maintain a canopy throughout the year. Also the majority of plants rely on animals for pollination and seed dispersal, producing flowers and fruits to attract insects,

reptiles, birds and mammals at various times during the year.

The structural organization of the plant community creates a variety of niches for these animals, which in turn are formed into distinct assemblages associated with each layer. Many insects and several vertebrates—monkeys, sloths, reptiles, and amphibians—confine themselves to the canopy or sub-canopy, rarely descending to the ground. A range of carnivores, including birds, feed here as well. Small pools of water held in the leaves of the upper layers or caught in the centre of epiphytic bromeliads support insect and amphibian communities—entire aquatic ecosystems, a few centimetres across, perhaps 30 m above the ground. Equally diverse animal, fungal and microbial communities wait on the forest floor for material arriving from above.

Rainforest is typical of areas where precipitation exceeds 200 cm per year and where no month has much less than 12 cm. If there is a prolonged dry season, forest gives way to open woodland, thorn scrub, and tropical grassland (**savanna**). The dry conditions favour trees and scrub with small leaves and with protection from browsers, such as the acacia trees (*Acacia* spp.) of the African plains (Plate 8.6) with their protective thorns. Even so, browsing checks the spread of woodland and scrub, so that elephant and giraffe, amongst others, are especially important in maintaining an open grassland.

Grazers are less selective, cropping the above-ground parts of the herbaceous (non-woody) plants. In the absence of grazers, the grasses of the savanna grow high—up to 3 m. Usually tussocked, they die back in the dry season, keeping their living tissues close to the ground, well protected from the grazers who crop closest. Fires are also important here; set by electrical storms, they will sweep through the savanna, in quick, short bursts. Together with the intense grazing, fires maintain an open landscape. Many flowers produce fire-resistant seed that germinate quickly when the rains come, while others sprout from deeply hidden tubers.

The soils are again ancient and infertile, with relatively little organic matter so that the animals grazing above are crucial for nutrient cycling. Savannas are typical of low, well-eroded flatlands, with depressions that become vital watering holes in the dry season. Their pronounced seasonality means the grass is grazed by successive waves of large mammals, cropping to different levels; in Africa, zebra is followed by wildebeest and thereafter gazelle. In attendance too are the carnivores, the big cats and hyaenas and also the scavengers who follow them—jackals and vultures.

The biomass represented by this hoof and claw testifies to the productivity of this biome. But in terms of biomass consumed, the most important herbivores are insects, primarily grasshoppers, locusts and ants. Equally important are the inconspicuous termites, obvious only by their sentinel mounds, yet whose activities are crucial in maintaining soil fertility.

Temperate biomes (Figure 8.14)

The stark contrast in temperatures between winter and summer has the most profound effect on temperate ecosystems. Cycles of biological activity anticipate and respond to the seasons. This clock times the flowering and fruiting of plants, while animals feed and breed, migrate or hibernate in tempo with the plants. Much of the discussion below focuses on the patterns in the northern hemisphere which has the most extensive temperate forests and grasslands.

The different temperate biomes reflect the abundance of water during the year. Forests dominate where water is plentiful, grasslands where there is a substantial dry season. We can distinguish several types of **temperate forest** according to their regimes of temperature and precipitation (Plate 8.7), from broad-leaved evergreen forest in the warmer latitudes to the conifer–broadleaf mixed forest further from the equator. Temperate forests have considerably fewer tree species than a tropical forest and are typically dominated by just three or four species, primarily determined by climate and soil. Often, there is a mix of coniferous and broad-leaved trees, with the balance shifting in favour of conifers, especially pine, where soils are nutrient-poor and there is a higher risk of forest fires. Many temperate coastal regions, with abundant rainfall or fogs throughout the year, support hemlock (*Tsuga* spp.), fir (*Abies* spp.) and redwood (*Sequoia* spp.), conifers growing in almost continuous stands.

The seasonal contrast in temperatures becomes more marked with distance from the equator. Where moisture is available all year round broadleaf forest develops. Its composition depends upon soil depth and drainage. Beech (*Fagus* spp.) dominates on drier, shallow soils and produces a dense shade with little under-storey development. Elsewhere, the forest floor beneath oak (*Quercus* spp.) is comparatively well-lit and these forests undergo a sequence of herbaceous growth each spring before the main canopy develops. There is also a distinct under-storey of saplings and low bushes as well as climbers (e.g. ivy *Hedera helix* and honeysuckle *Lonicera periclymenum*).

At very low temperatures, photosynthesis ceases and respiration is reduced. Average winter temperatures make the balance uneconomic for deciduous trees and so they reduce their metabolic costs by shedding leaves. Most conifers, which can photosynthesize well below 0 °C, retain their needles (Figure 6.4), but have a series of adaptations to reduce water loss, important when the soil is frozen. Some herbaceous plants over-winter as seeds; others store food in corms or tubers that will fuel growth early in the new season, before the forest canopy reforms.

The annual cycle of leaf-fall makes the soil an important repository of the nutrient wealth of the biome. Much of western agriculture lives off the organic capital remaining in these soils following the

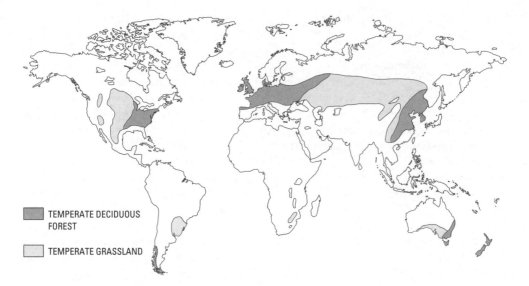

FIG. 8.14 Temperate biomes.

TEMPERATE DECIDUOUS FOREST

TEMPERATE GRASSLAND

clearance of the forest. The soils are deep and rich with a decomposer community stratified down their profile.

Animals also match their life cycles and activity to the times of plenty. Some over-winter as eggs or larvae, reducing their metabolic costs when food is scarce. Others hibernate or simply migrate to exploit more abundant resources. Again, insects are the dominant herbivores and these are, in turn, exploited by various migratory birds. In the northern hemisphere, mammals include squirrels, wild pigs and badgers with deer as the principal browsers. Before humans became the major influence on this biome, wolves, bears and several species of cat were the larger carnivores.

Where a dry season lasts for most of the summer, the forest gives way to **temperate grassland**. In the centres of large continental masses, away from moist coastal air streams, annual precipitation falls below 60 cm at these latitudes. Australia, South America (pampas), South Africa (veld), Eurasia (steppes), and North America (prairies—Plate 8.8) all have such zones. Once again, the grasslands are typically flat, gently rolling landscapes, which in the northern hemisphere are associated with the wind-blown sediments from recent glacial activity. In the south, they are again old and well-eroded plateaux.

Growth and flowering is distinctly seasonal, with grasses bearing their seed toward the end of the dry summer. Like the savanna, fires and grazing maintain the dominance of the grasses and the legumes. Thorn scrub and trees develop where grazing pressure is lifted. In the prairies of North America the dominance of the grasses and the variety of flowering plants were maintained by the bison. In Europe and Asia, various species of horse and antelope were the principal vertebrate grazers, but insects, above and below ground, are major herbivores in all temperate grasslands. Each region also has a range of burrowing animals, both vertebrate and invertebrate.

Soils are deep, often with a thatch of undecomposed vegetation overlying a thick humus layer. This is important protection against wind erosion when the soil is dry. Light grazing helps to speed decomposition processes and adds moisture. Today, most grasslands around the world have few wild grazers, but instead are used for meat and cereal production (Section 6.5).

Where temperatures range higher, and rainfall is confined to the winter months, mediterranean-type communities can develop (Plate 5.4). These have a very restricted distribution (Section 5.1) and the Mediterranean basin had lost much of the evergreen woodland before humans arrived. Where rainfall is very low, true desert will form (Plate 8.9, Figure 8.15).

FIG. 8.15 Mediterranean and desert biomes.

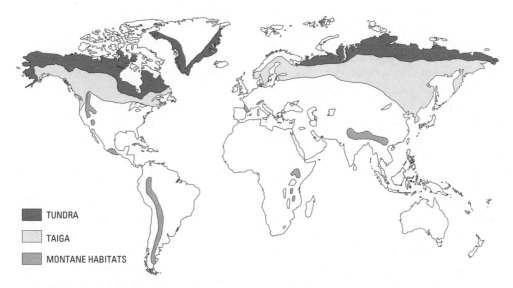

FIG. 8.16 Boreal and montane biomes.

Boreal biomes

Towards the Poles, average temperatures decline further and so does available water. Here, plants are adapted to long periods when little photosynthesis or nutrient uptake is possible.

Again, we can identify a forested region, the northern coniferous forest or taiga and a low-lying plant community where water is unavailable (because it is frozen for much of the year), the tundra. These communities circle the northern hemisphere where the land mass is almost continuous and together comprise the boreal (northern) biomes (Figure 8.16). Equivalent communities also occur at lower latitudes but at high altitudes, in montane areas where similarly severe climates exist (Plates 8.4, 8.12).

Where the temperate forest gives way to unbroken conifer forests, the **taiga** starts (Plate 8.10). Vast tracts of this dense forest are dominated by just two or three species—primarily pines, firs, spruces (*Picea* spp.), or hemlock—beneath which is a herbaceous layer of mosses, lichens and grasses. In some areas deciduous trees are found, mainly birch (*Betula* spp.) and poplars (*Populus* spp.). The fine leaves (needles) of conifers reduce water loss and are efficient for photosynthesis at low temperatures. By retaining their leaves they can make full use of the early spring sunlight, but they also have to support or shed heavy loads of winter snow. The trees are shaped accordingly with a conical growth habit.

In more northerly areas, the soils are permanently frozen at depth. All are thin, infertile, and covered with a thick layer of partially rotted needles. The nature of this litter and its slow decomposition under cold, semi-waterlogged conditions makes the soil acidic, so favouring fungi as the prime decomposers. A build-up of litter can promote forest fires, important to taiga ecology: fires create gaps which, once again, create opportunities for colonizing species. Some species (such as the jack pine—*Pinus banksiana*) take advantage of the gaps generated by these disturbances and shed their seeds only after a fire.

Relying solely on wind pollination, conifers have no direct need for insects and defend themselves against insect attack, primarily by using resins and their characteristically pungent terpenes (Section 4.4). Nevertheless, a variety of moths, sawflies and beetles burrow into the needles, buds and bark. Several migratory birds feed on these insects, while resident birds and squirrels take the conifer seeds. Other herbivores, including elk, caribou and deer, feed on the low-growing vegetation. Some remain active even when there is deep snow cover. Dogs (foxes and wolves), mustelids (martens, weasels, wolverines), birds (owls, eagles), bears and cats form the carnivore community. Many mammals hibernate for the winter.

In the summer, some of these herbivores and carnivores move into the **tundra** (Plate 8.11) to feed. This is the zone beyond the taiga, at the edge of the ice sheet. It was once heavily glaciated itself but is now uncovered for 2 or 3 months each year. Then the summer temperatures are high enough to melt the upper layers of the soil, down perhaps a metre or so. At depth, the soil never thaws and is called **permafrost**. Decomposition under these conditions is slow and organic matter accumulates as a dark, soggy, and compressed mass called peat. Similar conditions can exist at lower latitudes but high altitudes (Plate 8.12).

When the thaw comes, water, and the nutrients dissolved in it, becomes available. Plants bloom and grow—grasses, sedges and low shrubs, such as heathers (*Calluna* spp.), dwarf willows (*Salix* spp.) and birches—and produce seed in quick succession. Many will self-pollinate and spread by vegetative growth. Others flower to exploit the short burst of insect activity. Flies (especially mosquitoes) and butterflies that have spent the winter as larvae or pupae emerge to feed and mate. Birds migrate to exploit this pulse of plenty. Elk, reindeer and other herbivores also move to graze further north on exposed mosses and lichens. Some predators, from wolves to hawks, follow and feed especially on rodents, the lemmings and voles.

Across much of North America and Russia, the tundra presents a single unbroken expanse of a windswept and waterlogged landscape. Its flat monotony results from its repeated glacial scouring, as ice sheets have advanced and retreated with the repeated ice ages of the last 2.5 million years.

8.3   Climate change

The position of these biomes are far from fixed. With the last glaciations the tundra moved south in front of the advancing ice sheets, while during warmer interludes tropical conditions have extended well into regions that today we regard as temperate. Amongst the sediments deposited over the last two million years we find pollen, mollusc shells and beetle wing-cases that indicate communities and climates very different from today.

The Ice Ages are still fresh in the memory of the planet, so that we can still find well-preserved carcasses of mammoths and the remains of other recently extinct species. Sediments record a series of four major ice advances and retreats in the last million years, when cold- and warm-loving species replaced each other with each swing of the climate. When the ice retreated last time, the Ice Age ended abruptly. Temperatures rose extremely rapidly—averages in Europe climbed by about 7 °C in just 50 years and the Earth shifted from Ice Age to warm interglacial in the space of a human lifetime (Figure 8.17).

Most likely, this followed an increase in the energy received from the sun, perhaps as the relative position of the Earth changed because of variations in its orbit. Although there have been minor shifts in average air temperatures since then (including a 'mini-Ice Age' in the seventeenth century), overall the last 10 000 years have been relatively mild. Today we enjoy a climate substantially warmer than the average for the last 2 million years, though the difference between the Ice Age and the current interglacial is just 6 °C.

Now things seem to be changing again. Since industrialization started 200 years ago, humanity has been slowly altering the chemistry of the atmosphere. Climatologists are trying to model the consequences for the global climate and ecologists are being asked to predict the implications for the hierarchy of ecological processes.

Atmospheric composition and mean temperature

Orbital variations may be responsible for the oscillations in the ice advances (though see Chapter 9), but the amplitude of these temperature changes, the scale of their fluctuations, have been magnified by differences in the composition of the atmosphere (Figure 8.17). Samples of ancient atmospheres trapped in the air bubbles of ice cores dating back 160 000 years show that carbon dioxide levels have matched the changes in mean local temperature. Indeed, the same is true of another carbon-rich gas, methane and since methane is generated by microorganisms, this suggests that the level of biological activity has followed these temperature changes.

We know that the composition of the Earth's atmosphere is mediated by biological activity but the problem is to decide the extent to which the current warming is cause or effect: has the higher atmospheric temperatures increased biological activity thereby causing carbon dioxide concentrations to rise? Or, have the warmer temperatures induced by the increased carbon dioxide caused biological activity to rise? In fact, it is a combination of both. Along with water vapour, atmospheric carbon induces heating of the atmosphere via the greenhouse effect (Box 8.3). The fluctuations in temperature are 'amplified' because of the positive effect temperature has on biological activity. If nothing else changes, a warm climate increases the release of carbon by accelerating respiration and decomposition and this, in turn, promotes further atmospheric warming.

Carbon-rich gases also originate from non-biological but natural sources, such as volcanic activity. Carbon is removed from the atmosphere through non-biological absorption by the oceans. However, the carbon content of the atmosphere is primarily regulated by the biosphere—fixed through photosynthesis and released through respiration and decomposition (Section 6.1). That fixed in living tissues need not be released again directly, but may instead become incorporated into sediments (Figure 8.19). One important mediator of this process is the marine phytoplankton, whose photosynthesis and productivity rise with temperature. They remove significant amounts of carbon rapidly from the system if allowed to settle out, as does the death of marine invertebrates that combine carbon dioxide with calcium to build their shells and exoskeletons.

The scale of this sedimentation can be seen in the vast formations of limestone rocks over the planet. More significant perhaps, at least as far as our economic activity is concerned, are the coal and oil deposits derived primarily from terrestrial biomes of the geological past, especially the compressed forests of the carboniferous period.

Anthropogenic sources of atmospheric carbon

In the last 200 years, we have been releasing this fossilised carbon at an accelerating rate. On an annual basis, our inputs are relatively small (Box 8.4), but we have been burning coal and oil for a long time now. Overall, anthropogenic (human-derived) sources have led to a 30% increase in carbon dioxide over

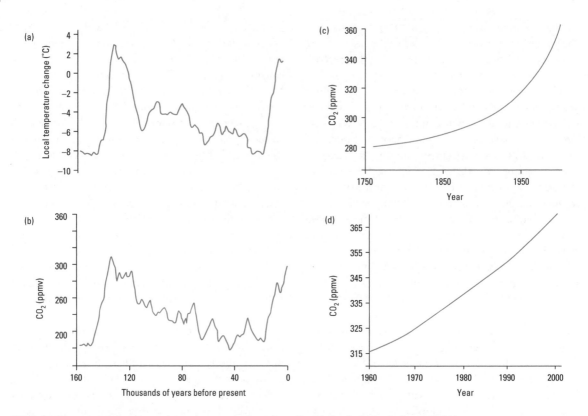

FIG. 8.17 Changes in temperature (a) and atmospheric carbon dioxide levels (b) in the last 160 000 years. Atmospheric methane levels show a similar pattern, indicating the rise in carbon-rich gases was due, in part, to increased biological activity as the global climate warmed. A rapid rise in temperatures marks the end of the last glaciation around 10 000 years ago. Carbon dioxide levels have risen consistently since the onset of industrialization in the eighteenth century (c), most rapidly with the increasing fossil fuel consumption and forest clearance in the last 40 years (d). (Note the change of scale in b, c, and d).

background levels since industrialisation began. In the 1980s, about one-fifth to one-fourth of the annual rise was attributable to the effects of forest clearance and changes in land use.

Burning forests (Plate 8.13) not only releases carbon oxides directly to the air, it also prompts an increase in decomposition in the exposed soil. Re-planting the forest is rapid way of fixing or sequestering carbon, but replacing it instead with agriculture will add further carbon to the atmosphere; 50 per cent of the carbon held in the organic matter of a soil is released under intensive cultivation. In addition, methane concentrations have increased 140 per cent, primarily from rice paddies and from the various emissions associated with the massive increase in cattle rearing. Developed nations also burn colossal amounts of fuel to sustain their agricultural production, from the manufacture of fertilizers and pesticides to the machinery used in planting and harvesting. Additional energy is then used to make the food more convenient—in food processing, packaging, transportation and refrigeration.

We are also responsible for a range of new gases, some with very powerful 'greenhouse' properties. **Chlorofluorocarbons** (CFCs) and their relatives were unknown before 1930 but until very recently were used extensively as aerosol propellants, as coolants in refrigerators, and in a variety of industrial processes. CFCs absorb infrared radiation in the one part of the spectrum where other greenhouse gases do not.

BOX 8.3   The greenhouse effect

The greenhouse effect is a natural and vital feature of our atmosphere. Without it, the mean atmospheric temperature of the Earth would be −17 °C, rather than its current average of 15 °C. The presence of water vapour, nitrogen, oxygen, and the carbon-rich gases (primarily carbon dioxide) all absorb the heat reflected off the Earth's surface. While the comparison with a greenhouse is inaccurate (there is no physical barrier, like the glass, to heat transfer), the name is at least evocative of the effect of heat capture.

Figure 8.18 shows how this works. Energy arrives from the sun primarily as short-wave radiation warming any surface it strikes. The heat re-radiated from these surfaces, is in the form of infrared radiation and these wavelengths are absorbed by different atmospheric constituents (Table 8.2). Besides water vapour, the principal greenhouse gases are CO_2, CH_4, N_2O, and CFCs. The contribution that each makes to global warming depends upon several factors—their concentration, the length of time they remain in the

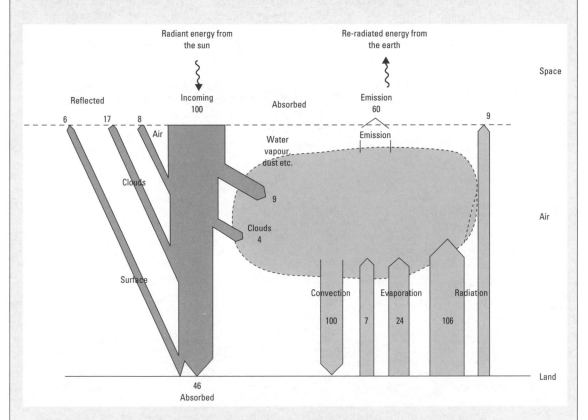

FIG. 8.18 An energy budget for the atmosphere and the greenhouse effect using arbitrary units. Solar radiation enters the upper atmosphere, where much of its shorter wavelength ultraviolet frequencies are reflected or absorbed. Longer wavelengths are absorbed as they pass the atmosphere, by water vapour and carbon dioxide. The radiation reaching the ground warms its surfaces and as their temperature rises they emit thermal radiation. These re-radiated infrared wavelengths are also absorbed by the water vapour and carbon-rich gases in the air. This is the main mechanism by which the temperature of the atmosphere is raised. Increasing the carbon and water content of the atmosphere increases this absorption, and its temperature rises as energy is lost more slowly than it is received.

BOX 8.3   Continued

atmosphere, and the wavelength at which they absorb re-radiated energy. CFCs are particularly potent because they absorb in part of the spectrum where the atmosphere was previously transparent (Table 8.2).

Any hot body loses its energy through thermal radiation and the hotter the body, the more infrared radiation it emits. Around 240 W/m^2 of solar radiation heats the earth's surface which then has to be lost, via the atmosphere, if the atmospheric energy budget is to remain in balance. Air temperatures are high because the re-radiated heat is retained by the atmosphere, captured on its return journey by gases which are not transparent to the infrared wavelengths. This energy is eventually lost by the atmosphere, at high altitude (around 5–10 km), but at this height the gases are cold and only lose their energy slowly. The overall effect of the greenhouse gases is thus to slow the passage of energy from the Earth's surface back into space and thereby raise the mean atmospheric temperature.

A planetary balance between energy input and output can be achieved at any atmospheric temperature. A higher concentration of greenhouse gases means this stabilisation is reached with more heat held in the atmosphere. Carbon dioxide added to the atmosphere will remain there for about 100 years and this means the pollution we add today will continue to have an effect at the beginning of the next century.

TABLE 8.2 The principal greenhouse gases

	CO_2	CH_4	N_2O	CFC_{12}
Concentration (ppmv)				
Pre-industrial	280	0.8	0.29	0
Now	370	1.72	0.31	0.00048
Rate of increase (% per year)	0.5	0.9	0.25	4
Lifetime (years)	50–200	10	150	130
Greenhouse effect per molecule over 100 years relative to CO_2	1.0	11	270	7100
Proportional contribution over 100 years*	72	18	4	(Highly variable)
Reduction in emissions needed to stabilize (%)	>60	15–20	70–80	75–85

* These estimates also allow for the indirect effects of different gases on CO_2 production.
CFCs have a very powerful greenhouse effect but their contribution depends (amongst other things) on their latitude—at higher latitudes they have a reducing effect due to their role in ozone reduction (which also has a greenhouse effect).

They are also long-lived (Table 8.2) and contribute significantly to ozone depletion (Box 8.5).

Some of our emissions help to ameliorate the greenhouse effect. Sulfate aerosols, themselves a product of fossil fuel combustion, can reduce short-term warming. They increase the reflectivity of the atmosphere, both by their presence and by inducing cloud formation. Sulfate aerosols are readily washed out of the atmosphere, so their cooling effect tends to be highly localized, close to the source of emissions. The sulfate pollution in industrial areas may be the main reason why the northern hemisphere has warmed less rapidly in the last 200 years than the southern hemisphere. Persistent sulfate emissions

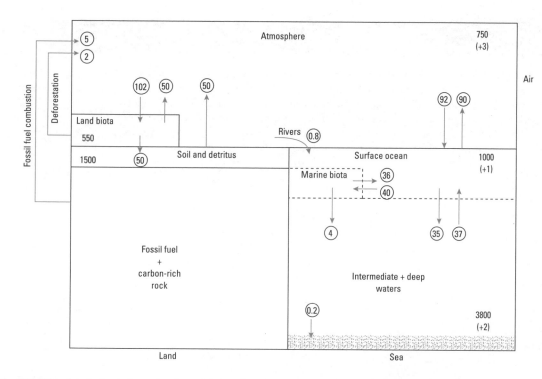

FIG. 8.19 The global circulation of carbon. Notice that the most important fluxes (circled) are between the land and air and the sea and air, and each are largely in balance. What has changed in the last two decades is the amount moving through the atmosphere and the reduction of the principal sinks on land, the forests. Fluxes are in 10^{12} kg per year and reservoirs in 10^{12} kg (gigatonnes).

BOX 8.4   The carbon balance and the missing sink

Without anthropogenic inputs, carbon added to the atmosphere is balanced by that removed. We add an additional 7 gigatonnes (10^9 tonnes or 10^{12} kg) each year, yet the atmosphere is accumulating only 3.4 gigatonnes each year and the oceans an additional 2 gigatonnes, so where does the missing 1.6 gigatonnes go?

One possibility is that our estimates for the oceans have been too low, but since there is little evidence of major changes in phytoplankton productivity with the rise in atmospheric carbon over the last century, this is perhaps unlikely. Most studies suggest that the global forests probably account for the deficit (Figure 8.20). About 31% of the Earth's land area is under forest and that retains 86% of total above-ground carbon. Around 73% of global soil carbon is found in forest soils.

By virtue of their size and the rapid growth of trees at higher temperatures, the tropical forests may seem to be the most likely long-term sink for the missing carbon, but this biome is suffering the most extensive logging.

Many argue that their scale of deforestation probably means the tropical latitudes are most likely to be net contributors to the carbon budget, others suggest their effect is broadly neutral.

In contrast, the temperate forests, especially those in the north, have been expanding. Commercial forestry and the abandonment of farmland have led to a spread in both the deciduous and boreal forests in the North, especially in the latter part of the last century. In addition, the production of the northern forests may have been heightened because of the fertilizing effect of nitrogen pollution from regional industrial activity as well as the widespread use of agricultural fertilizers. Also, sulfate deposition and the overall warming of the atmosphere would also have promoted the growth of established woodlands.

Roger Sedjo calculates that the increase in biomass in the northern forests, representing 0.7 gigatonnes each year of carbon, could account for a sizeable proportion of the missing carbon, and thus represent the

BOX 8.4 Continued

single most important sink. Pointing to evidence that nitrogen may not always limit carbon sequestration other ecologists suggest that a more integrated approach is needed, one which allows for soil processes as well as the photosynthesis and respiration of plants. More sophisticated analyses note that carbon and nitrogen ratios vary in both the growing plant and in the soil with heightened atmospheric carbon dioxide. There may not be one single major sink, but instead a more even distribution of carbon sequestration between temperate and tropical zones, and between forest and grassland.

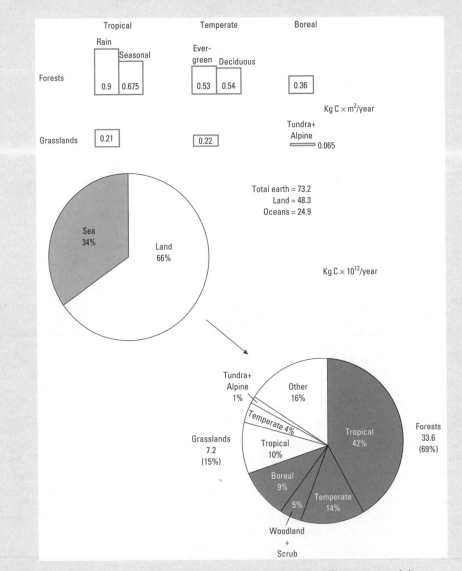

FIG. 8.20 The productivity of the planet measured as the carbon fixed by different parts of the biosphere: (a) shows the differences in mean net primary productivity per square metre between forests and grasslands and between the three major latitudinal biomes. Overall, terrestrial biomes are responsible for about two-thirds of the carbon fixed in the biosphere each year; (b) of which tropical forests (shaded) represent the largest proportion.

BOX 8.5   Ozone depletion

Changes in the chemical composition in the atmosphere can all be related to the radiation balance of the atmosphere—what is being let in and what is not being allowed to escape. The effect of one of these constituents, ozone (O_3) depends on its concentrations at different altitudes.

In the lowest layer of the atmosphere, the troposphere, ozone levels have been increasingly locally, as pockets of pollution around cities, generated through photochemical reactions involving car exhaust gases. High local concentrations may be responsible for forest dieback around major industrial regions. However, low-level ozone is also a major source of hydroxyl radicals (OH^-), important for reacting with methane, carbon monoxide, sulfur, and nitrogen oxides, and helping to reduce their impact on the atmospheric energy budget.

In the next atmospheric layer, the stratosphere, ozone is being depleted. Here, the gas absorbs harmful incoming ultraviolet radiation (UV-B), protecting living systems from energy at wavelengths that are strongly absorbed by a range of organic chemicals including DNA. Increased UV-B has been blamed for the rapid rise in skin cancers amongst the peoples of the higher latitudes. It also depresses plant productivity and therefore reduces the fixation of atmospheric carbon.

The depletion of stratospheric ozone is not uniform over the globe. It is most pronounced over the polar regions in the spring when low temperatures help produce a high altitude luminescent cloud within discrete circulation patterns. The ozone 'holes' recorded over Antarctica and the Arctic since 1977 follow from chemical reactions with sunlight and a number of gases, most especially CFCs. Ultraviolet radiation slowly breaks down these compounds to release chlorine monoxide, which, in turn reacts with ozone to produce molecular oxygen.

In addition, CFCs themselves account for perhaps 15% of the greenhouse effect. Because of their longevity and impact on UV levels, action to phase them out has been relatively swift. According to the Montreal Protocol agreed in 1987 and revised in 1992, large-scale use of all CFCs and related gases has largely ended in most industrialised nations. The success of the protocol demonstrates that international agreements can work; forecasts in 2002 suggest that the stratospheric ozone layer will be restored within 100 years.

from industrial areas are a major cause of acid rain (Box 7.2).

The role of ecological and biogeochemical processes in regulating the composition of the atmosphere/ ocean system is still poorly understood, mainly because we lack information on the role of microbial and other communities, especially in deep-ocean cycling. We often learn much from unexpected perturbations to the system. For example, the millions of tons of ash released by the eruption of Mount Pinatubo in the Philippines in June 1991 led to cool weather, globally, the following year (Box 9.2). Although global mean temperatures were lowered by 0.25 °C, there appears to have been no lasting effect on the long-term trend of global warming. Indeed, our understanding of the climate and oceanic system of the planet is improved by observing how it responds to such perturbations.

Predicting climate change

Of all the greenhouse gases, carbon dioxide is responsible for about half the greenhouse effect. The Intergovernmental Panel on Climate Change (IPCC), commissioned by the United Nations and the World Meteorological Organization, concluded in 1992 that carbon dioxide emissions would need to fall by 60 per cent to eventually stabilize at 1991 atmospheric levels. With no curbs on emissions and with current rates of increase in atmospheric carbon dioxide (the IPCC 'business as usual' scenario), average temperatures on the planet will rise by 0.25 °C every decade. As John Houghton points out, this is the fastest rate of change of any time over the last 10 000 years. We can expect carbon dioxide levels of around 450 ppm by 2030 compared to the current level of 370 ppm (Figure 8.17).

The first predictions of the panel have been substantially revised since they were reported in 1990. The models used to make the early predictions have been increasingly refined and now take greater account of the effect of cloud cover, deep ocean currents and, most recently, sulfate aerosols. The trend is still for a substantial rise—at least 1.5 °C by 2060 (Plate 8.14). Oceans may buffer changes over the longer term as higher temperatures lead to greater evaporation, but the time needed for stabilization, when the atmosphere reaches a new constant temperature, is still uncertain.

These predictions are based on the outputs from a number of *global circulation models* (GCMs). These are immensely complex mathematical models of the air/water system of the planet, originally derived from the weather forecasting simulations used on more local scales. There are several such models, but all require very powerful computers running for long periods. The current UK model (the Hadley Centre model) requires four months to run on one of the fastest supercomputers. Even so, these models are not complete descriptors of the system, primarily because they do not include ecological processes and the biological and chemical interactions that are mediated by the biotic components of the planet.

Predictions from the various models differ, primarily because of the different weights each attaches to the critical factors, especially deep ocean currents and cloud cover. These differences and the uncertainty of their predictions have been seized upon by their critics (Box 8.7). One way to test the predictive power of these models is to 'predict' past temperature movements, starting from a known set of conditions at some date in the past. If a GCM can then match the past observations, we can have some confidence that it captures the essential features of the system. Increasingly, models can do just this, and the latest output from the Hadley CM3 model manages to match both the scale of temperature change and the major fluctuations. It also demonstrates that the increase observed in the last half of the twentieth century is largely anthropogenic, that is, a result of human activity (Plate 8.14).

So here is the weather forecast for the rest of this century: warming will be uneven over the globe, increasing most rapidly in the northern hemisphere, perhaps more so as sulfate emissions are controlled in this region. Here, winters will be milder and in the middle of continents summers will be hotter and drier. The land will warm rather more than the oceans and the range in temperatures between day and night will become smaller. Continental margins will be wetter with higher precipitation at mid and high latitudes, though much less rainfall is forecast for the sub-tropical zones.

The Mediterranean basin is likely to experience longer and hotter summers, promoting more arid conditions throughout the year. Similarly, much of the western edge of North America will have higher temperatures and a reduction in summer rainfall. Particular models of the Sacramento basin suggest that runoff in the summer will decline by up to 50 per cent, should average temperatures rise by 4 °C. In the winter, there may be too much water—the Sierra Nevada mountains are predicted to receive more winter precipitation and they will also experience more days with extreme rain and snow falls. The consequence could be more winter floods in coastal California.

With persistently drier conditions, changes in the ecology of these catchments will become obvious much sooner. The rich flora of the Mediterranean basin will suffer major species loss as the drier top soil becomes more easily eroded. Agricultural productivity in these regions is also likely to fall.

Sustained warming over the long term, with loss of much of the polar ice caps will almost certainly slow the thermohaline circulation (Figure 8.11). This is a global circulation of oceanic water, driven by variations in salinity and temperature as different densities of water rise and fall. The massive amounts of water moving around the globe makes this a critical method for distributing heat. If this circulation slows the oceans will become increasingly stratified, with little mixing between depths for long periods—a change that has occurred with previous climatic oscillations. This is predicted to lead to twice as much warming of the waters in the northern hemisphere compared to the south and a reduction in winter sea ice across the Arctic. Sea levels will rise, primarily because of the thermal expansion of the water (2–4 cm per decade) and only partially because of melting ice (1.5 cm per decade). A number of countries face inundation during storms, including many low-lying Pacific islands, large areas of Bangladesh and Northern Europe.

Ecological changes likely with atmospheric warming

If the thermohaline circulation ceases, the carbon cycle and biogeochemistry of the planet will undergo profound changes. With increased oceanic stratification, nutrients released from the overturning of marine sediments will decline, and this in turn will limit primary productivity in the surface waters. One consequence of global warming could thus be dramatic changes to the communities in the upper waters of some oceans, with declines in both phytoplankton and zooplankton biomass. If such changes persisted, the larger animal community of these waters would also be reduced. The most productive fisheries occur where deep ocean upwellings bring nutrients to the surface waters. Rather like the impact of an El Niño off the coast of Ecuador and Peru, with fewer nutrients to support primary production, fisheries quickly collapse.

The distribution of terrestrial biomes will also shift as temperature and moisture regimes change. The extension of the monsoon rains polewards will increase the moist tropical forest cover in some areas, such as northern Australia. Some more southerly plant communities, most probably the grasses, will encroach polewards, as will the conifers of the taiga.

Deciduous hardwoods and grasslands will follow. Temperate mixed forests are expected to lose their conifers and become dominated by broad-leaf trees. Parts of Canada (Northern Alberta, Saskatchewan, and Manitoba) may lose their coniferous forests altogether. Current estimates suggest that the northern tree line will move 100 km northward for every 1 °C rise in mean temperature. The tundra will be warmer for longer and experience longer periods of biological activity and some predictions see the tundra and taiga shrinking by about one-third as carbon dioxide levels double and temperate forests extend their range.

Patterns of cultivation are likely to change in the same way. Growing seasons will be extended in northern latitudes, so that leaf and root vegetables could be grown in Central Alaska. Crop yields will increase in northern Europe, (perhaps by one-third in Denmark) but decline where water becomes limiting (down by one-third in Greece). More significantly, the drying of the prairies and the steppes in the southern part of their range will reduce cereal production. The overall impact will be increased food shortages in Russia and neighbouring countries. The Sahara will extend into the Sahel and increase its range in central Africa.

These are all large-scale changes that will take place over decades, as biomes move. In the short term, lower level changes in photosynthesis and respiration will determine carbon release and these in turn will influence our predictions about future warming.

Changes in primary productivity

Can we expect an increase in photosynthesis to mop up the abundant carbon dioxide and so offset the warming? It seems reasonable to expect photosynthesis to increase as atmospheric levels rise, helping to remove some of the excess carbon dioxide, but the devil once again is in the detail. The complication is that plants are also responding to higher temperatures, to a shortage of water in some areas, or varying levels of the major nutrients. Higher rates of photosynthesis demand more water. Equally additional nutrients, such as nitrogen and phosphates, would be needed to sustain both photosynthesis and growth if more carbon is to become locked in new plant tissues.

An absence of nutrients might thus be expected to limit any increase in photosynthesis in nutrient-poor ecosystems. Nitrogen is a key element in the chloroplast and one of the main enzymes responsible for photosynthesis, but different plants have different strategies for apportioning resources. For example, in the tundra nitrogen and phosphates are limiting for most plants, yet any increase in phosphate availability (by greater melting of the permafrost, or through pollution) often leads to greater seed production, rather than a general increase in biomass. The effects of elevated carbon dioxide in this biome are far from decided, but there is evidence that its plant community acclimates to the raised carbon dioxide and shows little increase in net primary productivity. By contrast, grasslands add large amounts of carbon to the soil each year, and the soil becomes a major site of carbon storage. The decaying plant material also fixes some of the available nitrogen, and retains it for the period that this organic matter persists. Ecosystem-level studies on calcium-rich grasslands suggest that nitrogen will once again become limiting as organic matter builds up in the soil, making such nitrogen unavailable. This is one form of higher-level buffering, constraining carbon fixation by individual plants.

At the level of the leaf, high carbon dioxide levels cause partial closing of stomata, so reducing transpiration and making the plant more efficient in its water usage. For some crop plants this is off-set by the plant producing more leaves, so any shortage of water remains limiting. Experiments on a variety of crop plants have shown that photosynthesis may rise initially with rising carbon dioxide but then some species adjust to the new level, and eventually show no overall increase in carbon fixation. One reason is thought to be the speed at which their photosynthetic enzymes can be regenerated, but other limiting factors include a build-up of the end-products of photosynthesis, such that starch granules impair the function of the chloroplasts. Together these represent a form of lower-level buffering.

But some plants do indeed fix more carbon, at least when measured as the weight of carbon fixed per unit of nitrogen in the leaf. In effect, these plants, especially some grasses, use their nitrogen more efficiently when carbon dioxide levels rise and appear to have no fixed ratio of carbon to nitrogen in their tissues. Also, many plants use the extra biomass to grow longer root systems, to forage in a larger volume of soil for the water and nutrients they need.

At the ecosystem level, the assumption that nutrient-poor plant communities may be unable to increase their carbon sequestration is perhaps too simplistic. It seems that some grasses and sedges may then be at a competitive advantage since they can increase their photosynthesis and carbon fixation even when nitrogen is in short supply. In this case, changes in the composition of the plant community are likely to follow, with less adaptable species being competitively displaced as the composition of the atmosphere changes.

Changes in decomposition

Higher temperatures are likely to accelerate decomposition in the colder areas of the planet and increase respiration more generally. Alone this would add to carbon release. Yet, experiments in the Alaskan tundra show carbon storage actually increases at elevated temperature and raised carbon dioxide concentrations. This is because the litter takes longer to degrade (due to its low nutrient levels) so additional growth leads to more carbon being added to the soil. In contrast, decomposition rates for older, nutrient-richer deposits will indeed increase as mean temperatures rise.

Other changes follow if the permafrost thaws to a greater depth. The higher temperatures allow the greater volume of waterlogged soil to generate more methane. Wetlands produce about five times as much methane as dry areas (whereas dry soils are an important sink for methane—particularly neutral woodland soils). Overall, microbial decomposition and respiration rates are expected to increase with high temperatures, releasing more carbon from soils rich in organic matter.

Another significant source of methane is rice paddies, which have expanded considerably in the drive to produce more food, often at the expense of tropical forests. A large proportion of their methane production is oxidized to carbon dioxide in the upper layers of the soil, but significant amounts still enter the atmosphere. Methane is also generated in large quantities by animal husbandry (the wastes, gaseous and otherwise, of cattle), by landfill sites (domestic refuse) and by mining and oil extraction.

Other interactions between species—competitive, exploitative and cooperative—may change as the balance shifts favouring some physiologies and metabolisms over others. As the nutrient status of their diet improves, some insect herbivores may well enjoy higher population growth rates. For others, the extended periods of higher mean temperatures may prolong their active phases, perhaps to the detriment of species they exploit. For example, some ecologists have suggested that the heathlands of North Western Europe may be threatened if a key herbivore, the heather beetle (*Lochmaea suturalis*) is able to increase its rate of reproduction, and produce two generations in a summer. In this case, the heather will suffer sustained attack and may well decline, and lose competitive battles with invading grasses.

More ominous perhaps is the evidence that, under the influence of climate change, the species composition of the topical forests of Amazonia have been undergoing long-term change. Oliver Phillips and his co-workers have shown that the proportion of lianas in the rain forest of the western Amazon has been rising over the last 20 years, at the expense, it seems, of the trees. These vines appear to benefit more readily from the raised carbon dioxide and have a faster response than the trees. Because lianas invest less in support tissues they fix less carbon in their stems and the total carbon storage in the forest declines as lianas increase.

Stabilizing the atmosphere: ecological methods

The main mechanism for controlling the rise in atmospheric carbon dioxide is to limit our emissions. Next, it would be to limit our deforestation programmes. Titus Bekkering suggests that together these two measures could reduce the growth in atmospheric carbon by 12 per cent by the year 2100. Yet even with a massive, perhaps unrealistic, programme of reforestation and forest regeneration extending cover by 865 million hectares, carbon levels in the atmosphere are still projected to be 54 per cent higher in a hundred years time.

A variety of schemes for increasing carbon fixation have been suggested. One is based on a proposition that oceanic phytoplankton communities are limited by a shortage of iron. Adding iron as an artificial fertilizer may be one means of increasing sequestration by the large oceans, hopefully leading to greater sedimentation to the ocean floor (Box 8.6). Although this has been seen as a potentially commercial operation, important questions remain about the wisdom of inducing large-scale change in marine ecosystems.

On land, forests are the principal means of fixing atmospheric carbon in terrestrial ecosystems, and the one component of these biomes over which we have the greatest control. Estimates in the IPCC report suggested that 370 million hectares planted at 10 million hectares per year over the next 40 years would eventually fix 80×10^{15} g carbon (80 gigatonnes), somewhere between 5 and 10 per cent of the

carbon derived from fossil fuel combustion. After this time, when tree death eventually matches tree growth, the forest ceases to be a sink for carbon. By then, the total area planted would be equal to about half of the Amazon basin.

Can planned reforestation be combined with a conservation strategy that protects the biological diversity of the tropics? This may only be realistic in those countries currently able to feed themselves. Bekkering suggests this comprises 15 countries, which between them have the potential to replant a total of 580 million hectares. The UN's Tropical Forest Action Plan (TFAP), begun in 1985, attempted to introduce forest management techniques into developing nations, bringing about an eightfold increase in reforestation in participating countries. Unfortunately the total area is small (1.9 million ha) and more carbon is released by these countries (282 million tonnes per year) than is fixed (12.4 million tonnes).

Hermann Behling has estimated the area by which the four major types of Brazilian forest have grown since their smallest extent at the last glacial maximum (18 000 years ago), and how much carbon they have fixed in the process (Figure 8.21). The area under forest in Brazil has increased by 2.4 million km^2 (55 per cent) and by 1.1 million km^2 (nearly 20 per cent) in the last 7000 years. This means the trees of Brazil now hold nearly 43 gigatonnes (26 per cent) more carbon than they did in the last Ice Age, and an additional 19.6 gigatonnes (10 per cent) since the Holocene. The carbon now fixed in these forests since the glacial maximum represents about 6 per cent of the current mass in the atmosphere.

Such calculations allow us to estimate the impact of deforestation on carbon release and storage. Today in Brazil, 90 per cent of the tropical rain forest remains, 9 per cent of the Atlantic rain forest, and 19 per cent of the *Araucaria* forests. Interestingly, Behling suggests that the most significant increase in carbon sequestration would occur if the *Araucaria* forests were encouraged to regenerate.

Rates of deforestation have begun to decline in Amazonia, though elsewhere in the tropics they are increasing. Recent evidence indicates that the scale of deforestation may have been over-estimated, because of a poor understanding of long-term land-use

---

BOX 8.6   Can the oceans be engineered to reduce atmospheric carbon dioxide?

Walker Smith, Virginia Institute of Marine Science

Most ecologists and geochemists now accept that carbon dioxide concentrations in the atmosphere are increasing as a result of man's activities, leading to changes in the earth's climate. We might moderate this increase by sequestering more carbon dioxide in the oceans and one suggestion is to fertilize the oceans with iron, the element limiting marine phytoplankton production in many areas.

This might be particularly effective in the Southern Ocean, surrounding Antarctica. The impact on atmospheric carbon dioxide is likely to be large here because of the extremely high nutrient concentrations present in the surface waters (a result of the massive convective overturn of water in this region). With iron enrichment this nutrient supply could support greater phytoplankton production, perhaps reducing the increase of atmospheric carbon dioxide by some 72 ppm over 100 years. In contrast, the same treatment applied to the nutrient-poor Pacific Ocean would lead to a reduction of just 3 ppm over the same period. However, attempts to model the impacts on the marine communities suggest that the implications are far from straight forward.

The principle is simple enough—the fixation of the nutrients by the phytoplankton would increase the rate at which organic matter represented by its dead cells sank to deeper water. According to early models this would produce a layer devoid of oxygen at depth, especially south of Africa. In this case, the entire ecosystem of the Southern Ocean would be changed, largely because the main herbivore in its food chain, krill

(*Euphausia superba*) would be destroyed—krill migrate to depths of 1000 m during their life cycle and would die under the anoxic conditions of the deeper waters. More recent models suggest that such anoxia would be unlikely, but depleting nutrients in the Southern Ocean could nevertheless reduce the nutrient supply in waters upwelling in the equatorial region. In this case the productivity would be greatly reduced within 100 years, impacting tropical fisheries such as tuna. Others have suggested that the phytoplankton community of the Southern Ocean would change under iron enrichment, perhaps favouring species that are not so readily grazed. One species *Phaeocystis antarctica*, a close relative of a species known to cause algal blooms in the North Sea, may well flourish because it is incompletely grazed, causing substantial changes in the local food web. This species is also important as a source of dimethyl sulfide, a key greenhouse gas, which could make the iron enrichment counter-productive.

Since much of the Southern Ocean is covered with ice for large periods of the year, adding iron to its waters is not without its practical difficulties. Nevertheless, substantial interest remains and some commercial organizations recognise that an industry which removes carbon dioxide can claim a potential 'energy credit', a licence to emit carbon which can be sold to other industries unable to reduce their carbon dioxide emissions. It may well be that the international trade in carbon emission credits starts between companies rather than between nations, even though the enrichment would be of ecosystems to which no nation or company could lay claim.

---

patterns and a misinterpretation of the patterns on the ground today. Too often, it seems, local farmers have been blamed for poor woodland management practices, when they are actually helping to sustain the forests at their margins.

Tropical reforestation can be a realistic prospect only where the needs of local people are taken into account. Forests can be exploited without decimating them. They have an economic potential from timber and fuel production, as well has higher value goods. If exploitation is managed with a proper regard to the natural processes which govern community structure, there is every indication that this is sustainable.

The real issue is not carbon fixation by forests but carbon release by fuel consumption. The solution depends on the will of industrialized nations to curb their fossil fuel usage. Perhaps, the major user nations should pay tropical countries to maintain their carbon-fixing potential—a service they rely on, but for which they currently do not pay. Costed on a *per capita* basis we as nations or we as individuals might begin to recognize the true price of cheap energy (Box 8.7). As in any hierarchy, we should all remember that our individual rates of consumption cumulatively determine the impact humanity has on the atmosphere.

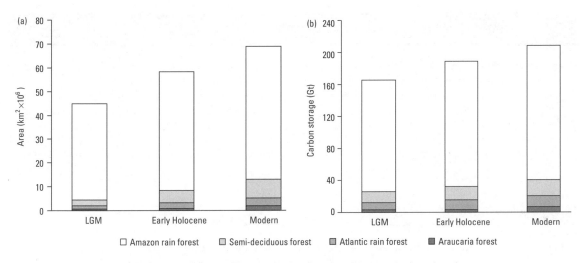

FIG. 8.21 Behling's estimate of the total carbon fixed in the forests of Brazil since the last glacial maximum (LGM, 18 000 years BP) and the Early Holocene (7000 years BP), based on pollen deposits in sediments. Using samples from 45 locations, Behling mapped the spread of four forest types. The 'modern' value is an estimate of the maximum extent of each type in the recent past (a) and the amount of carbon fixed in mature forests of each type (b). The latter use estimates ranging from 260 tonnes C per hectare for semi-deciduous forest to 300 tonnes C per hectare for Amazonian tropical rainforest.

BOX 8.7 So much hot air?

The uncertainty of predictions from the different GCMs, and a recognition that small-scale measures are unlikely to be effective, prompt many to argue against international proposals to cut carbon emissions. Further, the key role that cheap energy plays in industrial economies means that for some people, emission targets equate to checks on economic growth. Better, they argue, to let market forces constrain fuel use, through the costs of pollution and environmental degradation.

These commentators point out that the match between carbon dioxide levels and warming is not perfect. One hiccup occurred between 1940 and 1960 when, despite rising carbon dioxide levels, the atmosphere cooled slightly. This certainly demonstrated the weaknesses of our atmospheric models but was most likely caused by astronomical factors that do not feature in most GCMs. On another occasion, warming and gas concentrations appeared to halt in mid-1991 following the eruption of Mount Pinatubo, though both resumed their rise in the 1993. However, in late 1995, the UN IPCC finally accepted that global atmospheric warming was a reality and this has since been accepted by the vast majority of the scientific community.

Quite correctly, climatologists have been careful not to link any single event with climate change, but rather predict more extreme conditions generally. Higher energy levels in the atmosphere will drive larger movements of air and water around the planet. Carbon dioxide concentrations will continue to rise in the foreseeable future, with or without checks on human production, simply because of the time needed for atmospheric stabilization. The pressure to adopt strategies for developing non-fossil fuel energy systems was suggested by the majority of delegates at the Johannesburg summit of 2002 as a both a statement of principle and a fillip to alternative technologies (Box 9.4) —from expanding nuclear and solar-powered generation to newer ideas such as hot air based generators.

Its detractors argue that the Kyoto protocol will not reduce carbon emissions by a sufficient amount quickly enough to have any impact. Others argue that it puts an undue burden on the larger polluters—effectively those

BOX 8.7 Continued

nations who are profligate in their use of carbon-based energy. Most of its proponents acknowledge that it is primarily a gesture, symbolic of the sacrifice all peoples will need to make. Since there are few straightforward comparisons of its impact on different nations, perhaps the protocol will serve simply to demonstrate that we are all prepared to be good neighbours.

Eventually, the targets we agree will depend on a range of factors, including the scale of environmental change we expect, the compromises we are prepared to make and the technologies we have available. We need to be careful not to injure the economic prospects of many poor and malnourished peoples, wherever they are. The longer we delay action the longer stabilization will take.

SUMMARY

The global distribution of the major biomes reflects the climatic zones across the planet, most especially the mean temperature ranges and patterns of precipitation. Where water is abundant forests develop, where an extended dry season occurs grasslands are found. Between the tropics, forests are lush and with an immense diversity and the savanna grasslands support a variety of game. Temperate regions have communities adapted to distinct seasonal cycles. The taiga and tundra are limited by short growing seasons and long winters.

Landscape ecology provides an approach for studying the large-scale processes that determine the nature of component ecosystems. Hierarchy theory suggests that the organization of biological systems, from the biome to the individual cell, is a product of functional interactions between different levels of organization. Some of these mechanisms become apparent when we look at the relationship between the carbon balance of the atmosphere and the climate of the Earth. The two are linked primarily through biological activity and the evidence increasingly suggests that a large-scale warming, primarily due to human fossil fuel combustion and its effect on atmospheric carbon levels, is underway.

FURTHER READING

Archibold, O. W. 1995. *Ecology of World Vegetation*. Chapman & Hall, London. (A comprehensive account of the major biomes.)

Forman, R. T. T. 1995. *Land Mosaics. The Ecology of Landscapes and Regions*. Cambridge University Press, Cambridge, UK. (An exhaustive account of the principles and practice of landscape ecology.)

Houghton, J. 1997. *Global Warming. The Complete Briefing*, 2nd edn. Cambridge University Press, Cambridge (An excellent introduction to the science behind global warming from the Chairman of the first IPCC.)

WEB PAGES

The IPCC page, which gives the latest data and reports:
http://www.ipcc.ch/

as does the Pew Center on Climate Change:

http://www.pewclimate.org/

The World Resources Institute offers a comprehensive and up-to-date summary on global climate change:

http://earthtrends.wri.org/index.cfm

The World Conservation Union page covers many of the issues surrounding global conservation, beyond endangered species:

http://www.iucn.org/

A good general-purpose directory with directions to details on the major biomes and landscape ecology can be found at:

http://www.biologybrowser.org

EXERCISES

1 Select the correct answer for each of the following:

(i) Deserts are found where

(a) moist air is ascending on its way to the tropics,

(b) 40° either side of the equator,

(c) rainfall is less than 10 mm each year,

(d) dry air descends having shed its moisture in the tropics,

(ii) The global distribution of biomes is primarily a product of

(a) the effects of latitude on temperature and moisture,

(b) the effects of longitude on temperature and moisture,

(c) the effects of altitude on temperature and moisture,

(d) all of the above.

(iii) Grasslands are found where

(a) moisture limits forest growth,

(b) soils are thin and infertile,

(c) in flat areas,

(d) in the centre of continents.

(iv) Which of the following limits the supply of nutrients to tundra plants

(a) waterlogging of the upper soil in summer,

(b) the frozen soil in winter,

(c) grazing by vertebrates,

(d) lack of litter.

2 Assign the following to either tropical or temperate forests:

(a) deep rich soils with a high organic content,

(b) rich diversity of herbaceous plants,

(c) rapid rates of decomposition in the soil,

(d) a distinct season when respiration exceeds primary production,

(e) a high and continuous canopy,

(f) an associated community of plant creepers and climbers,

(g) most nutrient capital is held in the living components of the ecosystem,

(h) a resident population of frugivores.

3 Give two mechanisms by which changes in the water content of the atmosphere affects its mean temperature.

4 Much of the heat of the atmosphere is lost to space at high altitude. How does this contribute to the greenhouse effect?

5 Why might an increase in the concentration of carbon dioxide not lead to a corresponding increase in photosynthesis by a particular species of plant? What significance does this have for global atmospheric warming?

6 Using the simple diagram of a temperate agricultural landscape in Figure 8.1 as a prompt, suggest

(a) How nutrients held in the aquifer on the left side (shown as blocked rock strata) of the valley might find their way to the pond in the right foreground;

(b) the number of habitat patches available to a woodland bird that dislikes edge habitats;

(c) reasons for the distribution of woodland patches in this watershed.

Tutorial/seminar topics

7 Assume that tropical reforestation is unlikely to be used to mop up carbon from the atmosphere. Speculate on what possible technological fixes might be possible within the next 50 years, either to limit carbon releases or to remove it from the atmosphere.

8 What are the ecological arguments for paying less-developed nations for maintaining their natural ecosystems as a means of removing carbon added to the atmosphere, originally produced by the industrialized world?

9 Discuss the ways in which an understanding of landscape ecology might inform our efforts to conserve an endangered species.

9 CHECKS

'. . . the beauty of the cosmos derives not only from unity in variety, but also from variety in unity.'

Umberto Eco: The Name of the Rose

Humankind walked this way. Rubbish left by walkers visiting the Himalaya

CHECKS

The evolution of human beings, you will remember, is closely tied to that of the ice ages and the major climatic changes that began 7 or 8 million years ago. Back in Chapter 1 we should, perhaps, have asked what initiated the cooling and drying of the tropical African climate that prompted our evolution.

The short answer is we do not know. One suggestion is of a general cooling of the atmosphere caused by a shift in the Earth's orbit relative to the Sun. Another key event was the creation of the Gulf of Mexico around 3 million years ago, when the Panama isthmus closed. This deflected a major flow of warm water, the Gulf Stream, north towards Western Europe, and above it warm, moist air. Close to the pole this moisture fell as snow, allowing the arctic ice cap to build. An expanding ice sheet would have reflected more solar radiation back into space, and begin the major temperature oscillations that became the sequence of ice ages.

The Earth was perhaps more easily toppled into this instability by older and larger-scale continental movements. One theory says that global cooling followed when the Tibetan plateau and the Himalaya were formed as the Indian crustal plate pushed up into Asia, around 40 million years ago (Plate 9.1). This is thought to have induced vast amounts of precipitation as the atmospheric circulation was deflected over the new upland mass, depriving surrounding areas of moisture. In contrast to the drier conditions developing to the west (including the Mediterranean basin and East Africa—Section 1.1), massive rivers began draining and eroding the new upland, pouring minerals and nutrients into the seas. This fed the primary productivity of the oceans and resulted in an explosive growth of their phytoplankton. The increase in biological activity lowered atmospheric carbon dioxide levels, reducing the greenhouse effect, and so allowed global temperatures to fall. Primed in this way, it was easy for the cooler Earth to slip into the oscillations of the ice ages. If this was the ultimate cause of the climatic change in East Africa, then the birth of humanity was prompted by a collision of the continents. Literally, the Earth moved.

That new species arise following climatic change should come as no surprise to us. Nor that other species are lost. But it is spectacularly ironic that low levels of atmospheric carbon dioxide might be part of the explanation for our appearance on Earth, when our enhancement of the greenhouse effect threatens to tip global ecosystems into new states, making life increasingly difficult for many people living in more marginal habitats.

Rates of climate change have been particularly rapid over the last 100 years, primarily because of increased atmospheric pollution and large-scale deforestation (Section 8.3). As natural habitats disappear, so do species. Currently, extinctions are running at a rate perhaps 1000–10 000 times higher than estimated background rates. We are at the beginning of a mass extinction event, comparable to those documented in the fossil record. The major extinctions of 65 or 250 million years ago have been attributed to massive volcanic eruptions or the impact of large meteors. Today, that same scale of loss has only one principal cause, *Homo sapiens sapiens*.

How many species are there and how many should there be? We start this chapter by looking at the number of species that ecological theory predicts for an area and the rate at which species are going extinct. We see how this might be related to the pattern of species diversity across the planet. The rapid decline in biodiversity, measured as the loss of the big and obvious, is now at the centre of many calls for environmental action.

Here, we consider which species can be lost without impairing ecological processes, and, in particular, the significance of microbial and invertebrate diversity in maintaining soil fertility. The agricultural productivity of a region's soils is, along with the availability of water, critical for the quality of most people's lives. Yet, the species richness of soils goes largely uncounted and we have only a limited knowledge of the role these communities play in the stability of soil ecosystems. It may well be the activity of the small and inconspicuous species that determine how many humans the planet can support over the next century.

9.1 Predicting the number of species

Some patterns are easy to find but not always easy to explain. Here is a pattern of biodiversity you can find yourself: choose some way of sampling a large, uniform habitat using repeatable units—say the number of kinds of insect found on a leaf of a large bush or perhaps the number of plant species in a quarter square metre of lawn. Then double the size of your sample—two leaves or a half square metre and count again. Continue doubling your sample size until the count becomes impractical or your enthusiasm wanes (whichever comes later). Now plot the total number of species found against the area sampled. The chances are that you will produce a graph rather like Figure 9.1. As your sample size increases, so does the number of species found. However, the relationship is rarely linear—that is, S (the number of species) flattens out as the sample size get larger. New species are easy to find early on

but the rise in S does not keep pace with the increasing area. As Figure 9.1 suggests, the rate of addition of new species slows considerably until very large areas need to be surveyed to add the very rare and infrequent. If you are a birdwatcher you may have already experienced the difficulties of adding rarities to your tally and the frustrations this pattern can evoke.

This pattern is called the **species–area relationship** and is so common in ecology that we can make predictions based on its regularity. It can tell us the total number of species to expect in an area or how much sampling effort we need to detect the rarest species. It works best for a particular group—for example, herbivorous beetles or annual flowering plants—and shows that species number is some function of the available space, or a space-limited resource, such as food or water.

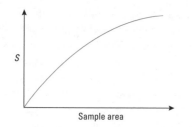

FIG. 9.1 The species–area relationship. As the area sampled increases, more species are counted, at least in the initial stages. *Within* a habitat, greater sampling effort will eventually find few new species so S tails off with area sampled. This is a very important relationship—it provides some indication of the size of sample we need to adequately sample an area and also tells us how many species of a particular group to expect in a habitat of a given size.

The relationship between species number and area can be summarized as

$$S = cA^z$$

where S is the number of species (often termed species richness), A is the area, c is a constant representing the number of species in a unit area (or sample) and z is the rate of increase of S with A. Notice that z is a power term, which means that there is no simple arithmetic increase of species number with area. In fact, it is logarithmic, so, for example, when $z = 0.3$ a 10-fold increase in the area will only double the species count:

For a 100 ha habitat with 10 species found in a unit sample

$$S = 10 \times 100^{0.3} = \text{about } 40$$

and for 1000 ha

$$S = 10 \times 1000^{0.3} = \text{about } 80$$

Calculating z using data for different groups of plants and animals shows that it typically ranges from 0.15 to 0.35. z is high for species that do not travel readily between patches: birds or reptiles are likely to have different abilities to colonize isolated patches and will have very different values of z.

When z is large the rise in S falls away sharply with increasing area, so z is usually higher for isolated habitats. When z is low, S flattens slowly with area. This should make sense: less isolated habitats are easily reached so most patches will share most species with most other patches and the number of species

(S) does not change greatly between one patch and another. When colonization is rare, patches will tend to have different collections of species. Consequently, z is high for oceanic islands and low for habitat patches within a landscape mosaic.

Reaching a habitat is only part of the problem. A species only colonizes a habitat patch if it can establish a reproducing population. This may require a male and female of breeding age (which are reproductively compatible) to be present at the same time, and to locate each other. For many animals, colonization begins only when they cease to be visitors, and successfully raise a new generation.

An island habitat accumulating species from a nearby mainland or 'source community' can also lose them. Species may become locally extinct even after establishing a breeding colony. Small islands with a small carrying capacity can only support small populations, so the chances of going extinct are high here compared to a larger habitat or the mainland (Section 3.7). However, if islands are close to the mainland, and new colonists arrive frequently, the local population may be sustained by new arrivals, as part of the movements within a metapopulation (Sections 3.7 and 8.1).

Between them, immigration and extinction means there is no fixed collection of species in a habitat, but instead a turnover of species, the comings and goings of colonists and failed colonists. Over time, the total number of species within a group—say the birds or the reptiles—will settle down to some long-term value, even though the list of which species are present will change with every immigration or extinction event. An island is, thus, said to reach a dynamic equilibrium for S. A simple model of this was first set out by Robert MacArthur and Edward Wilson in their **island biogeography theory**.

If a new island is large and close to the mainland (i.e. easily found and easily reached by the migrating group), it will rapidly acquire species from the source community. In its early days, habitats and resources on the island are unoccupied and new arrivals have little difficulty establishing themselves. Later, when most of the mainland species have colonized the island and the majority of its niches are occupied, there are fewer opportunities and rates of immigration decline (Figure 9.2a).

This competition for the increasingly scarce resources causes extinction rates to rise as S increases (Figure 9.2a). Individuals or species able to make best use of the limited resources are most likely to colonize. Since S is, in part, fixed by island size, area again must be a summary of those important resources that limit how many species can colonize—such as the range of food types or the variety of habitats.

With fewer migrants to supplement their numbers, populations on more distant islands will suffer higher extinction rates than those close to the mainland, and these will be higher still if the island is small (Figure 9.2b). So equilibrium S differs between near and far, and large and small islands; colonization rates will be higher on the near and extinction rates higher on the small (Figure 9.2b). When colonization is balanced by extinction the total count of species for a group stays more or less the same. This is a dynamic equilibrium because there is a turnover of species and the species list is continually changing.

Although we commonly think in terms of oceanic islands, the theory can be applied to any habitat isolated by hostile surroundings, much as we described habitat patches within the landscape mosaic of Chapter 8. Indeed, the model has been instrumental in the development of landscape ecology (Section 8.1) and underpins much of what we had to say about habitat patches and metapopulations. Since its proposal in 1967, the theory has been tested against real data for different groups on different islands and found to be a consistent and effective predictor of equilibrium S.

In its simplest form, island-biogeography theory makes no predictions about which species are found on the island and says little about the interactions between them as they establish themselves. Nor does it distinguish between species, even though some will be more important to a community than others: some may facilitate the arrival of later colonists, or perhaps preclude subsequent colonization by competitors (Section 5.3).

Over long time scales, new species, adapted to the local conditions, are also likely to arise, especially in isolated populations that become established on the islands. Close to an equilibrium S, when local niches are occupied, there is intense competition for resources and the pressure to differentiate, to define a new niche, would be high (Section 2.5). Indeed, a relatively high number of endemic species is a feature of many isolated islands.

The original model did not consider speciation, but in a new variation upon it, Stephen Hubbell includes this process, at least in the source community. His 'unified theory' also seeks to incorporate the relative abundance of species—the rarity or otherwise of each species. Which species are common and which are rare is some indication of the availability of resources demanded by these species and numbers will change from community to community, reflecting resource differences. Relative abundance is a key determinant of species diversity and is taken by ecologists to indicate how resources are partitioned between species within a community (Box 9.1).

By allowing for speciation in the source community (assumed to have its own equilibrium S, in this case where speciation is balanced by extinction),

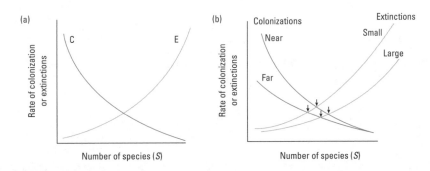

FIG. 9.2 The MacArthur–Wilson theory of island biogeography. (a) Rates of colonization (C) will decline as species are accumulated and rates of extinction (E) will rise as niches are occupied and competition for resources is intensified. (b) Colonization rates will also vary with island proximity and extinction with island size. When C is balanced by E there is a dynamic equilibrium of relatively fixed number of species but with a continuing turnover of species as local extinctions are replaced by new arrivals. For any particular group the value of S will reflect the habitat size and the ease with which it is reached from the mainland.

BOX 9.1 Measuring diversity

As with so many terms, the meaning of diversity depends very much on context. At different times diversity or even biodiversity refers to species richness (simply the number of species), to genetic diversity (the variety of genotypes), or to the variety of niches within a habitat. The term biodiversity may, thus, encompass everything from genes and nucleotides to biomes.

Community ecologists have concentrated their efforts on measuring species diversity as a way of comparing communities. Much effort has been devoted to find a universally acceptable measure or index, and these generally seek to combine the two components of diversity: **species richness** (the number of species) and **species equitability** (the proportion of individuals belonging to each species).

Two communities may share the same number of species and the same total number of individuals, but the one with the more even distribution of individuals (the highest equitability) is the most diverse (Figure 9.3). Species equitability is an indication of the resource space that each species occupies: a rare species with only a few individuals utilizes a small proportion of the resources available (Section 2.4). High equitability implies that resources are evenly shared between species and that the system is not dominated by just one or two abundant species. For example, surveys of the deep sea fauna of the North Atlantic show that equitability makes a significant contribution to the latitudinal gradient in the diversity of bivalve molluscs (i.e. there are few highly abundant species dominating the samples).

Some indices of diversity assume an underlying species distribution, that is, the relative abundance from the commonest to the rarest species, and so make assumptions about how resources are partitioned between them. That some of these patterns repeat themselves in very different ecosystems suggest similarities in the way these communities are organized (Section 5.2).

Other indices do not assume any distribution and as such treat all species as equivalent. This may seem rather arbitrary, especially since the taxonomic distance between groups is also rarely measured: two samples may have the same species richness, but the one with four species of ant would be seen by most people as less diverse than a sample with a fly, a wasp, a beetle and an ant.

Several indices are dependent on the size of the sample taken, and this can be problematic if we are unsure about our sampling efficiency. For these reasons, no single measurement of diversity has yet been adopted as being the most effective under all circumstances.

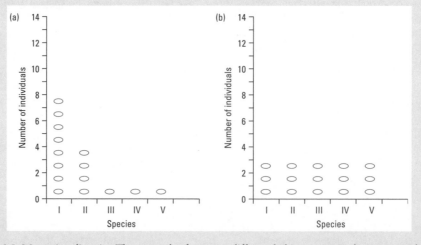

FIG. 9.3 Measuring diversity. These samples from two different habitats contain the same number of species ($S = 5$) and the same total number of individuals ($N = 15$). However, we would not take them to be equally diverse because community (a) is dominated by just two species, whereas (b) shows equal numbers of all species. (b) has high equitability. This would not be indicated by a simple ratio of S/N and so ecologists have sought more indicative measures of diversity.

Hubbell derives the value of S for a receiving island. For each island, S is a function of its size, the migration rate, and the combined effect of the total number of individuals in the source community and its speciation rate. Colonization is more likely for relatively abundant species, since they have more individuals who could make the journey to the island. Compared to their source community, islands will tend to have more common species and fewer rare species. Common species quickly dominate the resources on the island, whereas rare species, travelling less frequently, will tend to arrive later when resources are already occupied. Rare species on the mainland will thus be underrepresented on the island's species inventory.

The theory also provides insight into the larger picture—the number and relative abundance of species in the larger metacommunity, that is, across islands and mainlands or across a landscape of habitat patches. Based on his model, Hubbell concludes that the large-scale regional stability in S seen in many landscapes is a consequence of species mobility, rather than local species interactions and competitive battles for niche space. If the model is accurate, the assembly rules of a community are less important for its stability than its frequency of colonization and extinction. In simple terms, the more mobile the colonists the more stable the local communities. On the basis of this, Hubbell has predicted that coral reefs will show higher stability in S than tropical rain forests simply because species move more readily between reefs. The fossil record and evidence of recent studies of living reefs do indeed suggest they recover more rapidly from a disturbance than tropical forests.

Such models confirm the need for the frequent exchange of individuals and genetic material between patches for the long-term prospects of the populations within a metapopulation (Section 3.7). Refuges may be important for a species reduced to small numbers but on their own, small populations do not have long futures.

9.2 Extinction

Species come and go at different rates according to where you look in the fossil record. Based on what can be distinguished from the sequence in sedimentary rocks, we know that species number has built steadily over millennia. Currently, the oceans have a variety of animals about twice the average for much of the fossil record and today the Earth is at its most diverse. Robert May estimates that perhaps as many as 10 per cent of all multicellular species that ever lived are alive now.

This cranking up of S through geological time is despite several past episodes when there has been large-scale and rapid extinction of species. There have been at least five mass animal extinction events in the planet's history—the 'Big Five'—the result of major upheavals in the global environment, when a sizeable proportion of multicellular species were lost. Yet, on both land and sea there have been impressive recoveries from these immense disturbances. Up to 95 per cent of oceanic species were lost in the Permian extinction 250 million years ago, but species richness recovered quickly. The same recovery has happened after each event.

These extinctions are interesting not only because of the pace at which species disappear, but also the insight they provide into momentous changes in the Earth's biosphere. Even so, most extinctions occur outside these episodes—the fossil record suggests over 90 per cent of extinctions are part of the background rate of species turnover, outside of any major event. Unfortunately, our species has only ever known rapid rates of extinction and we are currently in the midst of the latest mass event. Indeed, with some methods of counting, extinctions are today happening faster than at any time in the Earth's history. At background rates, the average life of a species in the fossil record is estimated to be around 5–10 million years (though with great variation between groups and species), but Robert May estimates this has now dropped to just 10 000 years.

Extinctions are most obvious for the species we value and count most readily: Edward Wilson notes that one-fifth of all bird species have been lost in the last 2000 years and a further 1000 species are currently endangered. One estimate suggests that 3 per cent of the entire flora of the United States is endangered due to habitat loss from human activity. And it seems that our expanding sphere of influence is the principal cause of the current rate of species loss.

We can readily track human colonization in different continents or on islands by the disappearance of species. Europe, Asia and the Americas lost many of their large mammals soon after humans arrived. The arrival of humans in Australia around 40 000 years ago marks the loss of its large vertebrates including nearly all the large marsupials. Much of the mammal fauna and bird fauna of Madagascar was devastated following large-scale human settlement. We can see the same process operating today—for example, in the flowering plants, amphibians and reptiles—with the scale of human activity across the Americas threatening high numbers in these groups (Figure 9.4b).

The loss of habitat-sensitive, specialist species many years ago probably explains why the remaining flora of the Mediterranean basin is comparatively resilient today (Section 5.1). Just 0.15 per cent of its higher plants have been lost in the recent past, presumably because the survivors are those who have accommodated change. A much greater proportion have disappeared from the other mediterranean climatic regions over the same period, especially from the mallee of Western Australia (0.66 per cent or 54 species), the most recently exploited region. Once again, the scale of the threat across these regions shows some relation to the degree of human disturbance. While current losses are not yet significantly above the background rate, all of these areas face much higher losses in the near future as aridity increases.

Globally, the World Conservation Union estimates that 13 per cent of all plant species are threatened. Others think this a gross underestimate, taking no proper account of the under-recorded species in the tropical latitudes. In the late 1980s and early 1990s, the rain forests were declining at about 1.8 per cent each year. Using the normal range of z values (0.15–0.35), Edward Wilson calculated that this would amount to a loss of about 0.5 per cent of forest species going extinct each year. Recently, the Brazilian government has reported rapid declines in the rate of deforestation in Amazonia, though by 2001 somewhere between 13 and 20 per cent of the naturally forested area had gone.

Some losses are more important than others. In any inventory of diversity, losing the last member of a family must be a greater loss than losing a subspecies or variety. Yet, within a species every individual has a value and adds to the variation in the gene pool upon which the adaptability of future generations will depend. This genetic diversity, along with the variety of species (or higher taxa), and the diversity of habitats within an ecosystem are, collectively, what constitutes **biodiversity** (Box 9.1).

Some losses have greater ecological significance than others. Corals, for example, are keystone species upon which whole marine communities depend. Despite major fluctuations in sea level there have been relatively few extinctions within tropical reef communities over the last 2 million years. Instead, the animals associated with coral reefs have tracked the shift in the climatic regions, moving with them, probably, because, as Stephen Hubbell has noted, these species move freely between reef patches. In contrast, corals themselves are often highly localized and relatively immobile, and a species frequently disappears when its single population goes extinct.

Nevertheless, these communities have shown rapid diversification after each of the Big Five, requiring only 5–10 million years to re-establish their species richness. Indeed, a more rapid evolution rate and an early restoration of species number may be a feature of all tropical marine ecosystems. We find this elasticity, 2 million years ago, in the mollusc communities of the Caribbean. As sea temperatures fell during the glacial advances, mollusc extinction rates increased markedly, but this was more than offset by an increase in speciation rates.

Over the last 500 years, the reduction of manatees, turtles, jewfish, and conches through hunting appears to have had little effect on Caribbean reef communities, possibly because other species, primarily invertebrates, seem to have increased their abundance as a response. Unfortunately, human impact in the last 50 years may have tested the elastic limit of these ecosystems—increased sedimentation and pollution have contributed to widespread coral death in the region and now the reef systems are showing clear signs of permanent change. The loss of the corals means that a much wider range of marine species, especially in the shallow tropical waters of a reef, are now under threat.

The species and ecotypes going extinct take with them genetic variation that we might have one day used: the variation within wild types that we need to invigorate our domesticated species (Section 2.6).

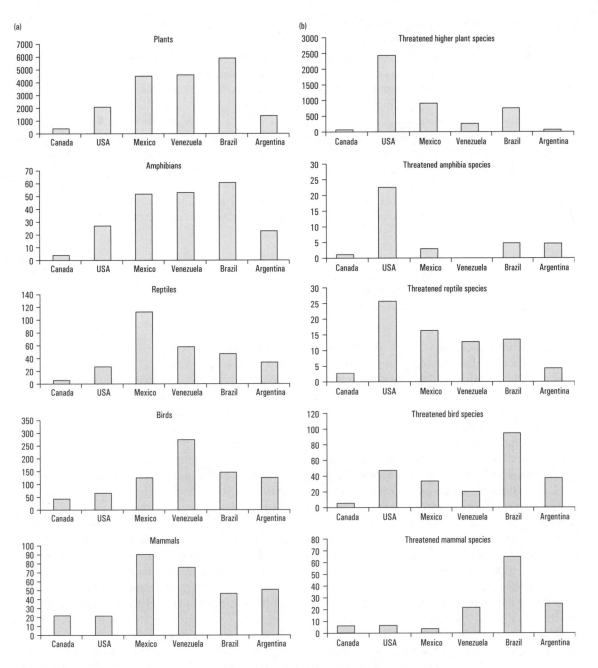

FIG. 9.4 Latitudinal diversity through the Americas. (a) This shows the number of species per 10 000 km², and thus standardizes the species count for countries of different sizes. Because some countries cover very large latitudinal ranges this will tend to obscure some of the latitudinal change in species richness, but, nevertheless, it does pick out the higher diversity for each group towards the tropics (the Equator runs through Brazil, close to its border with Venezuela). (b) The total number of species classified as threatened in each country. This is not standardized to area and is thus a simple count of those thought to be endangered.

Only in recent years have we begun to recognize the range of potential drugs, foods, fibres and other materials represented by unidentified and unclassified biodiversity, especially amongst tropical species. As it stands, much of our current economic activity is based on the small number of species that we found easy to domesticate many years ago.

On one hand, the numbers lost in a mass extinction are dramatic, but, on the other, they are not dramatic enough: it is often the small species, especially bacteria and invertebrates, which are critical for ecological processes. Yet, their diversity is not easily counted and not frequently recorded in the fossil record. Indeed, so difficult are they to type and to count, their contribution still goes unmeasured in many ecosystem studies carried out today. We assume prokaryotes are resilient species, able to survive cataclysm, and presume that few have been lost with their habitats.

Despite the scale at which species are being lost, most people do not witness an ecological collapse, say, on the scale of an Aral Sea (Section 8.1). It may seem that most ecosystems and their processes are resilient to small-scale species losses. Given that species are coming and going all the time what does it mean to say that an ecosystem shows stability?

9.3 Ecological stability

Strangely, when communities change we expect them to remain the same. Even as we prepare for winter, we expect spring to return and the luxuriant forest to re-establish itself, more or less as we saw it last year. Predictable change is incorporated into the biology and behaviour of plants and animals, and is even written into their genes. Squirrels hoard caches of food and plants swell their tubers to survive the times of shortage, banking on the times of plenty returning soon.

This continuity is what many people mean when they refer to a 'balance of nature'. It implies stability and the capacity of the system to restore itself. Beyond seasonal changes, natural ecosystems can withstand occasional shocks, so that even a major forest fire only resets the system (Section 8.1). Stability implies checks and control mechanisms to keep things the same from one year to the next, or which at least follow a predictable path of recovery.

Much of what we have to say about stability can be applied to any level in the ecological hierarchy, from an individual to an ecosystem. To encompass these different levels we use the term 'system', meaning two or more interacting elements that remain together. Ecologists have used a variety of measures to gauge stability in ecological systems, such as fluctuations in the population size of a key species, or coarse measures like decomposer biomass or rates of nutrient transfer. We concentrate here on populations and communities, where the components are individuals and species, respectively.

Stability is a property of systems that change little following a disturbance, or which return quickly to their previous condition. In fact, stability can be represented by various measures of a system's response to a disturbance, shown in Figure 9.5. A population or community might be judged stable if it resisted a disturbance, changed little, or quickly resumed its previous pattern or state after being deflected. Many disturbances cause no permanent shift and some, as we have seen, are anticipated by the system. Such disturbance is said to be incorporated, so that beyond the short term there is little lasting effect. A more sizeable or unpredictable disturbance may exceed the capacity of the system to absorb the change and then individuals or even species are lost. A very severe or extended winter can decimate squirrel populations if their food stocks become exhausted.

A good analogy for a stable system is the weighted spring (Figure 9.5). At rest, undisturbed, the spring has a particular length. Pull the weight gently (the disturbance) and it bounces around but eventually returns to its original position. The larger the disturbance the greater the amplitude of its fluctuations and the longer it takes to return. Exceed its elastic limit and the spring will be permanently stretched, coming to rest at a lower position.

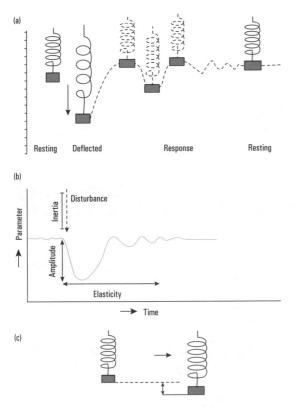

FIG. 9.5 A simple representation of stability in a system, shown by a weighted spring. In its resting position the spring and weight extend to a certain length. (a and b) If we disturb the system, the amount of force we need to deflect the spring is termed its inertial stability (or **resistance**). The scale of any deflection (its amplitude or **resilience**) is the increase in length. The time taken for the spring to settle down to its original position is its **elasticity** (or adjustment stability). (c) If we exceed the capacity of the system to return to its original position (we stretch it), then it will return to a new (slightly lower) equilibrium position. In ecological systems, we might measure such dynamics as changes in population size, biomass production, species richness, or a range of other measures.

For any organism to survive in a variable environment it must have mechanisms to maintain or restore itself after some sort of change. The internal environment of a cell or a tissue needs to be relatively constant to maintain metabolic efficiency, so that its complement of enzymes function close to their optimum. We are familiar with these homeostatic processes from observing our own biology—when we are hot or thirsty our behaviour and physiology changes to restore our poise and equilibrium.

Such mechanisms have been refined by natural selection to protect the individual and to allow them to function efficiently. It is easy to see why individuals would have evolved mechanisms to regulate internal systems. But how could such mechanisms have arisen in a community of species, each responding to different selective pressures? Communities, unlike individuals, do not have a heritable code that can be selected and passed on, so how does a collection of species come to operate together with any regularity and constancy?

In fact, any constancy we observe at this level will only be preserved across generations if traits directly or indirectly imparting these properties are selected in individuals. Community-wide regulation can arise only if genes are favoured that confer some advantage through their effect on the cooperative or competitive interactions of their owner. We have seen such finely tuned cooperative interactions between flowers and their pollinators (Section 2.2), in the competitive resource partitioning as niches becomes separated (Section 2.5), or in the evolutionary arms race of predator and prey (Section 4.4).

Stability at this level is, thus, a feature of highly integrated communities and ecosystems, emerging from the behaviour of its constituent individuals and species. An integrated, functioning ecosystem appears coherent in the same way that a flock of birds or a shoal of fish form organic shapes, a composite that emerges as individuals fly or swim together. In just the same way, their coherence flows from the selected traits of each individual, with each adapted to protect itself or to reduce its energy costs.

Some relations are very obviously constrained by selective forces—a parasite that kills its host before its eggs have been shed will not leave a long lineage and its genotype will soon be lost. Neither do most predators outstrip their prey; instead, most will shift to alternative prey species according to their relative abundance and the amount of effort needed to find and consume them. Similarly, symbiotic relations only persist while the selective advantage of one partner does not override the interests of the other. Change the relation and the selective pressure shifts for one or both species. Host and prey, symbiont and symbiont, are adjusted to each other and change in one is likely to induce change in the other.

Very often, the terms of the relationship is not simply decided by the adaptive capacity of the two interacting species. For example, a number of competing predators and their alternative prey species will collectively determine how closely the abundance of one particular predator follows that of one particular prey. Often, the fortunes of a species may be determined by species several trophic steps away. Even with a top predator like the polar bear, ecologists find it difficult to predict the impact of its loss on the marine community in the Arctic (Box 9.2). Deciding which interactions control the composition of a community is part of the 'rules of assembly' debate we discussed earlier (Section 5.1).

Some communities are more able to resist disturbance and some recover more rapidly than others. Stability clearly depends upon what you measure and how long you measure it for (Figure 9.6). However, we can learn much from comparing systems, finding

---

BOX 9.2   Making the connections

As we indicated in First Words, one of the major problems in ecology is establishing cause and effect. Although we may think some things are linked, proving that one environmental phenomenon causes another is often very difficult. Today, we see patterns that appear to indicate climate change but we rarely have sufficient data to prove that local conditions reflect a larger picture of long-term, global change. An example is given by the fortunes of the polar bears in western Hudson Bay.

Ian Stirling and his colleagues in the Canadian Wildlife Service have studied this population, again using radio collars to track the movements of 41 females. They asked whether a trend of reducing sea ice over this part of the bay over the last 50 years was reflected in the condition of the bears. There is considerable variability in sea ice extent and duration from one year to the next, so this trend is, as yet, poorly defined. The extent of sea ice cover has been declining in the Artic by between 3 and 5 % annually, though principally in the Siberian part of the arctic. Because of the distribution of land and air movements, the eastern Hudson Bay has, in fact, cooled since 1950, but the western bay has warmed by 0.2–0.3°C per decade. Since this area is a closed system, largely unaffected by ocean currents, the decline in sea ice in the western bay results from the rise in mean atmospheric temperatures.

The critical season is the late spring when females have to grow fat reserves for themselves, for their unborn cub, and for their nursing thereafter. As we have seen before, the bears hunt ringed seals (*Phoca hispida*) from the ice, and the longer the ice lasts, the longer they have to feed. Since the bears tend to stay close to a particular part of the coast, even when they are feeding from the sea ice, the population in the western bay is well defined (Box 3.4).

The weight of sedated male and female bears divided by the square of their length was used as an index of their condition—heavier and more well-fed bears have a higher index. Based on this measure, the condition of the bears deteriorated consistently since 1981. Studies between 1991 and 1998 showed a significant correlation between the condition of the females and the date at which the ice breaks up in each year, suggesting that a reduced feeding period on the ice caused their poorer condition.

In 1992, the ice lasted longer—3 weeks beyond the average—and the bears came ashore in good condition. This extended period of pack ice was almost certainly because of the cooling effects of two global weather events—the eruption of Mt Pinatubo in the Philippines in 1991 and an El Nino in 1991/92. Temperatures were reduced by around 2°C globally after the eruption, as they were in the bay. It may be a measure of the connectedness of this planet that a volcano erupting in the South Pacific may have meant, for 1 year, richer pickings for polar bears on the other side of the globe.

Thereafter, the bear's condition resumed its decline. There was also a decline in the proportion of yearlings (bears 1–2 years old) captured. Although the size of the population has remained more or less the same, its natality rate has showed a similar pattern of decline. Stirling, Lunn and Iacozza suggest that the condition of the bears in the Western Hudson Bay is responding to long-term global warming—Pinatubo briefly halted the warming and allowed 1 or 2 years respite, but the long-term trend has been resumed.

They point out that this population of polar bears is not under threat—compared to other populations they are in good condition, have a higher natality rate and a more stable population. But life is getting harder for them because they can spend less time feeding—if the duration of the sea ice continues to reduce, their means of feeding themselves is being cut from beneath their feet.

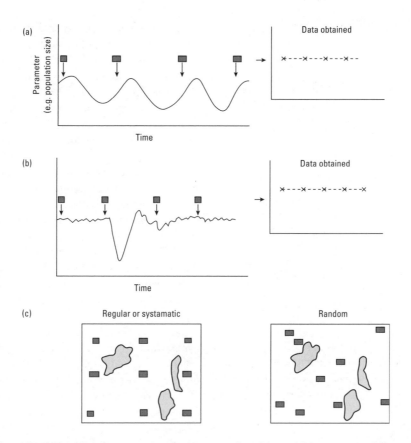

FIG. 9.6 The effect of sampling frequency on detecting temporal and spatial patterns. Sampling at regular intervals (shown by the rectangles) (a, b) or randomly (c) in time and space can fail to detect the dynamics of a system, a disturbance or its patchiness. This makes detecting significant change and stability in variable ecological systems very difficult. We have to use preliminary surveys to match our sampling programmes to the pace of processes or the distribution of features in the real world.

common features between communities, and describing the configurations that persist.

Our understanding of the link between species richness, community organization, and ecosystem stability is crucial for our capacity to predict changes over the long time scales and large spatial scales. Indeed, some ecologists doubt whether stability is actually a property of many ecological systems over the long term or even whether most species associations are especially long lasting. They ask instead whether any natural ecological system really does have an equilibrium configuration.

Stability and diversity

One of the most persistent ideas in ecology is that a diverse ecosystem is a stable ecosystem. A link between the number of species and stability was first suggested in the 1950s, based on comparisons of species-poor and species-rich communities. Pest outbreaks were thought to be more prevalent in species-poor agricultural systems and populations more variable in temperate and boreal biomes compared with tropical forests. Indeed, it is possible to find experimental evidence for such effects (Figure 9.7).

The stability of species-rich communities was thought to come from their **functional redundancy**, that is, more species allowed for more pathways through a food web (Section 6.4). If one species was lost others provided routes by which energy or nutrients could still flow, say when a predator switches to a different prey species. The greater the redundancy the greater the capacity for withstanding

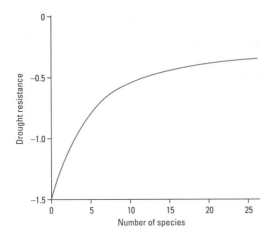

FIG. 9.7 David Tilman and his co-workers have shown that temperate grassland plots with more species have a greater resistance to the effects of drought (a smaller change in total plant biomass between a drought year and a normal year). However, there was a limit—each additional plant species contributed less and less, so that the most diverse plots showed only marginal increases in resistance. The reason for the greater stability seems to be that species-rich plots are more likely to contain some drought-resistant plants. Beyond a certain level, however, new species are less likely to differ in this ability from established species.

the disturbance of species loss. Populations in a diverse community would, thus, show less variability (more stability) than a species-poor community.

In the early 1970s, Robert May challenged this idea. He constructed mathematical models of food webs with different numbers of species and different numbers of connections between them, so varying the complexity of the system. A connection represented some interaction between two species (predation, competition, etc.) and was of variable strength (how much the population size of one species affected the size of the other). May was able to show that the link between diversity and stability was far from simple.

If everything else remained unchanged, adding more species would actually reduce the stability of the populations. In May's models more species could be accommodated only if the number of connections or the strength of these interactions declined: stability in its populations followed when the system's complexity was reduced. The evidence from 40 published food webs seemed to support this—connectance within a web falls with species number. A more elaborate model by Kevin McCann and co-workers suggests that it is the strength of these interactions that is critical for constancy in the web. Most of the complex webs seen in nature have a few strong links (where the abundance of one species has a major effect on another) but a large number of weak links. The McCann model shows these webs are stable because a hierarchy of relations involving many other species dampens oscillations of populations, even those that are partners in a strong interaction. Webs with many weak links serve to buffer major population swings and thereby check the oscillations of the larger system.

According to May, a highly diverse community could only persist if its interconnectedness was reduced and McCann and his colleagues conclude something similar. A food web divided into a series of **compartments** might achieve this. A compartment consists of a group of species sharing strong interactions with each other, but have more tenuous links with the rest of the community. The abundance of one species within a compartment can have major implications for the other members but not necessarily the rest of the community. John Moore and William Hunt found evidence of such compartments in their study of nitrogen transfers in the below-ground communities of North American grasslands. They describe compartments distinguished by their food source, and which connect with each other only further up the food web. For example, species clusters of soil nematodes (roundworms) feeding solely on bacteria had no connection with other nematodes feeding on fungi. Top predators (predatory mites) eventually linked these compartments, two trophic steps away. Moore and Hunt were also able to show a close relationship between connectance and species richness within each compartment.

Nigel Waltho and Jurek Kolasa also found evidence of compartments in their study of the fish populations in communities inhabiting Jamaican coral reefs. They describe a community structured according to habitat use, with specialist fish confined to narrow reef habitats and generalists occupying larger ranges. Not only were clusters of species (compartments) evident from the analysis, the population stability of each species also depended on its ecological range. Generalists, feeding high up in the hierarchy, were less variable.

Robert May concluded that complex communities are found only where the environment was unchanging and large-scale disturbance was unlikely. Their complexity was possible because members of the community had had a long history of evolution in each other's company. With time, new species evolved and, through competition and other interactions, defined their role precisely (Section 2.5). Long-lasting, unchanging environments would tend to accumulate more species and these would be the only circumstances where complex communities and their balancing act could develop. But such communities would also be most easily tilted from their precarious position by a disturbance.

There is evidence that these are indeed the characteristics of complex communities. As we shall see below, this may help to explain global patterns of species richness.

9.4   The big questions: latitudinal gradients in diversity

There are large and small gradients of species richness across the planet. Diversity declines with altitude up a mountainside. Down the length of a river the variety of invertebrates and plants increases, from highland stream to lowland river. Many open water (pelagic) marine animals show a maximum diversity some 1.5 km deep. Horizontally, across the oceans, their diversity also reflects the gyral circulation of water in each of the major oceans (Figure 8.10).

At the global scale, species richness for many groups increases from the poles to the equator so that the most diverse communities are found between the tropics. A range of benthic fish and invertebrate groups (especially molluscs and crustaceans) show this pattern, at least in the North Atlantic (Figure 9.8). The gradient is less distinct in the South Atlantic and Indo-Pacific oceans, where regional patterns dominate and there are 'hotspots' for some sea-floor invertebrates in the temperate latitudes. The greatest variety of marine isopods, for example, is found off the Argentinian coast, possibly reflecting ancient ocean currents. The latitudinal pattern is also confounded in shallow water communities, where local features, such as coastal configurations, are important.

Comparable processes create the patterns found in terrestrial ecosystems: regional contrasts superimposed on the global gradient because of local climate or landform, or the history of the region (Plate 8.4). One interesting example is the high diversity of mammals in the mountainous regions of North America, probably caused, in part, by the greater range of habitats in upland areas. However, despite such local hotspots, species richness of nearly all vertebrate groups rises toward the Equator in North America (Figure 9.9) and the pattern becomes more marked across North and South America (Figure 9.4a).

Globally, the distinction between tropical and temperate biomes could not be more dramatic: 6 per cent of the Earth's land surface is covered with tropical forests but it is home to perhaps 70 per cent of all multicellular species. Table 9.1 provides a checklist of the factors that might explain this gradient. A brief review of the long list of theories follows, though many of these are interlinked.

Ecologists have looked for some prime abiotic factor that changes with latitude, and that would apply in both marine and terrestrial communities. Temperature is the obvious candidate. Not only is it warmer near the equator, it remains so throughout the year. Perhaps a consistently warm climate allows for more specialization and greater differentiation of niches (Section 2.5). So, for example, tropical forests have birds and mammals that feed exclusively on fruit, possible because this resource is available all year round. There are no equivalent fruit-eaters in the seasonal forests of temperate regions and, similarly, most of their insectivorous birds have to migrate to avoid winter shortages.

In contrast, some argue that the marked seasonality of the temperate zones should promote greater niche differentiation. We know, for example, that changes in daylength are used to cue plant growth and flowering. In the early spring of an oak forest, bluebells and a host of other flowers bloom before the tree canopy closes and the competition for sunlight becomes limiting. The seasonal clock sequences the availability of a resource, allowing species to specialize and avoid competing with their neighbours. Even so, the temperate forests cannot match the net primary

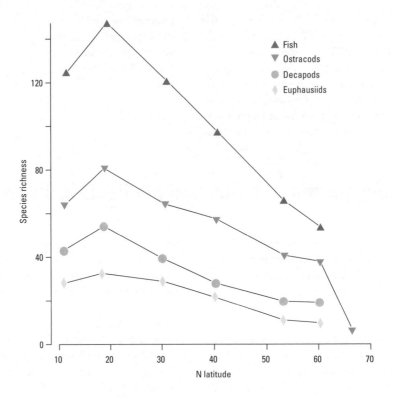

FIG. 9.8 Four latitudinal gradients in the diversity of a variety of marine vertebrates and invertebrates [Euphausiids (planktonic krill), Decapods (shrimps, crabs, etc.) and Ostracods] described by Martin Angel for the North Atlantic. Mean body size of benthic invertebrates also decreases toward the Equator. The length of food chains tends to be longer and cycling of nutrients is faster closer to the Equator.

productivity of the tropical forest, nor temperate grassland for savanna (Figure 8.20, Table 6.2), if only because their carbon fixation is turned off for several months when temperatures are too low for growth.

With the stark seasonality of the higher latitudes, all significant biological activity may cease for some part of the year. In this case, close associations that tie the fortunes of one partner to another can be a risky strategy when the association has to be re-established next year. Species in temperate areas also have to adapt to a wider range of conditions, favouring a few adaptable generalists, rather than narrow niche, specialist species.

The adaptations needed include the capacity to survive the seasonal cold—fur, feathers, or, in some cases, even anti-freeze in the body fluids. Such changes are so extensive that they can only be the result of a long period of selection, by which a species gradually colonizes a habitat. The latitudinal leakage

model suggests that the decline in S toward the poles is due to species only slowly invading the higher latitudes after the retreat of the ice sheets. Moving further from the Equator, a smaller leakage of species has occurred, roughly corresponding to the scale of adaptation required.

Based on sunlight received we might expect primary production to show a simple latitudinal gradient. In the oceans this pattern is masked by the availability of nutrients and local turbulence where turbid waters limit the radiation reaching the phytoplankton. The most productive areas are often at continental margins where sediments are stirred and nutrients are periodically mixed with the surface waters. The seas of the higher latitudes are distinctly seasonal and when nutrients and sunlight are abundant, they are highly productive. Because tropical waters have little overturning of surface waters, marine primary production shows only localized

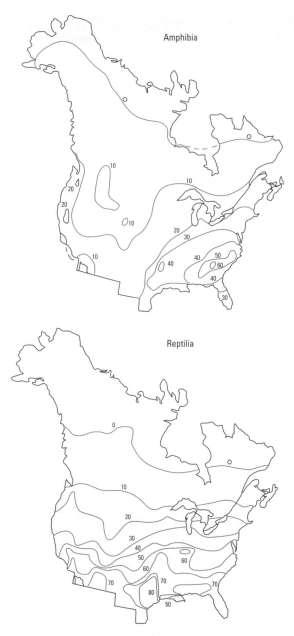

FIG. **9.9** The distribution of the number of species of amphibians and reptiles in North America. David Currie attributes this primarily as a response to solar radiation and its impact on water availability and temperature, both of which change with latitude and continentality.

maxima around the Equator. Nevertheless, the greatest diversity of animals is found on the tropical sea bed (Figure 9.8) surviving on the year-long 'rain' of organic matter from above. This high diversity is most probably a consequence of the reliability of the food supply rather than the amount arriving.

High productivity, therefore, does not guarantee more species, only more biomass. Many productive plant communities are dominated by one or two species occurring in very large numbers, often at the expense of others. Estuarine and wetland communities typically have just a small number of competitive, fast growing, and abundant plants (e.g. *Spartina* or *Typha*), whereas the nutrient-poor, semi-arid soils of the Mediterranean biomes have a prodigious variety of flowering plants (Section 5.1). Similarly, a shortage of the major nutrients is a feature of tropical forests and several studies point to a latitudinal gradient of soil fertility. The distribution of the trees often reflects the presence of key minerals in the soil. David Tilman suggests that low nutrients in tropical soils allows many species to coexist, preventing any one species becoming abundant and dominant (Section 4.3).

Much of the plant and animal richness of the tropical forests is attributable to the diversity of the trees and the opportunities they represent. Adding more species creates opportunities for other species—that is, each represents a niche or a collection of niches that can be exploited by other species. It may be that the diversity of tropical forests and ocean floors follows from the resources generated by their diversity. Species richness itself promotes species richness.

Some large areas of tropical forest have survived intact for millions of years, although much of it did change with the climatic shifts associated with the ice ages. Despite changing sea levels and habitats, most tropical areas were able to maintain large-scale forest refuges that today are represented by 'hotspots' of biodiversity. An unchanging environment causes fewer extinctions and, with time, greater specialization and niche differentiation. It may be that the tropics have simply had longer to accumulate species. In contrast, the higher latitudes have had a more changeable history and been subject to the migrations and extinctions associated with climatic change, especially the glacial advances. The deep ocean floors are buffered from significant change and the benthos of a tropical sea typically has a low density of individuals but high species richness. Amongst these are some of the oldest taxa.

TABLE 9.1 Possible factors contributing to global patterns of species richness

Physical conditions

Climate generally more benign and less variable in tropics, harsh and highly seasonal in higher latitudes

Highly contrasting seasons at the higher latitudes limit close associations between species

Shorter generation times promote higher speciation rates

Greater range of adaptations needed to exploit higher latitudes — major groups originate in tropics and 'leak' to higher latitudes

Limited nutrient supply allows coexistence of many tropical plants in a small area

Larger area

Greatest area of the globe found in the tropics, allows for larger populations and, therefore, fewer extinctions

Fragmentation of populations allows for higher speciation

Productivity

Higher net primary productivity in tropics, more elaborate food webs

Disturbance

Moderate levels of disturbance may create a greater range of opportunities for colonists in the tropics, more distinct regeneration niche for trees in tropical forests

Constancy and long history

Longer periods and larger areas undisturbed by climate change

Longer period of co-evolution of species in tropics without major change in biotic or abiotic parameters

Species accumulated since extinction rates are low

Greater complexity

Complex plant communities provide more niche opportunities for animals in the tropics

Greater competition and predation

High levels of competition, herbivory, and predation prevent any species from becoming dominant

Sexual reproduction is favoured to promote variation, to escape especially parasites

Greater diversity

Promotes greater diversity

The tropical biomes straddle the girth of the planetary sphere, where its land area is greatest. One possible effect of such large ranges is that populations may become discontinuous and develop local ecotypes that can go on to differentiate into new species (Section 2.6). This area has expanded (and contracted) in geological history, but as Michael Rosenzweig has pointed out, it nevertheless represents a vast region of relatively uniform climate that stretches back many millions of years. The size of this biome means many species can develop large populations over large areas, with consequently little risk of extinction. This is another reason for expecting the taxa to be older in the tropics and there is supporting evidence not only for the trees, but also for the birds. The diversity of trees in tropical forests suggest a long evolutionary history because they are represent different families and genera, with few species belonging to the same genus. The temperate zones have more tree species per genus, probably the result of recent speciation.

Animal species richness in tropical forests and coral reefs seems linked to the architectural complexity of these systems. These are termed three-dimensional habitats because their superstructure offers a variety of niches in space, in contrast to the limited vertical structure of a grassland or tundra. This distinction was found to be a good predictor of food web complexity by Frédéric Briand and Joel Cohen. Their survey of 113 webs from all types of ecosystems linked structural complexity and a relative constancy of the environment to longer food chains (Section 6.3). Without major disturbance over a long period, a complex community can assemble itself, and provide opportunities for later colonists.

Disturbance too has its part to play in maintaining diversity. Fugitive species rely on gaps to maintain a population and can only sustain themselves if the gaps appear with sufficient regularity (Section 5.3). Again, the scale of the disturbance is critical, since diversity will only be maximized if the adjustment stability of the system is not exceeded (Figure 9.5). We saw in our survey of Mediterranean-type communities how a certain frequency of minor fires was needed to maintain the scrubland mosaic (Figure 5.3), a scale of disturbance from which these communities could recover rapidly and upon which some species relied.

Intermediate levels of disturbance remove competitive species that would otherwise exclude new arrivals (Section 5.3). The intermediate disturbance hypothesis has been used to explain species richness in more variable environments, especially in coastal zones and temperate forests. For this to create a latitudinal gradient, however, disturbance rates would have to be close to optimum in tropical regions and minimized in the higher latitudes. In fact, turnover rates for trees in temperate and tropical forest biomes do not differ.

Robert Ricklefs argues it is the nature of the gaps, not their quantity, which is key. He points out that an opening in the tropical forest creates a habitat (a regeneration niche—Section 4.1) that contrasts sharply with the forest floor under a closed canopy, with a vast increase in light. A wide range of very different niches becomes available. In temperate forests the gaps remain in partial shade because of

the sun's angle at these latitudes (Figure 9.10). Ricklefs believes this explains the astounding variety of trees in tropical forests when their under-storey vegetation is surprisingly uniform.

With its vast array of animal species, predation and herbivory are probably more severe in tropical forests and this has important implications for the organization of these communities—a species consumed is a species constrained. At a high intensity, the consumed species is prevented from becoming dominant. That pressure, operating throughout a food web, itself promotes greater species diversity. Certainly, the intense seed predation observed around the parent tree would explain the large distances between individuals of the same species in tropical forests (Figure 9.11).

Finally, there seems to be more sex in the tropics. Sexual reproduction may be favoured here because it promotes variability, essential in the continual battle to escape the attentions of the large number of pathogens, parasites and predators. At higher latitudes, the frequency of asexual reproduction increases in some plant and animal groups, presumably because the selective advantage of sexual reproduction is less when there are fewer natural enemies. Again, species richness promotes species richness.

Despite its simple pattern, it seems that the latitudinal diversity gradient is probably the result of several ecological factors operating together. It may be that two or three abiotic factors set off a chain of

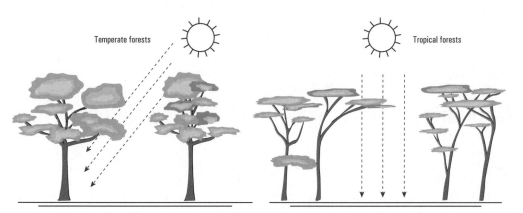

FIG. 9.10 Gaps in the canopy in forests are much more fully illuminated in the tropics because of the angle of the sun, producing a much greater contrast with the closed canopy. Robert Ricklefs suggests this creates very different niche conditions favouring a wider variety of trees, compared to temperate forests.

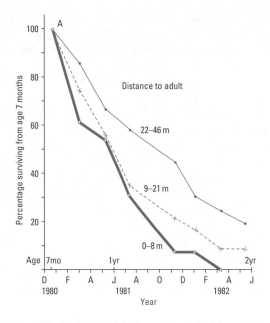

FIG. 9.11 The high diversity of trees in tropical forests may be due to high mortality of juveniles around the parent tree. Janzen and Connell proposed that the pressure of herbivory on seedlings would mean that distance between the parent tree and its offspring would be very large. Clark and Clark were able to demonstrate this for one canopy tree, *Dipteryx panamensis*, by following the survival of its seedlings. Although 80 per cent of the seeds fall within 13 m of the tree none survived within this zone after 2 years. The nearest saplings from this cohort were 36 m and 42 m away from their parent. The graph shows how seedling survival decreases closer to the parent tree with time. This provides some evidence that biotic interactions within the community, in this case herbivory or attack from pathogens, help to promote high diversity. In the same way, predation or competitive pressures may prevent a species from becoming dominant.

biotic reactions that accentuate the differences between latitudes. Perhaps the greater energy influxes and primary productivity of the tropics provides the resources for a greater variety of herbivores, which, in turn, provide for more carnivores. As the community builds, it amplifies the opportunities for other species. In each case, the consumers serve to constrain species they feed on, preventing any one species becoming dominant. The contrast between latitudes then follows from this potential for positive feedback, so adding more species provide more opportunities, which, in the tropics at least, can be supported by their higher productivity.

But can we keep adding species? Highly diverse communities may be more fragile, yet through natural selection, evolution will try to add new species. Perhaps only the tropics provide the stable conditions necessary for complex communities to assemble, and where, because of their long history, many 'trials'—different combinations of species—have taken place. What we see today is the current result, the current balancing act.

But can we keep removing species? Diverse ecosystems may remain stable only if there are redundant species whose removal does not cause the community to topple. We have seen how the Mediterranean-type communities have, under human influence, lost species in the past and developed relatively stable configurations (Section 5.1). As more are lost, however, the role played by the remainder is increasingly critical to the stability of the whole system. Lose too many and collapse becomes inevitable—a threat facing areas of several Mediterranean communities today.

9.5 The big questions: stability and sustainability

The question of how many species are needed for an ecosystem to function is critical, both to ecological theory and to political practice. Perhaps we could afford to be unconcerned about species going extinct if their ecosystems continued to work in their absence. If ecosystems were unaffected by their loss we might be able to live without them and the economic argument for protecting large and talismanic species would be weaker.

However, we rely on natural ecosystems to purify our water, replenish our oxygen supply and moderate our climate. We have to know whether the species going extinct are likely to affect the delivery of these, and other, essential services. It is a cliché but we do take these services for granted and they are rarely part of any accounting procedure, properly quantifying the costs of maintaining environmental quality.

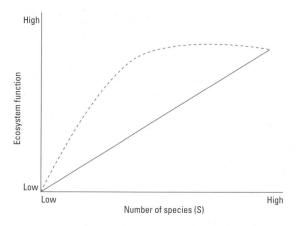

FIG. 9.12 Functional redundancy in ecological communities. If every species is important in maintaining some ecological function then we would expect a linear relationship between S and the process (the solid line). If, instead, some species are more important—that is other species are to some degree redundant—the ecological function will not decline linearly as each species is lost (the broken line).

Rather more parochially, ecologists are interested in the principle and the stability of ecosystem function as species are lost. If all species in a community were essential to a process, we might expect its performance to decline progressively as species were lost (Figure 9.12). Alternatively, with a large measure of functional redundancy, species could be lost with little loss of performance or stability in the system. Then a major reduction in the function only occurs when non-redundant species disappear.

M. W. Schwartz and colleagues have looked at both the experimental and theoretical evidence for such patterns. Most observational and experimental work showed that ecosystem function did indeed fall as species number reduced, but only in three out of 19 cases was there a simple linear response; that is, when function declined in direct proportion to the number of species. We saw this pattern before with the effects of drought on temperate grasslands (Figure 9.7), suggesting a high level of redundancy.

A nonlinear response means the less common species are not critical for the function of the larger ecosystem. Simple theoretical models suggest why a linear response is rare and why many communities can remain stable even when losing species. Once again, it seems that a large number of weak interactions appear to be critical for dampening major fluctuations. In addition, species often respond in different ways to an ecosystem change, with some species increasing their abundance when others decline. This pulling in opposite directions may also help to dampen the response in the larger system.

It is often the little things that count. We saw that a large proportion of the productivity of the savanna (Section 8.2) was processed by termites, and their activity, more than any other multicellular species, is critical to nutrient cycling in this biome. The same is true of soil invertebrates, bacteria and fungi in all soils of the world. The soil bacterial community is essential to nutrient fixation and turnover (Section 7.1), and as a consequence determines the composition of higher plant community. This interaction is an example of the checks and regulation within a community working both ways—the microbial community that develops in a soil will itself be determined in large measure by the plant community growing above it.

There are formidable technical difficulties in measuring the microbial and microinvertebrate diversity of a soil and relatively few studies have quantified the species richness or relative abundance of its decomposer community. Even experiments using laboratory soil microcosms have concentrated on the easily manipulated species—the larger invertebrates and plants.

We do know that intensive agriculture often simplifies soil communities. Unsympathetic agricultural practices have been washing away the soils of the Mediterranean basin for 3000 years. In the years of the Kansas dustbowl, the 1930s, the United States was losing 3 billion tonnes of sediment in surface runoff annually. Estimates from the midwest United States suggest that arable lands have lost about 7 cm of topsoil, equivalent to 900 tonnes per hectare since cultivation began. It is in these upper layers where most of the soil's organic matter and decomposer community resides.

The way we treat the soil often determines its microbial health. For example, dramatic change follows when the grazing of a pasture ceases, heralded by changes in the plant community. Roy Neilson and his co-workers have shown that stopping sheep grazing leads to creeping buttercup (*Ranunculus repens*) replacing perennial rye grass (*Lolium perenne*) as the dominant plant on upland pastures in Scotland. Because of its effect on the above-ground plant community, grazing shifts the below-ground community from one dominated by bacteria to a fungal-based economy, with consequent changes in the invertebrates. In contrast, stopping the use of artificial fertilizers caused little change.

Grazing induces the production of root exudates by the grasses and these promote bacterial growth near the roots. Without this interaction, fungi dominate the soil and a different community of soil invertebrates—dominated by fungivorous nematodes and springtails—prevails. Springtails are one of the main fungivores in grasslands, but interestingly, some species only become functionally redundant if one of the larger species—*Folsomia*—is present. On its own this species can determine rates of litter decomposition and soil respiration. When it is present other species have no major impact on these processes.

It seems that much of the immense diversity of bacteria and protozoa in soils is redundant. Even the coarsest of manipulations of bacterial and fungal communities (or the food chains above them), appear to have little effect on soil functioning. Mira Liiri and co-workers in Finland found no significant change in biomass or in drought resistance of coniferous woodland soils when a sizeable part of the invertebrate community was removed from field plots. Even when changes are induced in the community by adding a pulse of nutrients (wood ash), the soil's capacity to resist disturbance and maintain its stability (measured as the loss of nitrogen from soil microcosms) was unaltered.

At the microscopic level, a soil is not one microbial community, but several—a series of habitat patches—with bacteria or fungi flourishing where opportunities present themselves. Perhaps this degree of compartmentation helps to buffer the response of the larger system. Many high latitude and temperate soils appear to show this resilience, in part, because they have incorporated major seasonal changes, including pulses of nutrients, into their annual cycle. These are also young soils, where many key plant nutrients, such as phosphates, are relatively abundant. Elsewhere, nutrients are not so available, and soils are more easily pushed beyond their capacity to restore themselves.

Gradients of sustainability

The global gradient of diversity also tells us something about the agricultural capacity of the different biomes. Generally, low latitudes have old and infertile soils and nutrients are rapidly reclaimed from these by the intense competition amongst microbial and invertebrates decomposers (Section 8.2). When we remove the forest we remove much of the ecosystem's nutrient capital and its supply to the decomposer community. Without the protection of the trees, the soil is readily eroded by the intensity of direct rainfall and sunshine, as the glue of its organic matter quickly disappears. Michael Huston suggests this alone should give us every incentive to stop the wholesale destruction of the tropical forests, since these serve a more useful role fixing carbon from the atmosphere and preserving biodiversity.

Yet, these forests are predominantly in regions where agriculture is still the principal means of wealth creation. Huston describes the global gradient of declining GNP (Gross National Product, a measure of the economic wealth of a nation) toward the equator. These are the nations where human population growth is often the most rapid, but which are relying on ecosystems unsuited to the large-scale disruption (Section 7.2). The indigenous practice of the slash and burn of small clearances, followed by cultivation for 2–5 years, is an appropriate means of exploiting the short-lived fertility, without overcoming the elasticity of the forest ecosystem. Indeed, the disturbance may be important for the gaps it creates, promoting diversity through regeneration. Commercial logging operations exceed the adjustment stability of the system so that in many areas of Amazonia recovery is non-existent and a different plant community becomes established, typically a semi-arid grassland.

The loss of tropical forest leads to local climatic change, in turn, causing soils to become drier and more easily eroded. Somewhere between 100 and 200 tonnes of soil per hectare are lost annually from cultivated slopes in tropical areas. Even where there has been commercial reforestation (for rubber or paper pulp production) the poor soil causes yields to be low. Up to 20 per cent of the Amazonian forest (600 000 km^2) has been lost through human activity in recent decades and overall it now has about half of its pre-human forest cover.

Drought and periodic famine are written into the history of many tropical and semi-arid areas, but the scale of habitat degradation today and, with it, the rate of species loss, indicate more permanent

BOX 9.3 The growth of the human population

As far as we can tell, the human population has been relatively stable for much of its recorded history. Various methods have been used to measure the size of past populations in different parts of the world and the overall picture is one of a slow rise for much of its history, with one or two noticeable declines due to disease.

However, the global rate of population growth showed a marked increased in the twentieth century. Most nations began a phase of rapid growth at some time during the last 100 years, primarily as they adopted modern medical techniques and standards of hygiene, and as nutritional standards improved. Together these have led to major reductions in death rates.

At its current rate of growth (1.3 % per year), the global population will double in the next 50 years (Figure 9.13). Today, the global population is around 6 billion people with an age structure characteristic of rapid growth. However, the increase is not uniform across the planet: the United Nations gives annual growth rates of 0.3 % in the developed world and 1.6% in the developing nations. At the beginning of the twenty-first century the overall rate of growth has started to slow and fertility rates have begun to decline in developing countries—from 4.7 births per female in 1975–80 to 3.0 for 1995–2000. Nevertheless, these rates operate on much larger populations, so the increase in absolute numbers is set to continue for some decades to come.

The total global population is expected to be close to 10 billion by the middle of this century. As Tuckwell and Koziol (Figure 9.13) put it '. . . there are apparently logistic-type regimes which persist until some major event occurs'.

This is not news. Predictions about the consequences of large human populations have been around for centuries (and perhaps longer). The rates of increase in the 1950s and the 1960s provided momentum for the early

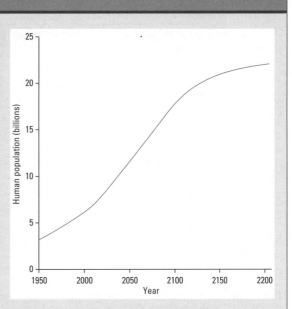

FIG. 9.13 The growth in the world population, modelled according to the logistic model described in Section 3.2. Henry Tuckwell and James Koziol found that this simple model gave the best fit to the data between 1950 and 1985 and that it accurately predicted the population in 1992, of 5.48 billion. If it continues to be accurate, a carrying capacity of around 24 billion, four times the current size, will be reached in the year 2200.

environmental movement, and their forecasts of imminent environmental collapse. These were somewhat premature, though numbers do outstrip resources, especially in the developing world. In particular, water shortages are expected to affect two out of three people by 2025 and the struggle for limited regional supplies may provide, in the words of Kofi Annan, 'the seeds of violent conflict'.

change. Much of the developing world relies upon wood as its principal fuel and their growing populations have stripped the land of its trees, especially in sub-Saharan Africa, India and the Himalaya. As the trees disappear, so does the soil. Henry Kendall and David Pimental describe how, in the latter half of the twentieth century, most nations ceased to be self-sufficient in food. Despite the advances in agricultural technology—the 'green revolution' based on new crop strains, artificial fertilizers and pesticides (Section 6.5)—per capita food production has slowed down. Food production in 70 per cent of developing nations can no longer match their population growth.

BOX 9.4 Final thoughts

'The rescue of biodiversity can only be achieved by a skilful blend of science, capital investment and government: science to blaze the path by research and development; capital investment to create sustainable markets; and government to promote the marriage of economic growth and conservation.'

Edward Wilson

Human economic development seems inevitably to lead to environmental degradation—so often rapid economic and industrial development has been followed by pollution, the loss of natural habitats and of species. Few of us are guiltless and few of us see the consequences of our actions and our choices. Quite naturally, we tend to react to the local and the obvious rather the global and the insidious. In the developed nations we are insulated from the environmental costs and these changes have yet to impinge on our lives. Elsewhere people are reminded daily of the quality of their soil and the availability of their water.

At some stage, a limit to human population growth will be part of the resolution of the global environmental deterioration. Many people argue now for greater population control and more sustainable use of the existing resources. Others suggest that rather than constrain economic activity it is better to promote economic growth to provide the financial means to address to the problems of famine, poverty, and poor sanitation amongst developing nations.

Finding technological fixes may be part of the answer, but many of these solutions once again neatly divides the world into two: those nations with the economic resources to support advanced technologies, and those without. Even then, some of these fixes have different efficacies either side of the North–South Divide—for agricultural economies working poor or marginal soils,

the high-energy solutions used to maintain soil productivity in the higher latitudes do not work. Typically, only the simplest innovations cross the divide.

South of the divide the connection between economic well-being and the state of the environment is immediate and obvious. In sub-Saharan Africa, drought means famine and solutions which work in San Diego rarely work here. Technical fixes in the Sahel cannot demand highly refined materials or large energy subsidies. Fixes here have to be sensitive to regional ecological processes, the nature of the soil, the annual regime of rainfall, and so on. Local systems of resource conservation and food production have to become the priority.

Sustainability is becoming an issue in the developing nations too. Our profligate use of water is becoming problematical in Mediterranean climates and a potential check on economic growth (Section 7.2). Today, with or without the Kyoto protocols, most industrial economies hope their market-based approach will produce the technological solutions before their energy-use makes the atmospheric system shift into a new climatic regime. Whole nations submit the rest of the globe to passive smoking, and then, through the haze, ask 'what's the problem?'

The long-term trends are thought by policy-makers to demand technological fixes, rather than a change in behaviour by the consumer. To others, a technological fix will be just that—a fix. It would extend the sleight of hand that our technology has performed since we first warmed ourselves by the fire—insulating some from the environmental consequences but reserving the costs for the next generation or for those without the means to pay. Properly applied, our ingenuity does not have to mean some losing out or environmental degradation, but it does requires us to look beyond the local and the here-and-now.

The usual response to food shortages is to bring more land into production, causing more natural habitats to be lost. The land available is inevitably marginal and cannot sustain agriculture over the long term. Fifty years ago much of the central highlands of Ethiopia were covered in relatively lush forest. Since then, the once nomadic peoples of sub-Saharan Africa have increasingly settled around

centres where medicine or water is available. In 1984, Ethiopia had its driest year since records began and the combination of a civil war, a large settled population, and a severely depleted ecosystem based on unprotected soils, conspired to produce one of the most severe famines of the twentieth century. Now, following the failure of the short and the long rains in 2001/2, warnings are being sounded about the largest famine yet to hit Ethiopia, perhaps affecting 15 million people.

The consensus today is that famine relief does not provide a long-term solution, that a simple reliance on food aid undermines local agriculture. Whatever political and economic changes are needed in Africa, large areas will need to be reforested and the population helped to achieve densities that are compatible with a functional soil. Carl Jordan argues that tropical agriculture needs to avoid adopting western methods with their dependence on fossil fuels. Better, he argues, to use practices that harness the natural processes in the local environment, alongside more imaginative cultivation practices. This was recognized by most world leaders at the Rio Summit of 1992 and the commitment of Agenda 21 to find sustainable strategies for all economic development. Even though the financing of the initiative remains problematical, it does at least recognize the link between human welfare and environmental quality (Box 9.3).

So we end up where we started—in sub-Saharan Africa with a receding forest (Section 1.1). Today, most people living south of the great desert suffer some form of malnutrition. What was the cradle of humanity has become the place where large-scale human death is most prevalent. In this part of the developing world, more than any other, agriculture has failed to grow as rapidly as the population.

Much of the world population is hungry and perhaps one-fifth suffer malnutrition. Yet, even as our numbers have grown, a smaller proportion today suffers from a lack of food than at the start of the twentieth century. That encourages some to argue that human ingenuity could eventually solve shortages of food and water. We will inevitably look for a technological fix (Box 9.4). Perhaps genetic manipulation might improve the disease resistance of key crops, or equip cereals with the capacity to fix their own nitrogen. Even assuming that this technology is freely shared (or that genetically modified organisms pose no environmental risks), we will still need to use sustainable agricultural practices to preserve the complex ecological interactions that makes food production possible.

Functional redundancy may seem an abstract idea, of interest to only some ecologists. But as we seek to restore the productivity of the degraded agricultural soils it assumes immense practical significance. In famine-prone areas, water management and irrigation schemes will be part of the solution, but we will still need to understand how to sustain the soil community that will continue to feed us. In the end, the solution will have to be ecological, but the problem is far larger. Change will follow when we address the cultural, political, social and economic factors that underlie environmental degradation.

Perhaps the extinctions that humankind have set in train are just the 'froth', the functionally redundant species the Earth has accumulated over the last 65 million years, and which it can afford to lose. Comforting as such a notion might seem, this fails to recognize that species loss is too rapid and too wholesale for the selectivity this implies. It would be difficult to show that Polar bears were critical to the functioning of the polar ecosystem. We might speculate that if this top predator were lost, seal populations would rise and there would be a general shift in the trophic structure of the seas around the Arctic ice sheet, perhaps impacting the fish stocks we exploit. But as yet we cannot make the connection.

It is too simplistic to suggest that all environmental degradation can be traced back to the explosion of the human population. Instead, it is more the number of individuals multiplied by the amount of resources we each demand, set against the fragility of the environments we exploit. What we each consume and demand of our environment varies greatly across the globe and is shaped by our culture, tradition and the political constraints on our activity. Nevertheless, the number of *Homo sapiens sapiens* is a prime factor in determining the rate of habitat and species loss. We have used our technology and ingenuity to bring us this far, now the question is whether we have the wisdom not to cut the ground from underneath our feet.

SUMMARY

The number of species increases with area and allows us to predict how many species a habitat or island might contain when colonization and extinction rates are balanced. At this point there is a dynamic equilibrium and the total species count (or species richness) stays more or less the same. Currently, the loss of habitats, largely through human activity, is the principal cause of a mass extinction event, possibly the most rapid in the history of the Earth. The loss of some species is more important than others, and ecologists are seeking to describe the proportion that can be described as functionally redundant, which can be lost without affecting the stability of the system.

An ecosystem is stable if it resumes its original state after disturbance. Stability can be divided into inertial stability (the capacity to resist deflection) and adjustment stability (the capacity to return to the undisturbed state). The association between the diversity of a community and its stability is not simple; often, the most diverse communities are the most fragile, and are typically a feature of long-standing environments that have suffered little disturbance. This may account for the global gradient of diversity toward the tropics, in which the oldest, least disturbed biomes around the equator show the highest diversity, both on land and in the sea. A range of other factors, including their benign climate and high productivity, may be important, and also the positive feedback of high species richness that creates further niche opportunities.

Tropical and more marginal soils appear to be more fragile, most probably because they have characteristically smaller reserves of nutrients and faster rates of decomposition. Much of the uncounted biodiversity at this level—of microorganisms and small invertebrates—is critical for agricultural productivity.

FURTHER READING

Jordan, C. F. 1995. *Conservation*. John Wiley & Sons, New York. (This provides an excellent overview of the ecological, social, and economic issues surrounding conservation, especially in tropical regions).

May, R. M. and Lawton, J. H. (eds). 1994. *Extinction Rates*. Oxford University Press, Oxford. (A useful collection of essays providing our current understanding of extinction rates in different ecosystems.)

Wilson, E. O. 1992. *The Diversity of Life*. Penguin, London. (A wonderfully written biography of life on Earth.)

WEB PAGES

The latest Red List of threatened species is available at the IUCN site.
www.iucn.org
and also
www.redlist.org/

Attempts at measuring global biodiversity are described at
www.all-species.org/

Images of the scale of tropical deforestation and forest fires from earth orbit can be
   seen at
http://earthobservatory.nasa.gov
and especially for images of human impact on the Earth:
http://images.jsc.nasa.gov/

Useful and up-to-date data on various global environmental parameters from the World
   Resources Institute:
http://earthtrends.wri.org/index.cfm

EXERCISES

1 Offer some reasons why some of the oldest taxa are found on the deep ocean floor.

2 Choose the *best* answer. In the species–area relationship z is low when

(a) most habitat patches share the same species,

(b) most habitat patches are large,

(c) most species do not move freely between habitat patches,

(d) most species have large populations,

(e) most habitat patches are close to a source community.

3 Decide which of the following groups you would expect to have high or low values of z for
oceanic islands. For each provide *one* reason for your decision.

Pines

Frogs

Mayflies

Dandelions

Butterflies

Rats

4 Which of the following statements has *not* been offered as a possible explanation for the
latitudinal gradient in diversity across the planet?

(a) more benign and more constant conditions promote greater speciation in tropical
   latitudes,

(b) the sharp seasonal contrasts in temperate zones mean that fewer tropical species have
   been able to adapt to these conditions,

(c) sexual reproduction and faster rates of speciation are favoured because of higher rates of
   parasitism and predation in the tropics,

(d) the rapid cycling of nutrients favours greater speciation in the tropics,

(e) the larger land area means that fewer species go extinct and species tend to accumulate
   in tropical forests.

5 Consider Figure 9.2. Explain why four values for equilibrium S are arrowed in the second
graph. Are there only likely to be four possible values for equilibrium S in reality?

6 Explain why more endemic species maybe found on isolated islands.

7 Consider Figure 9.3 and assume that the measure of stability in your experiment is N for
each population of polar bear. How, in real terms, would you decide that

(a) one population had greater elasticity than another,

(b) one population showed greater inertia than another.

8 Using the data provided in the following table, calculate for each ecosystem the S/N ratio and then attempt to decide which ecosystem has the highest diversity in terms of S, S/N, and species equitability. Comment on the differences you observe.

The number of individuals in each of nine species in seven ecosystems

Ecosystem	Sp. 1	Sp. 2	Sp. 3	Sp. 4	Sp. 5	Sp. 6	Sp. 7	Sp. 8	Sp. 9
A	11	36	23	2	4	97	23	12	79
B	0	4	78	14	0	123	34	34	32
C	34	11	93	8	0	56	56	13	9
D	23	29	39	6	2	92	78	19	11
E	21	11	16	9	1	48	43	26	12
F	13	56	12	2	2	55	55	34	85
G	4	78	39	0	5	96	89	62	12

Tutorial/seminar topics

9 review the data in Figure 9.4. To what extinct does the number of threatened species match their species richness across the Americas? You could prepare for this tutorial by

(a) calculating (roughly) the proportion of endangered species in each country, for each group of organisms,

(b) choose one group and look at the particular threats in two or three countries from this range,

(c) offer reasons why simple comparisons between countries are not strictly possible.

Suggest ways in which you could improve the comparability of these data.

10 To what extent could we use the polar bear as a sentinel for the state of the planet?

11 'It will be the diversity of species in the soil which will determine the carrying capacity of the human population on Earth.' Discuss.

Adaptation The collection of characters that enable an organism to survive. An organism may adapt in the short term, within its lifetime, by adjustments of its behaviour, physiology, or development. A species may adapt over generations by changes to its genetic code, through **natural selection** (*qv.*).

Admixture The flow of genes between two populations.

Age structures The proportion of individuals in each age class of a population. A stable age distribution is dominated by the younger age classes and is characteristic of a rapidly growing population. Closer to the carrying capacity a stationary age distribution is approached and the distribution is far more even.

Allele The various forms of a gene found at a particular position (**locus**) on a chromosome, and which thereby code for a particular character.

Allelopathy The inhibition of one organism by another using chemcial means. Some plants may inhibit the growth of others (especially those of the same species) by the chemical nature of their litter or by special secretions.

Allopatric speciation See **speciation**.

Amino acid sequencing The chemical analysis of proteins by determining the sequence of individual amino acids along the polypeptide chain. This method can be used to measure the degree of relatedness between two species by comparing the similarity between the sequences.

Anaerobic (anoxic) An environment that is devoid of or has very low concentrations of oxygen; **Aerobic** metabolism uses oxygen, anaerobic metabolism does not.

Annual A species that completes its life cycle within a year.

Antibody A protein produced by some animals in response to a foreign substance (antigen) as part of their immune system.

Apomict (micro-species) A species that reproduces without fertilization.

Asexual reproduction Reproduction in which there is no swapping of genetic material between individuals.

Archaeobacteria A primitive form of bacteria typically from extreme environments and with a distinctive metabolism.

Assimilation efficiency A measure of the efficiency by which an organism assimilates energy from the food it consumes (this is given by the amount of energy assimilated divided by the amount of energy consumed).

Autecology The study of the relationship between a single species and its environment.

Autotroph An organism able to obtain its energy either from sunlight (cf. photoautotroph) or from chemicals (cf. chemoautotroph).

Authority The person who first describes and names a particular species. The abbreviated name or initials of the Authority occur after the specific name (e.g. Linnaeus abbreviated to L.).

Batesian mimicry A form of mimicry in which a harmless species resembles a harmful or poisonous species to deter potential predators.

Benthic communities Those found on the bed of an aquatic ecosystem. **Pelagic communities** are found in the open water.

Biennial A plant that lives for 2 years, with growth and establishment in the first followed by maturation fruiting and death in the second.

Bipedalism Walking on two legs.

Biological control Pest control that makes use of a species' natural enemies.

Biological species concept Species defined as a group of individuals able to breed together and produce viable fertile offspring.

Bioaccumulation An increase in the concentration of a pollutant from water and diet by a living organism.

Bioconcentration An increase in the concentration of a pollutant from water by a living organism (qv. bioaccumulation, biomagnification).

Biodiversity Often taken to be simply the number of species (cf. species richness), but also used to encompass biological variety at all levels, from the genetic diversity between the individuals of a species to the total number of species on the planet.

Biomagnification An increase in the concentration in the tissues of a living organism either from the soil (in plants) or from the diet (animals) (qv. bioaccumulation).

Biomanipulation Changing the structure of an ecological community to bring about long-term changes in ecosystem structure and processes. This technique is used especially with eutrophic freshwaters (*qv.*).

Biomass The weight of living material.

Biome The characteristic vegetational community associated with a particular latitude or biogeographical region, such as tropical forests or temperate grasslands, and primarily defined by climate.

Binomial system The system used to formally name species, using a **generic** name followed by the **specific name** (e.g. *Canis rufus*).

Biosphere The part of the Earth that supports living organisms; the global ecosystem.

Biochemical oxygen demand (BOD) A measure of the amount of organic pollution in water. This is measured as the amount of oxygen used in the oxidation of the organic matter in a sample of water over a 5-day incubation at 20°C in the dark.

Biotic indices The use of the species assemblage of a community as a measure of its habitat quality. (qv. **Indicator species**).

Brood parasite A parasitic organism that lays its eggs in a nest of another species or individual that then rears the young.

Catchable stock Individuals sufficiently large to be harvested, such as fish being caught in nets with a certain mesh size.

Chamaephyte Low growing woody shrub, typical of maquis- or chaparral-type habitats.

Character displacement The divergence of a trait over several generations and as a result of genetic change, between two or more species competing for the same resource.

Chemoautotroph Microorganisms that derive their energy from oxidizing inorganic compounds and use carbon dioxide as their carbon source. These include the nitrifying bacteria.

Chimera An organism that consists of two genetically distinct cell types.

Chromosome The main site of the genetic material in living cells, composed of DNA (*qv.*) and protein.

Climax community A community of species found towards the end of a successional process, when species turnover is minimized and when the plant community assumes a persistent structural form.

Cohesion species concept The theory that discrete gene complexes are the result of adaptation and can be used as a means of defining separate species. (See also **Evolutionary Significant Unit.**)

Cohort A group of individuals of the same age in a population.

Commensalism An association between two species of benefit to one partner only, but of no detriment to the other.

Community A collection of plants and animals that live in the same habitat and interact with each other.

Compartmentation The extent to which a community, or more specifically a food web, is divided into discrete units each with strong internal interactions between its members, but weaker links to others.

Compensation point Condition in plants when energy fixed in photosynthesis is balanced by energy used in respiration. The extinction of light down a water column means this is represented by a certain depth in aquatic ecosystems.

Competition The fight for a limited resource between individuals of the same species (**intraspecific competition**) or between different species (**interspecific competition**). Organisms can compete for food, water, or space, and such competitive battles are one of the major driving forces behind natural selection.

Competitive exclusion principle Two species cannot co-exist in the same niche at the same time in the same place. Of two species with identical resource requirements, one will eventually be ousted.

Concentration factor The ratio of the pollutant concentration in an organism to that of its diet or the soil in which it grows.

Connectance The number interactions between different species within a food web expressed as a proportion of the total number of possible interactions. This is a measure of the complexity of the system.

Connectivity In landscape ecology, this refers to the extent to which habitat patches are connected by corridors.

Consumer A species that feeds on another species or organic matter—that is, herbivores, carnivores, parasites, detritivores, and others.

Convergent evolution The evolution of two or distinct species towards a similar adaptive solution under equivalent selective pressures, often resulting in similar morphologies or appearances.

Cultivar A distinct form of a cultivated species that has been produced as a result of artificial selection.

Cytochrome An important protein within the electron transport chain that is involved in the conversion of energy within chloroplasts and mitochondria.

Decomposer An organism that feeds on dead organic matter. **Detritivores** feed on **detritus**, fragmented organic matter.

Decomposer food chain A food chain made up of **decomposer** organisms that feed on dead organic matter.

Definitive host See **Parasite**.

Denitrification The conversion of nitrate to nitrite and nitrite to molecular nitrogen by a range of anaerobic bacteria.

Density-dependence Populations that grow in a density-dependent way increase their numbers in relation to the density of the existing population; usually, the rate of population increase falls as the habitat becomes more and more crowded. Other organisms have **density-independent** growth.

Diversity Species diversity simply refers to the number of different species in a habitat, or species richness (*qv.*). Other measurements of diversity include the relative proportions of different species, or equitability. Genetic diversity refers to the totality of the variation in the gene pool (*qv.*).

Detritus Dead and decaying organic matter.

Detritivore See **Decomposer**.

DNA Deoxyribose nucleic acid, the molecule that carries the genetic code.

DNA sequencing A technique used to determine the sequence of bases along a piece of DNA.

Ecology The science that studies the relationship between living things and their environment.

Ecological niche The totality of adaptations of a species to the biotic and abiotic factors in its environment; a species role in its community. **Niche breadth** measures the range of a resource exploited by a species. **Niche overlap** is the range of a resource used by two species occurring together. **Fundamental niche** is the niche breadth that a species would use (occupy) in the absence of competitors or predators; **realized niche** is the niche breadth a species is restricted to in the presence of competing species.

Ecoengineering The use of vegetation in the construction and stabilization of structures.

Ecotype A variety with sufficient genetic differences from other members of its species to be recognized as a separate race.

Ecosystem A community of living organisms and its physical environment.

Ectotherms Animals which, when active, attempt to regulate their body temperature by gaining heat from, or losing heat to, the environment, primarily by their behaviour (cf. endotherms).

Ecotone The boundary between two adjoining communities.

Effective population size The size of the breeding population at a particular time and location, taking into account the breeding behaviour of the species.

Emergent property A characteristic of a population or community that could not be detected by looking at individuals, which only emerges from seeing the assemblage operating together.

Endemic species Native to and only found in an area.

Endosymbiont theory The theory that suggests that a number of organelles within the eukaryotic cell have their origins in primitive prokaryotes that evolved a symbiotic relationship with the host cell.

Endotherms Animals that regulate their body temperature by generating heat internally through their metabolism (cf. ectotherms).

Energy subsidy The supplemental energy used in agriculture to aid cultivation (in the form of fuel, fertilizers and pesticides).

Epiphyte A plant that grows entirely on another plant. Typical epiphytes include the Bromeliads of tropical forests.

Eukaryotes Organisms whose cells have a discrete nucleus within a membrane. Eukaryotes include all algae, fungi, and all multicellular plants and animals.

Eutrophication Changes in the structure of an ecological community as a result of nutrient enrichment, most often phosphate and nitrogen enrichment.

Evolution A gradual, directional change in the characters of an organism. The process by which one species might arise from another (qv. natural selection).

Extinction vortex A descriptive title given to the three factors (genetic, population and habitat) that can act in combination to drive a population extinct.

Evolutionary Significant Unit (ESU) A population considered to be on a separate evolutionary trajectory than other populations of that species and which, if left, may well give rise to a new species. (See also **Cohesion species concept**.)

Extremophile An organism adapted to extreme conditions (frequently applied to microbes).

Facilitation An effect where the activity or presence of one species allows other species to colonize a habitat, as part of the process of **succession** (*qv.*).

Flux rate The rate of turnover of an element as it passes from inorganic reservoirs through living systems and return to the reservoirs.

Food web The trophic or feeding relations between different members of a community. A **food chain** consists of a series of trophic levels—primary producer, herbivore, carnivore.

Fossil record The entire catalogue of fossilized plants and animals that have been dated in geological time.

Founder effect The genetic differentiation of a small population isolated from a larger (parent) population. In a small gene pool, some alleles may have different frequencies compared to the main population and these differences become more pronounced as the isolated population interbreed.

Fugitive species Species that maintain a population by colonizing ephemeral gaps. Usually applied to plant species that persist in habitats with a predictable frequency of disturbance.

Functional redundancy The presence of a number of species within a community whose activity has no significance for its ecological processes. Such species can be lost without any major impact on the nature of the community.

Functional response A response of a predator to prey numbers in which it consumes more prey as they become available.

Gaia hypothesis The theory that the atmosphere of the Earth is maintained in its present form by the biological activity in the biosphere.

Gametes The sex cells, eggs (ova), or sperm, which are haploid, having a single set of chromosomes. The fertilization of an ovum by a sperm produces a diploid zygote with two sets of chromosomes.

GCM (Global Circulation Models) Computer models used to describe and predict the behaviour of the climate under the influence of the land, the oceans, and the chemistry of the atmosphere.

Gel electrophoresis A technique that separates proteins and other chemicals using an electrical current passing through agar gel. The molecules separate out according to their electrical properties and size.

Gene A sequence of nucleotides on the DNA molecule that are inherited as a unit and which codes for a specific RNA molecule or a polypeptide for which that RNA molecule is responsible.

Gene pool/Genome The totality of genetic material of a species.

Generalist species A species that is **r-selected** and has an opportunistic **life-history** *strategy* (*qv.*).

Generation The individuals belonging to one age class within a population.

Generic name See **Binomial system**.

Gene sequencing See **DNA sequencing**.

Genet An organism that grows from a fertilized egg and is therefore genetically distinct. In contrast a **ramet** is an individual that has arisen by asexual reproduction.

Genetic drift The genetic differentiation of a small, isolated population through random changes (and independent of any selective pressure).

Genotype The genetic code of an individual, or sometimes, rather confusingly, all individuals that share that code. (cf. phenotype).

Genotypic variation Variation between individuals that is wholly attributable to differences in their genetic code. Genetic diversity refers to the totality of the variation in the gene pool (*qv.*).

Genus A category between species and family in which a number of closely related species are grouped.

Geomorphology The study of landforms. **Topography** is the study of surface features of an area.

Grazing food chain A food chain involving plant material as the primary producer and herbivores as the primary consumers.

Green revolution A term used to describe high-tech agricultural innovation in plant breeding and cultivation.

Greenhouse effect The capacity of the Earth's atmosphere to absorb long-wave (infrared) radiation, while being largely transparent to the incoming short-wave radiation.

Gross primary production (GPP) The total amount of energy fixed by plants within a given area.

Growth efficiency A measure of efficiency by which an organism converts the energy it has assimilated into its tissues (as given by the amount of energy fixed in tissues divided by the amount of energy assimilated).

Guild A group of organisms sharing a similar set of adaptations and carrying out a similar role or exploiting the same resource, within a habitat.

Habitat The place where an organism lives; sometimes used in the general sense to refer to the type of place in which it lives.

Herbivore An animal that feeds on plants.

Heterotroph An organism that obtains its energy from other species; consumers.

Heterozygote An individual that, for a particular character, has different genes on each of its paired chromosomes. A **homozygote** individual has the same alleles (*qv.*) on each chromosome.

Hierarchy theory The proposal that large and complex systems are self-organizing by virtue of the constraints and limitations imposed by the interactions between different levels. A level is defined by the spatial and temporal scale over which key processes operate. Levels are nested, or formed into a hierarchy, in which a process occurs fastest at the smallest scales; change is slowest at the largest scales.

Holoparasite A plant parasite that is totally dependent on its host and cannot photosynthesize (see also **Parasite**).

Homoplasy A structural resemblance arising out of the convergent evolution of two different species or groups.

Homozygous With the same allele on each chromosome (See **heterozygote**).

Hunter-gatherer A non-settled form of agricultural activity in which people harvest animals and plants from the wild.

Hybrid A cross-bred individual derived from gametes from different populations, species, or genera.

Hybrid breakdown A series of mechanisms that prevent hybrids from breeding.

Hydrological cycle The movement of water from oceans onto land and back again.

Immunological methods Techniques that use antibodies to detect the similarity or difference between proteins and that operate on the basis on the specificity of the reaction between antibodies and antigen.

Inbreeding depression The reduced reproductive potential of a population in which individuals share much of the same genetic code. Conversely, **outbreeding depression** occurs when the genetic differences between individuals are too large to produce viable offspring.

Indicator species Species that are characteristic of certain environments (soil, climates etc. or can be used to measure habitat quality. *Sentinel* species are used as measures of pollution levels in a habitat).

Infield–outfield A form of agricultural activity where grazing and cultivation takes place in enclosed areas in around a settlement whilst the surrounding areas are used for grazing.

Inhibition The capacity of one species to inhibit the colonization of another, especially in a successional sequence.

***In situ* conservation** Conservation effort to protect an endangered species in its natural habitat. **Ex situ conservation** refers to captive breeding or culturing of an endangered species in a botanic or zoological garden.

Integrated pest management A method of pest control that includes the use of biological, chemical and cultural techniques.

Intermediate disturbance hypothesis The suggestion that maximum diversity occurs in communities subject to disturbance which is frequent enough to prevent competitive species becoming dominant but which is not so frequent to depress colonization rates by immigrant species.

Interspecific competition See **competition**.

Intraspecific competition See **competition**

Island-biogeography theory A model predicting a dynamic equilibrium of species number in a habitat, as a result of the opposing processes of colonization and extinction. This is allied to the **species–area relationship** (*qv.*)

Joule The SI unit of energy (equivalent to 0.2389 calories).

K The carrying capacity (or maximum sustainable size) for a population in a limited environment.

K-selection Applied to an organism whose reproductive and life history strategies are primarily adapted to life in an unchanging and limited environment where there is intense competition for resources (cf. r-selection).

Keystone species An organism whose abundance or activity is central to maintaining the nature of a habitat.

Kinetic energy Energy of a body due to its motion.

Kingdom Five broad taxonomic categories that are divided up between the Superkingdoms—**Prokaryota** and **Prokaryota**.

Landscape ecology The study of ecological processes across several ecosystems united by a shared landform, climate and regime of disturbance. Adjacent landscapes under an equivalent climate or topographical area may be collected together as a region.

Leaching The loss of nutrients when they are washed out of the soil.

Life history strategy The allocation of time and resources that an organism makes between different stages of its life cycle so as to maximize its reproductive potential.

Life table A summary of the rates of mortality and survivorship of different age groups in a population.

Locus The site of an individual gene on a chromosome.

Macrophyte An aquatic plant that is either rooted or attached to a submerged substrate (e.g. sand, mud, or rock) and with a stem that holds the leaves close to or above the water surface.

Maximum sustainable yield (MSY) The largest number of individuals or mass that can be harvested from a population year on year without damaging its reproductive potential.

Meiosis The division of the paired chromosomes necessary to form a gamete. In contrast, *mitosis* divides the nucleus but not the chromosomes, as a preclude to cell division.

Metapopulation A population of populations. A series of populations that may swap individuals with each other, but are usually divided into discrete patches.

Microcosm An ecosystem or part of an ecosystem isolated in the laboratory for experimental purposes, for example, an aquarium. A **mesocosm** is part of a real ecosystem partitioned for the same reason, sometimes called enclosures.

Mitochondria Organelles, the site of cellular respiration in eukaryotes.

Mitochondrial DNA Non-nuclear DNA that is contained within the **mitochondria** essential for maintaining their structure, function and replication.

Monera The taxonomic Kingdom that includes bacteria and cyanobacteria.

Mortality rate (m) The death rate.

Mullerian mimicry Warning signals (patterns and coloration) adopted by potential prey that signal danger to predators (*e.g.* the yellow and black stripes of the wasp).

Mutualism A form of symbiotic relationship between two species to their mutual benefit.

Mycorrhiza A symbiotic (*qv.*) association between plant and fungus, where the plant root acts as a host to a network of fungal threads. The plant benefits by having an enlarged 'root' system to collect more phosphates and water and the fungus benefits from the supply of carbon from the plant.

Natality rate (b) The birth rate.

Natural selection The reproductive success of different individuals under the constraints placed upon them by their environment. Less fit individuals fail to reproduce and their genes are lost from the gene pool (*qv.*). The persistence of particular adaptive traits through the generations produces the evolutionary change that may produce new species.

Net primary production (NPP) The amount of energy fixed within plant tissues after metabolic and photosynthetic energy costs have been met.

Neutral variation Variation within a population or species that confers no selective advantage.

Niche See **ecological niche**.

Niche breadth See **ecological niche**.

Niche overlap See **ecological niche**.

Nitrification A process carried out by a range of soil microbes by which nitrite is oxidized to nitrate.

Nomadic-pastoralist A semi-settled form of agricultural involving the management of grazing herds by setting up camp and then moving along with them (*e.g.* the Laplanders who herd reindeer).

Nucleus In eukaryotic cells, the organelle in which chromosomes are held.

Numerical response To the increase in numbers of predator as a response to the increased abundance of its prey.

Oligotrophic Nutrient-poor; oligotrophic ecosystems have low primary productivity (*qv.*). **Eutrophic** ecosystems are nutrient rich and have high primary productivity.

Ombrotrophs Plants that obtain their nutrients from precipitation.

Omnivore An animal that feeds on plants and animals.

Parapatric speciation See *speciation*

Parasite An organism that is metabolically dependent on another, at the expense of the host. A **definitive host** is one in which the parasite matures (and may reproduce sexually); an intermediate host serves as a **vector** in transmitting immature stages between definitive hosts.

Parasitoid Insect parasites (primarily Hymenoptera and Diptera) that lay an egg within a host insect and which eventually leads to its death. Often seen as a specialised form of parasitism and predation.

Pathogen An organism or virus that can cause disease.

Perennial A plant that lives for more than 2 years.

Permafrost The soil layer found at depth beneath tundra vegetation that remains frozen throughout the year.

Pest A species whose presence causes a nuisance and results in some economic cost.

Pheromone A specially produced chemical used by one individual to alter the behaviour of another, usually of the same species, perhaps to attract a mate. These are especially important among social insects for communication and control.

Phenotype The characteristics of an individual, the product of the expression of its genetic code and its interaction with the organism's environment.

Phenotypic variation Variation between individuals that is wholly attributable to physiological or developmental (i.e. non-genetic) differences.

Photoautotroph An organism that obtains its energy from sunlight through the process of photosynthesis.

Phylogeny The evolutionary relationships and history of an organism.

Physiognomy The characteristic features of a plant community, reflecting their adaptation to life in that environment.

Plagioclimax A succession that has been deflected from its normal climax state (*qv.*), often a result of human activity, and with a different plant community as a result.

Polymorphism Variation within a species where individuals may take different forms.

Polyploidy Organisms that possess multiple copies of the entire genome.

Population Individuals of the same species living in a defined area at a defined time.

Predator An animal that kills another animal to feed.

Prezygotic barriers A range of mechanisms that prevent the formation of a hybrid zygote from gametes from two different species or varieties.

Primary producer An autotroph (*qv.*). A secondary producer derives its energy from a primary producer.

Primary production The synthesis of complex organic molecules from simple inorganic ones by **autotrophs**, using either sunlight (**photoautotrophs**) or chemical energy (**chemotrophs**).

Prokaryotes Cells without a well-defined nucleus; the bacteria and the blue-green algae.

Protista The taxonomic Kingdom that includes protozoa, primitive algae and fungi.

Production efficiency A measure of efficiency by which an organism converts the energy from its food into tissues (the ratio of the amount of energy fixed in its tissues to the amount consumed).

Productivity The energy fixed in the biomass of a individual, population, species, or trophic level.

Postzygotic barrier Some biological failure in the subsequent development of the fertilized egg, the zygote, which makes it non-viable, typically in hybrids.

Potential energy Energy that is stored (e.g. stored chemically as starch and oils in plants, or fat and glycogen in animals).

$r(r_m)$ The rate of change in a population per individual. r_m is the maximum possible rate of population growth per individual under ideal conditions.

r-selection Organisms that are *r*-selected are primarily adapted to life in changeable, short-lived habitats, where rapid population growth is favoured (cf. **K-selection**).

Recombination The swapping of fragments of genetic code between chromosomes during meiosis and which may therefore produce a new genotype (*qv.*). Such variation is essential for the evolutionary process.

Recruitment The addition of new individuals to the catchable stock (*qv.*) from growth or immigration.

Reproductive rate (R_0) The average number of offspring per individual in a given time. The **net reproductive rate** (R_N) is the average number alive per individual in the next generation, allowing for births, deaths and survivors.

Resource spectrum The range of a resource that is available to organisms in a habitat. Different species may use different parts of the spectrum—say food items of different sizes.

Realised niche See **ecological niche**.

Reclamation The process of restoring a degraded area to a state which is more ecologically sustainable and aesthetically pleasing.

Rehabilitation The restoration of an ecosystem part way towards an intended endpoint, which allows the site to continue developing naturally through succession.

Replacement An alternative endpoint to a restoration programme, selected either because the original ecosystem is impossible to recreate or alternatively had little ecological value.

Ribosome Site of protein synthesis in the cell.

RNA Ribonucleic acid. A polynucloetide molecule that exists in several forms in the cell and carries out various functions, especially in protein synthesis.

Ruderal Plant species characteristic of disturbed and temporary habitats.

Rules of assembly The suggestion that there are certain ways in which a community can be configured, particularly regarding the structure of food webs. If there are community rules of assembly, we expect to find similar community configurations under equivalent abiotic conditions.

Saprotrophs Organisms that derive their nutrients from dead and decaying organisms.

Sclerophyllous Used to describe individual plants or a plant community with small, tough and leathery leaves, adapted for drought. Typical of Mediterranean-type communities.

Secondary consumer An organism that is dependent on primary consumers (herbivores) for its energy needs.

Secondary plant metabolites Biologically active compounds produced by plants that also have a defensive role against herbivores.

Sexual selection The selection of individuals according to particular traits by a potential mate. This often includes courtship behaviour and is commonly taken to be a means of selecting 'fit' partners.

Species A collection of individuals able to breed with each other and produce viable (fertile) offspring. Note this is a functional definition: the simple concept of a species as the basic unit of biological classification is now regarded as untenable, and is used largely as a convenient unit in taxonomy.

Speciation The formation of new species as a result of evolution by natural selection. Speciation can be **allopatric** in which species become genetically distinct because they are geographically isolated, or **sympatric** when genetic isolation occurs between populations with overlapping distributions. *Parapatric speciation* is where adjacent populations diverge, say along some environmental gradient or by exploiting different resources. Species with short generation times and under strong selective pressure to specialise on a particular part of a resource spectrum show parapatric speciation.

Specialist species A *K*-selected species is closely adapted to its habitat and a particular part of a resource spectrum. Invariably highly competitive species.

Species–area relationship The common pattern of declining species number as area or sample size increases.

Specific name See **Binomial system**.

Species richness, species diversity The number of species in a habitat. **Species equitability** is the proportion of individuals in each species, sometimes used together to measure **diversity** (*qv.*).

Stability The capacity of a system to resist change (inertial stability) or to return to its original position or original rate of change following a disturbance (elasticity or adjustment stability).

Sustainability In the simplest terms, maintaining the capacity of an ecological system over the long term to carry out a particular function or to provide a given level of productivity.

Succession A directional sequence of changes in the species composition of a community. A primary succession is where there has been no previous life on a site; a secondary succession re-establishes life on a site that may have some nutrients or organic matter from its previous occupants.

Symbiosis The association of two species living together. A variety of associations are possible. **Mutualism** (*cv.*) describes where each species gains an advantage from the association.

Systematics A method of classifying organisms according to their evolutionary relationships.

Stress tolerators A category of plants that have a **life history strategy** adapted to life in hostile conditions (e.g. cacti).

Superkingdom The broadest taxonomic groups. Two Superkingdoms encompass the five Kingdoms.

Sympatric speciation See **Speciation**.

Taxonomy The description, naming, and classification of organisms.

Tolerance (i) the ability of a species to survive adverse or toxic conditions; (ii) in a succession, the capacity of many late successional plants to persist in poor growing conditions, typically low nutrient levels.

Transcription The creation of a molecule of messenger RNA (*qv.*) using part of a DNA molecule as a template. See also **Translation**.

Transhumance A form of agriculture involving the movement of grazing herds between summer and winter pastures.

Translation The conversion of the code of the messenger RNA (*qv.*) to create a chain of amino acids.

Translocation The transfer of a habitat from a donor to a receptor site. This is an extreme form of restoration that is undertaken to rescue a site whose destruction might otherwise be irrevocable.

Transpiration The evaporation of water vapour from the aerial parts of a plant.

Trophic level Position along a food chain, from primary producer through a sequence of consumers.

Variety A distinct form within a species that occurs naturally.

Vector An organism that carries the parasite from one host to another (e.g. malarial mosquitoes of the genus *Anopheles*).

Xerophyte A plant adapted to live in an arid environment.

REFERENCES

First words

Bhatt, R. V. 2000. Environmental influence on reproductive health. *International Journal of Gynecology and Obstetrics* 70, 69–75.

Carlsen, E., Giwercman, A., Keiding, N., and Skakkebæk, N. E. 1992. Evidence for decreasing quality of semen during past 50 years. *British Medical Journal* 305, 609–613.

Cobb, G. P., Houlis, P. D., and Bargar, T. A. 2002. Polychlorinated biphenyl occurrence in American alligators (*Alligator mississippiensis*) from Louisiana and South Carolina. *Environmental Pollution* 118, 1–4.

Kucklick, J. R., Struntz, W. D. J., Becker, P. R., York, G. W., O'Hara, T. M., and Bohonowych, J. E. 2002. Persistent organochlorine pollutants in ringed seals and polar bears collected from northern Alaska. *Science of the Total Environment* 287, 45–59.

Pajarinen, J., Penttila, A., Laippala, P., and Karhunen, P. J. 1997. Incidence of disorders of spermatogenesis in middle aged Finnish men 1981–91: two necropsy series. *British Medical Journal* 314, 13–18.

Polischuk, S. C., Norstrom, R. J., and Ramsay, M. A. 2002. Body burdens and tissue concentrations of organochlorines in polar bears (*Ursus maritimus*) vary during seasonal fasts. *Environmental Pollution* 118, 29–39.

Suominen, J. and Vierula, M. 1993. Semen quality of Finnish men. *British Medical Journal* 306, 1579.

Wiig, O., Derocher, A. E., Cronin, M. M., and Skaare, J. U. 1998. Female pseudohermaphrodite polar bears at Svalbard. *Journal of Wildlife Diseases* 34, 792–796.

1 Origins

Section 1.1

Darwin, C. 1872. *The Origin of Species*, 6th edn. John Murray, London.

Isaac, G. 1978. The food-sharing behaviour of protohuman hominids. *Scientific American 238*, 90–108.

Leakey, R. 1995. *The Origin of Humankind*. Weidenfeld & Nicholson, London.

Leakey, R. and Lewin, R. 1992. *Origins Reconsidered*. Abacus, London.

Lewin, R. 1984. *Human Evolution*. Blackwell, Oxford.

Parker, S. 1992. *The Dawn of Man*. Crescent, New York.

Pough, F. H., Heiser, J. B., and Mc Farland, W. N. 1996. *Vertebrate Life*, 4th edn. Prentice Hall, New Jersey.

Trinkaus, E. and Howells, W. W. 1979. The Neanderthals. *Scientific American 241*, 118–133.

Walker, A. and Leakey, R. E. F. 1978. The hominids of east Turkana. *Scientific American 239*, 54–66.

Wood, B. 1992. Origin and evolution of the genus *Homo*. *Nature 355*, 783–790.

Sections 1.2–1.5

Campbell, N. A. and Reece, J. B. 2002. *Biology*, 6th edn. Benjamin/Cummings, Redwood, CA.

Dawkins, R. 1982. *The Extended Phenotype*. Oxford University Press, Oxford.

Dawkins, R. 1995. *River Out of Eden*. Weidenfeld & Nicholson, London.

Dennett, D. C. 1996. *Kinds of Minds*. Weidenfeld & Nicholson. London.

Lincoln, R. J., Boxshall, G. A., and Clark, P. F. 1982. *A Dictionary of Ecology, Evolution and Systematics*. Cambridge University Press, Cambridge.

Mace, R. 1995. Why do we do what we do? *Trends in Ecology and Evolution 10*, 4–5.

Purves, W. K., Sadava, D., Orians, G. H., and Heller, H. C. 2001. *Life: The Science of Biology*, 6th edn. Sinauer, Sunderland, MA.

2 Species

Section 2.1

Camp, W. H. 1951. Biosystematy. *Brittonia 7*, 113–127.

Ford, E. B. 1940. Polymorphism and taxonomy. In J. Huxley (ed.), *The New Systematics*. Clarendon Press, Oxford.

Jeffrey, C. 1989. *Biological Nomenclature*, 3rd edn. Edward Arnold, London.

Linnaeus, C. 1753. *Species Plantarum*. Laur Salvius, Holmiae.

Linnaeus, C. 1758. *Systema Naturae*. Laur Salvius, Holmiae.

Mayer, E. 1942. *Systematics and the Origin of Species from the Viewpoint of a Zoologist*. Columbia University Press, New York.

Minnelli, A. 1993. *Biological Systematics: The State of the Art*. Chapman & Hall, London.

Stace, C. A. 1989. *Plant Taxonomy and Biosystematics*, 2nd edn. Edward Arnold, London.

Wagner, W. H. 1970. Biosystematics and evolutionary noise. *Taxon 19*, 146–151.

Section 2.2

Margulis, L. 1992. Biodiversity: molecular biological domains, symbiosis and Kingdom origins. *BioSystems 27*, 39–51.

Margulis, L. and Schwartz, K. V. 1982. *Five Kingdoms*. W.H. Freeman, San Francisco, CA.

Whittaker, R. H. and Margulis, L. 1978. Protist classification and the kingdoms of organisms. *BioSystematics 10*, 3–18.

Section 2.3

Altschul, S. F. and Lipman, D. J. 1990. Equal animals. *Nature 348*, 493–494.

Carr, S. M., Ballinger, S. W., Derr, J. N., Blankenship, L. H., and Bickham, J. W. 1986. Mitochondrial DNA analysis of hybridisation between sympatric whitetailed and mule deer in West Texas. *Proceedings of the National Academy of Science of the United States of America 83*, 9576–9580.

Crandell, K. E., Bininda-Emonds, O. R. P., Mace, G. M., and Wayne, R. K. 2000. Considering evolutionary processes in conservation biology. *Trends in Ecology and Evolution 15*, 290–295.

Dowling, T. E., Moritz, C., and Palmer, J. D. 1990. Nucleic acids II. Restriction site analysis. In D. M. Hillis and C. Moritz (eds), *Molecular Systematics*. Sinauer Associates, Sunderland, MA.

Grant, V. 1985. *Plant Speciation*, 2nd edn. Columbia University Press, New York.

Gregg, K. B. 1983. Variation in floral fragrances and morphology: incipient speciation in *Cycnoches? Botanical Gazette 144*, 566–576.

Hewitt, G. M. 1988. Hybrid zones—natural laboratories for evolutionary studies. *Trends in Ecology and Evolution 3*, 138–147.

Hutchinson, G. E. 1957. Concluding remarks. *Cold Spring Harbour Symposium on Quantitative Biology 22*, 415–427.

Hutchinson, G. E. 1959. Homage to Santa Rosalia, or why there are so many kinds of animals? *American Naturalist 93*, 145–159.

Jansen, R. K. and Palmer, J. D. 1988. Phylogenetic implications of chloroplast DNA restriction site variation in the Mutisieae (Asteraceae). *American Journal of Botany 75*, 753–766.

Janzen, D. H. 1977. What are dandelions and aphids? *American Naturalist 111*, 586–589.

Janzen, D. H. 1979. Reply. *American Naturalist 114*, 156–157.

Janzen, D. H. 1979. How to be a fig. *Annual Review of Ecology and Systematics 10*, 13–51.

Knobloch, I. W. 1972. Intergeneric hybridization in flowering plants. *Taxon 21*, 97–103.

Lewis, K. B. and John, B. 1968. The chromosomal basis of sex determination. *International Review of Cytology 23*, 277–379.

Mallet, J. 1995. A species definition for the modern synthesis. *Trends in Ecology and Evolution 10*, 294–299.

Moritz, C. 1994. Defining 'evolutionary significant units' for conservation. *Trends in Ecology and Evolution 9*, 373–375.

Moritz, C., Dowling, T. E., and Brown, W. M. 1987. Evolution of animal mitochondrial DNA: relevance for population biology and systematics. *Annual Review of Ecology and Systematics 18*, 269–292.

Moyle, P. B. and Davis, L. H. 2000. A list of freshwater, anadromous, and euryhaline fishes of *California. California Fish & Game 86*, 244–258.

Nylander, W. 1866. Circa novum in studio lichenum criterium chimicum. *Flora 49*, 98.

Poulton, E. B. 1904. Presidential address. *Proceedings of the Entomological Society of London 73*–116.

Sarich, V. M., Schmid, C. W., and Marks, J. 1989. DNA hybridisation as a guide to phylogenies. A critical analysis. *Cladistics 5*, 3–32.

Sterck, A. A., Groenhart, M. C., and Mooren, J. F. A. 1983. Aspects of the ecology of some microspecies of *Taraxacum* in the Netherlands. *Acta Botanica Neerlandica 32*, 385–415.

Ryder, O. A. 1986. Species conservation and systematics: the dilemma of subspecies. *Trends in Ecology and Evolution 1*, 9–10.

Wagner, D. F., Furnier, G. R., Saghai-Maroof, M. A., Williams, S. M., and Allard, R. W. 1987. Chloroplast DNA polymorphism in lodgepole and jack pines and their hybrids. *Proceedings of the National Academy of Science of the United States of America 84*, 2097–2100.

Walters, S. M. 1986. The name of the rose: a review of ideas on the European bias in angiosperm classification. *New Phytologist 104*, 527–540.

Wayne, R. K. and Jenks, S. M. 1991. Mitochondrial DNA analysis implying extensive hybridzation of the endangered red wolf *Canis rufus. Science 351*, 565–568.

Young, J. P. W. 1988. The construction of protein and nucleic acid homologies. In D. L. Hawksworth (ed.), *Prospects in Systematics*. Clarendon Press, Oxford.

Zuckerkandl, E. 1987. On the molecular evolutionary clock. *Journal of Molecular Evolution 26*, 34–46.

Section 2.4

Begon, M., Harper, J. L., and Townsend, C. R. 1990. *Ecology: Individuals, Populations, and Communities*, 2nd edn. Blackwell, Oxford.

Inhorn, M. C. and Brown, P. J. 1990. The anthropology of disease. *Annual Review of Anthropology 19*, 89–117.

Thornton, R. 1997. Aboriginal North American population and rates of decline, ca AD 1500–1900. *Current Anthropology 38*, 310–315.

Wirsing, R. L. 1985. The health of traditional societies and the effects of acculturation. *Current Anthropology 26*, 303–322.

Section 2.5

Grubb, P. 1977. The maintenance of species richness in plant communities: the importance of the regeneration niche. *Biological Reviews 52*, 107–145.

Pianka, E. R. 1976. Competition and niche theory. In R. M. May (ed.), *Theoretical Ecology Principles and Applications*. Blackwell, *Oxford*.

Pianka, E. R. 1988. *Evolutionary Ecology*, 4th edn. Harper and Row, New York.

Schoener, T. W. 1989. The ecological niche. In R. M. May (ed.), *Theoretical Ecology Principles and Applications*. Blackwell, Oxford.

Vandermeer, J. H. 1972. Niche theory. *Annual Review of Ecology and Systematics 3*, 107–132.

Section 2.6

Antonivics, J., Bradshaw, A. D., and Turner, R. G. 1971. Heavy metal nutrient tolerance in plants. *Advances in Ecological Research 7*, 1–85.

Ammerman, A. and Cavalli-Sforza, L. L. 1971. Measuring the rate of spread of early farming in Europe. *Man 6*, 674–688.

Baker, A. J. M. 1987. Metal tolerance. *New Phytologist 106* (Suppl.), 93–111.

Beringer, J. E. 2000. Releasing genetically modified organisms. *Journal of Applied Ecology 37*, 207–214.

Chapman, G. P. (ed.). 1992. *Grass Evolution and its Domestication*. Cambridge University Press, Cambridge.

Ehrendorfer, F. 1970. Evolutionary patterns and strategies in seed plants. *Taxon 19*, 185–195.

Harlan, J. R. 1992. Origins and processes of domestication. In G. P. Chapman (ed.), *Grass Evolution and its Domestication*. Cambridge University Press, Cambridge.

Hellmich, R. L., Siegfried, B. D., Sears, M. K., Stanley-Horn, D. E., Daniels, M. J., Mattila, H. R., Spencer, T., Bidne, K. G., and Lewis, L. C. 2001. Monarch larvae sensitivity to *Bacillus thuringiensis*—purified proteins and pollen. *Proceedings of the National Academy of Sciences of the United States of America 98*, 11925–11930.

Hunt, P. F. 1986. The nomenclature and registration of orchid hybrids at specific and generic levels. In B. T. Sykes (ed.), *Infraspecific Classification of Wild and Cultivated Plants*. The systematics association special volume, Clarendon Press, Oxford.

Letourneau, D. K. and Burrows, B. E. 2002. *Genetically Engineered Organisms: Assessing Environmental and Human Health Effects*. CRC Press, Boca Raton, FL.

Hull, D. L. 1976. Are species really individuals? *Systematics and Zoology 25*, 174–191.

Lai, J. and Messing, J. 2002. Increasing maize seed methionine by mRNA stability. *The Plant Journal 30*, 395–403.

Larson, A. 1989. The relationship between speciation and morphological evolution. In D. Otte and J. A. Endler (eds), *Speciation and its Consequences*. Sinauer Associates, Sunderland, MA.

Losey, J. E., Rayor, L. S., and Carter, M. E. 1999. Transgenic pollen harms monarch larvae. *Nature 399*, 214.

Lupton, F. G. H. 1987. *Wheat Breeding: Its Scientific Basis*. Chapman & Hall, London.

Macnair, M. R., Cumbes, Q. 1987. Evidence that arsenic tolerance in *Holcus lanatus* L. is caused by an altered phosphate uptake system. *The New Phytologist 107*, 387–394.

Majerus, M., Amos, W., and Hurst, G. 1996. *Evolution: The Four Billion Year War*. Longman, Harlow.

Richards, A. J. 1986. *Plant Breeding Systems*. Allen and Unwin, London.

Salamini, F., Ozkan, H., Brandolini, A., Schafer-Pregl, R., and Martin, W. 2002. Genetics and geography of wild cereal domestication in the near east. *Nature Reviews Genetics 3*, 429–441.

Schilewen, U. K., Tautz, D., and Paabo, S. 1994. Sympatric speciation by monophyly of crater lake cichlids. *Science 368*, 629–632.

Schluter, D. 1994. Experimental evidence that competition promotes divergence in adaptive radiation. *Science 266*, 798–801.

Scriber, J. M. 2001. *Bt* or not *Bt*: Is that the question? *Proceedings of the National Academy of Sciences of the United States of America 98*, 12328–12330.

Sears, M. K., Hellmich, R. L., Stanley-Horn, D. E., Oberhauser, K. S., Pleasants, J. M., Mattila, H. R., Seigfried, B. D., and Dively, G. P. 2001. Impact of *Bt* corn pollen on monarch butterfly populations: a risk assessment. *Proceedings of the National Academy of Sciences of the United States of America 98*, 11937–11942.

Traverse, A. 1988. Plant evolution dances to a different beat. Plant and animal evolutionary mechanisms compared. *Historical Biology 1*, 277–301.

White, M. J. D. 1978. *Modes of Speciation*. Freeman, San Francisco, CA.

Zangeri, A. R., McKenna, D., Wraight, C. L., Carroll, M., Ficarello, P., Warner, R., Berenbaum, M. R. 2001. Effects of exposure to event 176 *Bacillus thuringiensis* corn pollen on monarch and black swallowtail caterpillars under field conditions. *Proceedings of the National Academy of Sciences of the United States of America 98*, 11908–11912.

3 Populations

Section 3.3

Cushing, D. H. 1981. *Fisheries Biology*. University of Wisconsin Press, Madison, WI.

Section 3.4

Holmes, R. 1994. Biologists sort the lessons of fisheries collapse. *Science 264*, 1252–1253.

Jedrzejewska, B., Okarma, H., Jedrzejewska, W., and Milkowski, L. 1994. Effects of exploitation and protection on forest structure, ungulate density and wolf predation in Bialowieza Primeval Forest, Poland. *Journal of Applied Ecology 31*, 664–676.

Section 3.5

Lampert, K. P. and Linsenmair, K. E. 2002. Alternative life cycle strategies in the West African reed frog *Hyperolius nitidulus*: the answer to an unpredictable environment? *Oecologia 130*, 364–372.

Section 3.6

Mauritzen, M., Derocher, A. E., Wiig, O., Belikov, S. E., Boltunov, A. N., Hansen, E., and Garner, G. W. 2002. Using satellite telemetry to define spatial population structure in polar bears in the Norwegian and western Russian Arctic. *Journal of Applied Ecology 39*, 79–90.

Section 3.7

Ashley, M. V., Melnick, D. J., and Western, D. 1990. Conservation genetics of the Black Rhinocerus (*Diceros bicornis*). 1. Evidence from the mitochondrial DNA of three populations. *Conservation Biology 4*, 71–77.

Caughley, G., Dublin, H. T., and Parker, I. 1990. Projected decline of the African elephant. *Biological Conservation 54*, 157–164.

Cohn, J. P. 1990. Elephants: remarkable and endangered. *BioScience 40*, 10–14.

Dinerstein, E. and McCracken, G. F. 1990. Endangered greater one-horned rhinoceros carry high levels of genetic variation. *Conservation Biology 4*, 417–422.

Douglas-Hamilton, I. 1987. African elephants: population trends and their causes. *Oryx 21*, 11–24.

Dublin, H. T., Sinclair, A. R. E., and McGlade, J. 1990. Elephants and fire as causes of multiple stable states in the Serengeti-Mara woodlands. *Journal of Animal Ecology 59*, 147–164.

Haynes, G. 1991. *Mammoths, Mastodonts and Elephants: Biology, Behaviour and the Fossil Record*. Cambridge University Press, Cambridge.

Hoare, R. 2000. African elephants and humans in conflict: the outlook for coexistence. *Oryx 34*, 34–38.

IUCN Status list; CITES Appendix 1, 2001. available at **www.worldwildlife.org/spec**

Lombard, A. T., Johnson, C. F., Cowling, R. M., and Pressey, R. L. 2001. Protecting plants from elephants: botanical reserve scenarios with Addo National Park, South Africa. *Biological Conservation 102*, 191–203.

4 Interactions

Bronowski, J. 1973. *The Ascent of Man*. BBC Publications, London.

Section 4.1

Krebs, C. J. 2001. *Ecology: Experimental Analysis of Distribution and Abundance*, 5th edn. Benjamin Cummings, London.

Section 4.2

Addicott, J. F. 1986. Variation in the costs and benefits of mutualism: the interaction between yuccas and yucca moths. *Oecologia 70*, 486–494.

Beattie, A. J., Turnbull, C., Knox, R. B., and Williams, E. G. 1984. Ant inhibition of pollen function: a possible reason why ant pollination is rare. *American Journal of Botany 71*, 421–426.

Bond, W. and Slingsby, P. 1984. Collapse of an ant–plant mutualism: the Argentinian ant (*Iridomyrmex humulis*) and myrmecochorous proteaceae. *Ecology 65*, 1031–1037.

Giraud, T., Pedersen, J. S., and Keller, L. 2002. Evolution of supercolonies: The Argentine ants of southern Europe. *Proceedings of the National Academy of Sciences of the United States of America 99*, 6075–6079.

Goguen, C. B. and Mathews, N. E. 2001. Brown-headed cowbird behaviour and movements in relation to livestock grazing. *Ecological Applications 11*, 1533–1544.

Janzen, D. H. 1966. Coevolution of mutualism between ants and acacias in Central America. *Evolution 20*, 249–275.

Munro, J. 1967. The exploitation and conservation of resources by populations of insects. *Journal of Animal Ecology 36*, 531–547.

Wigglesworth, V. B. 1964. *The Life of Insects*. Weidenfeld & Nicholson, London.

Section 4.3

Brown, V. K. and Southwood, T. R. E. 1987. Secondary succession: patterns and strategies. In A. J. Gray, M. J. Crawley, and R. J. Edwards (eds), *Colonisation, Succession and Stability*. Blackwell, Oxford.

Case, T. J., Bolger, D. T., and Richman, A. D. 1992. Reptilian extinctions: the last ten thousand years. In P. L. Fiedler and S. K. Jain (eds), *Conservation Biology*. Chapman & Hall, New York.

Crombie, A. C. 1946. Further experiments on insect competition. *Proceedings of the Royal Society of London, Series B 133*, 76–109.

Grime, J. P. 1979. *Plant Strategies and Vegetation Processes*, John Wiley, Chichester.

Grime, J. P., Hodgson, J., and Hunt, R. 1988. *Comparative Plant Ecology: A Functional Approach to Common British Species*. Unwin Hyman, London.

Law, R. and Watkinson, A. R. 1989. Competition. In J. M. Cherrett (ed.), *Ecological Concepts*. Blackwell, Oxford.

Muller, C. H. 1966. The role of chemical inhibition (allelopathy) in vegetational composition. *Bulletin of the Torrey Botanical Club 93*, 332–351.

Southwood, T. R. E. 1977. Habitat, the templet for ecological strategies. *Journal of Animal Ecology 46*, 337–365.

Southwood, T. R. E. 1988. Tactics, strategies and templates. *Oikos 52*, 2–18.

Tilman, D., Mattson, M., and Langer, S. 1981. Competition and nutrient conditions along a temperature gradient: and experimental test of a mechanistic approach to niche theory. *Limnology and Oceanography 26*, 1020–1033.

Section 4.4

Beeby, A. and Richmond, L. 2001. Intraspecific competition in populations of *Helix aspersa* with different histories of exposure to lead. *Environmental Pollution 114*, 337–344.

Beeby, A. and Richmond, L. 2002. Lead reduces shell mass in juvenile garden snails (*Helix aspersa*). *Environmental Pollution 120*, 283–288.

Belsky, A. J. 1986. Does herbivory benefit plants? A review of the evidence. *American Naturalist 127*, 870–892.

Blum, M. S., River, L., and Plowman, T. 1981. Fate of cocaine in the lymantriid *Elonia noyesii* a predator of *Erythroxylum coca*. *Phytochemistry 20*, 2499–2500.

Cox, F. E. G. (ed.). 1982. *Modern Parasitism*. Blackwell, Oxford.

Dawkins, R. and Krebs, J. R. 1979. Arms race between and within species. *Proceedings of the Royal Society of London, Series B 205*, 489–511.

De Gregorio, E., Lemaitre, B. 2002. The mosquito genome: the post-genomic era opens. *Nature 419*, 496–497.

Eisener, T. and Meinwald, J. 1966. Defensive secretions of arthropods. *Science 153*, 1341–1350.

Eisener, T., Alsop, D., Hicks, K., and Meinwald, J. 1978. Defensive secretions of millipedes. In S. Bettini (ed.), *Arthropod Venoms, Handbook of Experimental Pharmacology, 48*. Springer, Berlin.

Feeny, P. P. 1970. Seasonal changes in oak leaf tannins and nutrients as a cause of spring feeding by winter moth caterpillars. *Ecology 51*, 565–581.

Fitzgerald, T. D., Jeffers, P. M., Mantella, D. 2002. Depletion of host derived cyanide in the gut of the eastern tent caterpillar, *Malacosoma americanum*. *Journal of Chemical Ecology 28*, 257–268.

Gardener, M. J., Hall, N., Fung, E., White, O., Berriman, M., Hyman, R. W., Carlton, J. M., Pain, A., Nelson, K., Bowman, S., Paulsen, I. T., James, K., Eisen, J. A., Rutherford, K., Salzberg, A. C., Kyes, S., Chan, M.-S., Nene, V., Shallom, S. J., Suh, B., Peterson, J., Angiuoli, S., Pertea, M., Allen, J., Selengut, J., Haft, D., Mather, M. W., Vaidya, A. B., Martin, M. A., Fairlamb, A. H., Fraunholz, M. J., Roos, D. R., Ralph, S. A., McFadden, G. I., Cummings, L. M., Subramanian, G. M., Mungall, C., Venter, J. C., Carucci, D. J., Hoffman, S. L., Newbold, C., Davis, R. W., Fraser, C. M., and Barrell, B. 2002. Genome sequence of the human malaria parasite *Plasmodium falciparum*. *Nature 419*, 498–511.

Hammill, M. O. and Smith, T. G. 1991. The role of ecology in the predation of the ringed seal in Barrow Strait, Northwest Territories. *Canadian Marine Mammal Society 7*, 123–135.

Harborne, J. B. 1982. *Introduction to Ecological Biochemistry*, 2nd edn. Academic Press, London.

Hassell, M. P. and Anderson, R. M. 1989 Predator–prey and host–pathogen interactions. In J. M. Cherrett (ed.), *Ecological Concepts*. Blackwell, Oxford.

Jones, D. A. 1973. Co-evolution and cyanogenesis. In V. H. Heywood (ed.), *Taxonomy and Ecology*. Academic Press, New York.

Krebs, J. R. and Davies, N. B. 1987. *An Introduction to Behavioural Ecology*. Blackwell, Oxford.

LaPage, G. 1963. *Animals Parasitic in Man*. Dover, New York.

Owen, D. F. 1980. *Camouflage and Mimicry*. Oxford University Press, Oxford.

Payne, C. C. 1977. The ecology of brood parasitism in birds. *Annual Review of Ecology and Systematics 8*, 1–28.

Peterson, S. C., Johnson, N. D., and LeGuyader, J. L. 1987. Defensive regurgitation of alleochemicals derived from cyanogenesis by eastern tent caterpillars. *Ecology 68*, 1268–1272.

Pierkarski, G. 1962. *Medical Parasitology*. Bayer, Leverkusen.

Pimental, D. and Stone, F. A. 1968. Evolution and population ecology of parasite-host systems. *Canadian Entomologist 100*, 655–662.

Potts, G. R. and Aebischer, N. J. 1989. Control of population size in birds: the grey partridge as a case study. In P. J. Whittaker and J. B. Grubb (eds), *Towards a More Exact Ecology*. Blackwell, Oxford.

Pundir, Y. P. S. 1981. A note on the biological control of *Scurrula cordifolia* (Wall.) G. Don. by another mistletoe in the Sivalik Hills (India). *Weed Research 21*, 233–234.

Rosenthal, G. A. and Janzen, D. H. (eds). 1979. *Herbivores: Their Interaction with Secondary Plant Metabolites*. Academic Press, New York.

Silvertown, J. W. 1980. The evolutionary ecology of mast seeding in trees. *Biological Journal of the Linnean Society of London 14*, 235–250.

Stirling, I., Øritsland, N. A. 1995. Relationships between estimates of ringed seal (*Phoca hispida*) and polar bear (*Ursus maritimus*) populations in the Canadian Arctic. *Canadian Journal of Fish and Aquatic Science 52*, 2594–2612.

Section 4.5

Boudouresque, C. F. and Verlaque, V. 2002. Biological pollution in the Mediterranean Sea: invasive versus introduced macrophytes. *Marine Pollution Bulletin 44*, 32–38.

Bram, R. A., George, J. E., Reichard, R. E., and Tabachnick, W. J. 2002. Threat of foreign arthropod-borne pathogens to livestock in the United States. *Journal of Medical Entomology 39*, 405–416.

Christian, C. E. 2001. Consequences of a biological invasion reveal the importance of mutualism for plant communities. *Nature 413*, 635–639.

Furukawa, A., Shibata, C., and Mori, K. 2002. Synthesis of four methyl-branched secondary acetates and a methyl-branched ketone as possible candidates for the female pheromone of the screwworm fly, *Cochliomyia hominivorax*. *Bioscience, Biotechnology & Biochemistry 66*, 1164–1169.

Mack, R. N. and Lonsdale, W. M. 2001. Humans as global plants dispersers: getting more than we bargained for. *BioScience 51*, 95–102.

Meinesz, A., Belsher, T., Thibaut, T., Antolic, B., Mustapha, K. B., Boudouresque, C.-F., Chiaverini, D., Cinelli, F., Cottalorda, J.-M., Djellouli, A., Abed, A. E., Orestoano, C., Grau, A. M., Ivesa, L., Jaklin, A., Langar, H., Massuti-Pascual, E., Peirano, A., Leonardo, T., de Vauelas, J., Zavodnik, N., and Zuljevic, A. 2001. The introduced green alga *Caulerpa taxifolia* continues to spread in the Mediterranean. *Biological Invasions 3*, 201–210.

Mooney, H. A. and Cleland, E. E. 2001. The evolutionary impact of invasive species. *Proceedings of the National Academy of Sciences of the United States of America 98*, 5446–5451.

Novak, S. J. and Mack, R. N. 2001. Tracing plant introduction and spread: genetic evidence from *Bromus tectorum* (Cheatgrass). *BioScience 51*, 114–122.

Thomas, J. A., Knapp, J. J., Akino, T., Gerty, S., Wakamura, S., Simcox, D. J., Wardlaw, J. C., and Elmes, G. W. 2002. Parasitoid secretions provoke ant warfare. *Nature 417*, 505.

Williamson, M. and Fitter, A. 1996. The varying success of invaders. *Ecology 77*, 1661–1666.

5 Communities

Section 5.1

Archibold, O. W. 1995. *Ecology of World Vegetation*. Chapman & Hall, London.

Begon, M., Harper, J. L., and Townsend, C. R. 1990. *Ecology: Individuals, Populations, and Communities*, 2nd edn. Blackwell, Oxford.

Bottema, S., Entjes-Nieborg, G., and Van Zeist, W. (eds). 1990. *Man's Role in the Shaping of the Eastern Mediterranean Landscape*. Balkema, Rotterdam.

Boucher, C. and Moll, E. J. 1981. South African Mediterranean Shrub-lands. In F. Di Castri *et al.* (eds), *Mediterranean-type Shrublands*. Elsevier, Amsterdam, pp. 233–248.

Cody, M. L. and Mooney, H. A. 1978. Convergence versus nonconvergence in Mediterranean-climate ecosystems. *Annual Review of Ecology and Systematics 9*, 265–351.

Cowling, R. M., Rundel, P. W., Lamont, B. B., Arroyo, A. K., and Arianoutsou, M. 1996. Plant diversity in Mediterranean-climate regions. *Trends in Ecology and Evolution 11*, 362–366.

Di Castri, F. 1981. Mediterranean-type shrublands of the world. In F. Di Castri *et al.* (eds), *Mediterranean-type Shrublands*. Elsevier, Amsterdam, pp. 1–52.

Di Castri, F. and Mooney, H. A. (eds). 1973. *Mediterranean Type Ecosystems: Origin and Structure*. Springer, New York.

Di Castri, F. and Vitali-Di Castri, V. 1981. Soil fauna of mediterranean-climate regions. In F. Di Castri *et al.* (eds), *Mediterranean-type Shrublands*. Elsevier, Amsterdam, pp. 1–52.

Di Castri, F., Goodall, D. W., and Specht, R. L. 1981. Ecosystems of the World 11. *Mediterranean-type Shrublands*. Elsevier, Amsterdam.

Greuter, W. 1995. Extinctions in mediterranean areas. In J. H. Lawton and R. M. May (eds), *Extinction Rates*. Oxford University Press, Oxford.

Hanes, T. L. 1981. California chaparral. In F. Di Castri *et al.* (eds), *Mediterranean-type Shrublands*. Elsevier, Amsterdam, pp. 1339–1373.

Le Houreau, H. N. 1981. Impact of man and his animals on Mediterranean vegetation. In F. Di Castri *et al.* (eds), *Mediterranean-type Shrublands*. Elsevier, Amsterdam.

Mills, J. M. 1983. Herbivory and seedling establishment in post-fire southern California chaparral. *Oecologia 60*, 267–270.

Naveh, Z. 1990. Ancient man's impact on the Mediterranean landscape in Israel—ecological and evolutionary perspectives. In S. Botteman *et al.* (eds), *Man's Role in the Shaping of the Eastern Mediterranean Landscape*. Balkema, Rotterdam.

Polunin, O. and Smythies, B. E. 1973. *Flowers of South-West Europe. A Field Guide*. Oxford University Press, Oxford.

Quezel, P. 1981. Floristic composition and phytosociological structure of sclerophyllous matorral around the Mediterranean. In F. Di Castri *et al.* (eds), *Mediterranean-type Shrublands*. Elsevier, Amsterdam, pp. 107–122.

Rundel, P. W. 1981. The matorral zone of central Chile. In F. Di Castri *et al.* (eds), *Mediterranean-type Shrublands*. Elsevier, Amsterdam, pp. 175–201.

Specht, R. L. 1981. Mallee ecosystems in Southern Australia. In F. Di Castri *et al.* (eds), *Mediterranean-type Shrublands*. Elsevier, Amsterdam, pp. 203–231.

Tomaselli, R. 1981. Main physiognomic types and geographic distribution of shrub systems related to mediterranean climates. In F. Di Castri *et al.* (eds), *Mediterranean-type Shrublands*. Elsevier, Amsterdam.

Trabaud, L. 1981. Man and fire: impacts on Mediterranean vegetation. In F. Di Castri *et al.* (eds), *Mediterranean-type Shrublands*. Elsevier, Amsterdam.

Van Andel, T. H. and Zanger, E. 1990. Landscape stability and destabilisation in the prehistory of Greece. In S. Botteman *et al.* (eds), *Man's Role in the Shaping of the Eastern Mediterranean Landscape*. Balkema, Rotterdam.

Weiher, E. and Keddy, P. (eds). 1999. *Ecological Assembly Rules. Perspectives, Advances, Retreats*. Cambridge University Press, Cambridge.

Section 5.2

Bach, C. E. 2001. Long-term effects of insect herbivory and sand accretion on plant succession on sand dunes. *Ecology* 85, 1401–1416.

Clements, F. E. 1936. Nature and structure of the climax. *Journal of Ecology* 24, 252–284.

Connell, J. H. 1978. Diversity in tropical rainforests and coral reefs. *Science* 199, 1302–1310.

Connell, J. H. and Slatyer, R. O. 1977. Mechanisms of succession in natural communities and their role in community stability and organisation. *American Naturalist* 111, 1119–1144.

Connell, J. H., Noble, I. R., and Slatyer, R. O. 1987. On the mechanisms of producing successional change. *Oikos 50*, 136–137.

Diamond, J. and Case, T. J. (eds). 1986. *Community Ecology*. Harper and Row, New York.

Duckworth, J. C., Kent, M., and Ramsay, P. M. 2000. Plant functional types: an alternative to taxonomic plant community descriptions in biogeography? *Progress in Physical Geography* 24, 515–542.

Gleason, H. A. 1926. The individualistic concept of the plant association. *Bulletin of the Torrey Botanical Club* 53, 7–26.

Gray, A. J., Crawley, M. J., and Edwards, P. J. (eds). 1987. *Colonization, Succession and Stability*. Blackwell, Oxford.

Harper, J. L. 1977. *The Population Biology of Plants*. Academic Press, London.

Horn, H. S. 1976. Succession. In R. M. May (ed.), *Theoretical Ecology: Principles and Applications*. Blackwell, Oxford.

Janzen, D. H. 1986. Keystone plant resources in the tropical forest. In M. E. Soule (ed.), *Conservation Biology: The Science of Scarcity and Diversity*. Sinauer Associates, Sunderland, MA.

Keeley, J. E. and Bond, W. J. 1997. Convergent seed germination in South African fynbos and Californian chaparral. *Plant Ecology 133*, 153–167.

Kent, M., Owen, N. W., Dale, P., Newnham, R. M., and Giles, T. M. 2001. Studies of vegetation burial: a focus for biogeography and geomorphology? *Progress in Physical Geography 25*, 455–482.

Krebs, C. J. 1985. *Ecology, the Experimental Analysis of Distribution and Abundance*, 3rd edn. Harper & Row, New York.

Kutiel, P., Eden, E., and Zevelev, Y. 2000. Effect of experimental trampling and off-road motorcycle traffic on soil and vegetation of stabilised coastal dunes, Israel. *Environmental Conservation 27*, 14–23.

van der Meulen, F. and Salman, A. H. P. M. 1996. Management of Mediterranean coastal dunes. *Ocean & Coastal Management 30*, 177–195.

Moore, P. D. 1986. Site history. In P. D. Moore and S. B. Chapman (eds), *Methods in Plant Ecology*. Blackwell, Oxford

Paine, R. T. 1969. A note on trophic complexity and community stability. *American Naturalist 100*, 65–75.

Rubio-Casal, A. E., Castillo, J. M., Luque, C. J., and Figueroa, M. E. 2001. Nucleation and facilitation in salt pans in Mediterranean salt marshes. *Journal of Vegetation Science 12*, 761–770.

Whitaker, R. H. 1975. *Communities and Ecosystems*. Macmillan, London.

Section 5.3

Braun-Blanquet, J. 1932. *Plant Sociology: The Study of Plant Communities* (trans. G. D. Fuller and H. S. Conrad). McGraw Hill, New York.

Franklin, J. 2002. Enhancing a regional vegetation map with predictive models of dominant plant species in chaparral. *Applied Vegetation Science 5*, 135–146.

Rodwell, J. S. (ed.). 1989. *British Plant Communities*, Vol. 1, *Woodlands and Scrub*. Cambridge University Press, Cambridge.

Rodwell, J. S. (ed.). 1991. *British Plant Communities*, Vol. 2, *Mires & Heaths*. Cambridge University Press, Cambridge.

Rodwell, J. S. (ed.). 1992. *British Plant Communities*, Vol. 3, *Grasslands & Montane Communities*. Cambridge University Press, Cambridge.

6 Systems

Section 6.1

Cavicchioli, R., Thomas, T. 2000. Extremophiles. In J. Lederberg (ed.), *Encyclopedia of Microbiology*, 2nd edn.

Section 6.2

Cooper, J. P., 1975. *Photosynthesis and Productivity in Different Environments*. Cambridge University Press, Cambridge.

Davis, G. W., Richardson, D. M., Keeley, J. E., and Hobbs, R. J. 1996. Mediterranean-type ecosystems: the influence of biodiversity on their functioning. In H. A. Mooney, J. H. Cushman, E. Medina, O. E. Sala, and E.-D. Schulze (eds), *Environmental Physiology of Plants*. John Wiley & Sons Ltd, New York.

Fitter, A. H. and Hay, R. K. M. 1987. *Physiological Plant Ecology*, 2nd edn. Academic Press, London.

Hay, R. K. M. and Walker, A. J. 1989. *An Introduction to the Physiology of Crop Yield*. Longman, Harlow.

Humphreys, W. F. 1979. Production and respiration in animal populations. *Journal of Animal Ecology 48*, 427–474.

Jones, H. G. 1983. *Plants and Microclimate*. Cambridge University Press, Cambridge.

Orshan, G. 1963. Seasonal dimorphism of desert and mediterranean chamaephytes and its significance as a factor in their water economy. In A. J. Rutter and F. H. Whitehead (eds), *The Water Relations of Plants*. Wiley, New York.

Osmond, C. B., Winter, K., and Zeigler, H. 1981. Functional significance of different pathways of CO_2 fixation in photosynthesis. In A. Pirson and M. H. Zimmermann (eds), *Encyclopaedia of Plant Physiology*, New Series. Springer Verlag, Berlin.

Salisbury, F. B. and Ross, C. 1992. *Plant Physiology*, 4th edn. Wadsworth, Belmont, CA.

Schultz, E. D. 1970. Der CO_2-Gaswechsel de Buche (*Fagus sylvatica* L.) in Abhangigkeit von den Klimafaktoren in Feiland. *Flora Jena 159*, 177–232.

Schultz, E. D., Fuchs, M., and Fuchs, M. I. 1977. Spatial distribution of photosynthetic capacity and performance in a mountain spruce forest in northern Germany. I. Biomass distribution and daily CO_2 uptake in different crown layers. *Oecologia 29*, 43–61.

Schultz, E. D., Fuchs, M., and Fuchs, M. I., 1977b. Spatial distribution of photosynthetic capacity and performance in a mountain spruce forest in northern Germany. III. The significance of the evergreen habit. *Oecologia 30*, 239–248.

Whittaker, R. H. 1975. *Communities and Ecosystems*, 2nd edn. Macmillan, New York.

Whittaker, R. H. and Likens, G. E. 1973. The primary production of the biosphere. *Human Ecology 1*, 299–369.

Woodward, F. I. 1987. *Climate and Plant Distribution*. Cambridge University Press, Cambridge.

Zscheile, F. P. and Comar, C. L. 1941. Influence of preparative procedure on the purity of chlorophyll components as shown by absorption spectra. *Botanical Gazette 102*, 463–481.

Section 6.3

Begon, M., Harper, J. L., and Townsend, C. R. 1990. *Ecology: Individuals, Populations, and Communities*, 2nd edn. Blackwell, Oxford.

Colinvaux, P. 1980. *Why Big Fierce Animals are Rare*. Penguin Books, Harmondsworth, Middlesex.

Cousins, S. 1987. The decline of the trophic level. *Trends in Ecology and Evolution 2*, 312–316.

Golley, F. B. 1960. Energy dynamics of an old-field community. *Ecological Monographs 30*, 187–200.

Heal, O. W. and Maclean, S. F. 1975. Comparative productivity in ecosystems- secondary productivity. In W. H. Van Dobben and R. H. Lowe-McConnell (eds), *Unifying Concepts in Ecology*. Dr W Junk, The Hague.

Lawton, J. H. 1989. Food webs. In J. M. Cherret (ed.), *Ecological Concepts*. Blackwell, Oxford.

Moore, J. C., de Ruiter, P. C., Hunt, H. W., Coleman, D. C., and Freckman, D. W. 1996. Microcosms and soil ecology: critical linkages between field studies and modelling food webs. *Ecology 77*, 694–705.

Odum, P. 1989. *Ecology and Our Endangered Life-support System*. Sinauer Associates, Sunderland, MA.

Odum, H. T. 1957. Trophic structure and productivity of Silver Springs, Florida. *Ecological Monographs 27*, 55–112.

Phillips, J. 1966. *Ecological Energetics*. Edward Arnold, London.

Swift, M. J., Heal, O. W., and Anderson, J. M. 1979. *Decomposition in Terrestrial Ecosystems*. Blackwell, Oxford.

Varley, G. C. 1970. The concept of energy flow applied to a woodland community. In A. Watson (ed.), *Animal Populations in Relation to their Food Resource*. Blackwell, Oxford.

Section 6.4

Briand, F. 1983. Environmental control of food web structure. *Ecology 64*, 253–263.

Briand, F. and Cohen, J. E. 1984. Community food webs have scale-invariant structure. *Nature 307*, 264–266.

Cohen, J. E. and Briand, F. 1984. Trophic links of community food webs. *Proceedings of the National Academy of Sciences of the United States of America 81*, 4105–4109.

Jarman, S. N., Gales, N. J., Tierney, M., Gill, P. C., and Elliott, N. G. 2002. A DNA-based method for identification of krill species and its application to analysing the diet of marine vertebrate predators. *Molecular Ecology 11*, 2679–2690.

Kucklick, J. R., Struntz, W. D. J., Becker, P. R., York, G. W., O'Hara, T. M., and Bohonowych, J. E. 2002. Persistent organochlorine pollutants in ringed seals and polar bears collected from northern Alaska. *The Science of the Total Environment 287*, 45–59.

Pimm, S. L. 1982. *Food Webs*. Chapman & Hall, London.

Pimm, S. L., Lawton, J. H., and Cohen, J. E. 1991. Food web patterns and their consequences. *Nature 350*, 669–674.

Pimm, S. L. and Kitching, R. L. 1987. The determinants of food chain length. *Oikos 50*, 302–307.

Price, P. W. 1975. *Insect Ecology*. Wiley, New York.

Polischuk, S. C., Nortstrom, R. J., and Ramsay, M. A. 2002. Body burdens and tissue concentrations of organochlorines in polar bears (*Ursus maritimus*) vary during seasonal fasts. *Environmental Pollution 118*, 29–39.

Tilman, D. 1982. *Resource Competition and Community Structure*. Princeton University Press, Princeton, NJ.

Section 6.5

Ammerman, A. and Cavalli-Sforza, L. L. 1971. Measuring the rate of spread of early farming in Europe. *Man 6*, 674–688.

Beaufoy, G. 1998. *The reform of the CAP Olive-Oil Regime: What are the Implications for the Environment*. EFNCP Occasional Publication Number 14(En), European Forum for Nature Conservation and Pastoralism, Peterborough.

Beaufoy, G. 2001 *EU Policies for Olive Farming*. WWF Europe/Birdlife, Brussels.

Breznak, J. A. 1975. Symbiotic relationships between termites and their intestinal biota. In D. H. Jennings and D. L. Lee (eds), *Symbiosis*. Symposium 29, Society for Experimental Biology, Cambridge University Press, Cambridge.

Bronowski, J. 1973. *The Ascent of Man*. BBC Publications, London.

Caraveli, H. 2000. A comparative analysis on intensification and extensification in mediterranean agriculture: dilemmas for LFAs policy. *Journal of Rural Studies 16*, 231–242.

Duckham, A. N. 1976. Environmental constraints. In A. N. Duckham, J. G. W. Jones, and E. H. Roberts (eds), *Food Production and Nutrient Cycles*. North Holland Publishing Company, Amsterdam.

Evans, L. T. 1993. *Crop Evolution, Adaptation and Yield*. Cambridge University Press, Cambridge.

Fernandez Ales, R., Angle, M., Ortega, F., and Ales, E. E. 1992. Recent changes in landscape structure and function in a mediterranean region of SW Spain. *Landscape Ecology 7*, 3–18.

Giller, K. E., Beare, M. H., Lavelle, P., Izac, A.-M. N., and Swift, M. J. 1997. Agricultural intensification, soil biodiversity and agroecosystem function. *Applied Soil Ecology 6*, 3–16.

Goudie, A. 1993. *The Human Impact on the Natural Environment*, 4th edn. Blackwell, Oxford.

Hungate, R. E. 1975. The rumen microbiological ecosystem. *Annual Review of Ecology and Systematics 6*, 39–66.

Morgan, R. P. C. 1995. *Soil Erosion and Conservation*, 2nd edn. Longman, Harlow.

Nature Conservancy Council 1984. *Nature Conservation in Great Britain*. Nature Conservancy Council, Peterborough.

Polidori, R., Rocchi, B., and Stefani, G. 1997. Reform of the CMO for olive oil: current situation and future prospects. In M. Tracy (ed.), *CAP Reform: The Southern Products*. Agricultural Policy Studies, Belgium.

Richards, B. N. 1974. *An Introduction to the Soil Ecosystem*. Longman, Harlow.

Rodin, L. E. and Bazilevich, N. I. 1967. *Production and Mineral Cycling in Terrestrial Vegetation*. Oliver & Boyd, Edinburgh.

Simmons, I. G. 1989. *The Changing Face of the Earth: Culture and Environment*, 4th edn. Blackwell, Oxford.

Slesser, M. 1975. Energy requirements of agriculture. In J. Lenihan and W. W. Fletcher (eds), *Food, Agriculture and the Environment*. Blackie, Glasgow and London.

Smith, D. F. and Hill, D. M. 1975. Natural agricultural ecosystems. *Journal of Environmental Quality 4*, 143–145.

Spedding, C. R. W. 1975. *Biology and Agricultural Systems*. Academic Press, London.

Swift, M. J., Vandeermeer, J., Ramakrishnan, P. S., Anderson, J. M., Ong, C. K., and Hawkins, B. A. 1996. Biodiversity and agroecosystem function. In H. A. Mooney, J. H. Cushman, E. Medina, O. E. Sala, and E.-D. Schulze (eds), *Environmental Physiology of Plants*. John Wiley & Sons Ltd, New York.

Tivy, J. 1975. Environmental impact of cultivation. In J. Lenihan and W. W. Fletcher (eds), *Food, Agriculture and the Environment*. Blackie, Glasgow and London.

Tivy, J. 1990. *Agricultural Ecology*. Longman, Harlow.

Tivy, J. 1993. *Biogeography: A Study of Plants in the Ecosphere*, 3rd edn. Longman, Harlow.

Tivy, J. and O'Hare, G. 1981. *Human Impact on the Ecosystem*. Oliver & Boyd, Glasgow.

Usher, M. B. and Thompson, D. B. A. (eds). 1988. *Ecological Change in the Uplands*. Blackwell, Oxford.

7 Balances

Section 7.1

Barry, R. G. and Chorley, R. J. 1970. *Atmosphere, Weather and Climate*. Holt, Rinehart and Winston, New York.

Bettinetti, A., Pypaet, P., and Sweerts, J.-P. 1996. Application of an integrated management approach to the restoration project of the lagoon of Venice. *Journal of Environmental Management 46*, 207–227.

Boli, B. and Cook, R. B. (eds). 1983. *The Major Biogeochemical Cycles and their Interactions*. John Wiley, Chichester.

Brown, C. M., McDonald-Brown, D. S., and Meers, J. L. 1974. Physiological aspects of inorganic nitrogen metabolism. *Advances in Microbial Physiology 11*, 1–52.

Fochte, D. D. and Verstraete, W. 1977. Biochemical ecology of nitrification and denitrification. *Advances in Microbial Ecology 1*, 135–214.

Freedman, B. 1995. Environmental Ecology: The Ecological Effects of Pollution, Disturbance, and Other Stresses, 2nd edn. Academic Press, San Diego, CA.

Ghassemi, F., Jakeman, A. J., and Nix, H. A. 1995. *Salinisation of Land and Water Resources*. New South Wales Press, Sydney.

Jordan, C. F. and Kline, J. R. 1972. Mineral cycling: some basic concepts and their application in a tropical rain forest. *Annual Review of Ecology and Systematics 3*, 33–49.

Lee, J. 1988. Acid rain. *Biological Sciences Review 1*, 15–18.

McNaughton, S. J. 1988. Mineral nutrition and spatial concentrations of African ungulates. *Nature 334*, 343–345.

Miller, R. M. 1987. Mycorrhizae and succession. In W. R. Jordan, M. E. Gilpin, and J. D. Aber (eds), *Restoration Ecology: A Synthetic Approach to Ecological Research*. Cambridge University Press, Cambridge.

Moss, B. 1988. *Ecology of Fresh Waters: Man and Medium*, 2nd edn. Blackwell, Oxford.

Nature Conservancy Council 1990. *On Course Conservation: Managing Golf's Natural Heritage*. Nature Conservancy Council, Peterborough.

Odum, E. P. 1989. *Ecology and Our Endangered Life-support System*. Sinauer Associates, Sunderland, MA.

Reinoso, J. C. M. 2001. Vegetation changes and groundwater abstraction in SW Donana. Spain. *Journal of Hydrology 242*, 197–209.

Ricklefs, R. E. 1990. *Ecology*, 3rd edn. Freeman, New York.

Serrano, L. and Serrano, L. 1996. Influence of groundwater exploitation for urban water supply on temporary ponds from the Donana National Park (SW Spain). *Journal of Environmental Management 46*, 229–238.

Sprent, J. I. 1983. *The Biology of Nitrogen-fixing Organisms*. McGraw Hill, New York.

Section 7.2

Earle, S. 1992. Assessing the damage one year later. *National Geographical 179*, 122–134.

Edmondson, W. T. 1979. Lake Washington and the predictability of limnological events. *Archiv für Hydrobiologie, Beiheft 13*, 234–241.

European Investment Bank 1990. *The environmental program for the Mediterranean: preserving a shared heritage and common resource*. Report number 8504. International Bank for Reconstruction and Development/World Bank and European Investment Bank. Luxemburg.

Lehman, J. T. 1986. Control of eutrophication in Lake Washington. In *Ecological Knowledge and Environmental Problem-solving, Concepts and Case Studies*. National Academy Press, Washington, DC.

Piatt, J. F. and Lensink, C. J. 1989. *Exxon Valdez* oil spill. *Nature 342*, 865–866.

Ritchie, W. and O'Sullivan, M. (eds). 1994. *The Environmental Impact if the Wreck of the Braer*. The Scottish Office, Edinburgh.

Schindler, D. W. 1977. Evolution of phosphorus limitation in lakes. *Science 195*, 260–262.

SEO/BirdLife 2003. *The Disaster of Prestige Oil Tanker and its Impact on Seabirds*. BirdLife International, Madrid.

Van Donk, G. and Gulati, R. D. 1991. Ecological management of aquatic ecosystems: a complementary technique to reduce eutrophication-related perturbations. In O. Ravera (ed.), *Terrestrial and Aquatic Ecosystems: Perturbation and Recovery*. Ellis Horwood, Chichester.

Ward, D. M., Atlas, R. M., Boehm, P. D., and Calder, J. A. 1980. Microbial degradation and chemical evolution from the Amoco spill. *Ambio 9*, 277–283.

Section 7.3

Baker, A., Brooks, R., and Reeves, R. 1988. Growing for gold . . . and for copper . . . and zinc. *New Scientist 117*, 44–48.

Bradshaw, A. D. 1987. Restoration: an acid test for ecology. In W. R. Jordan, M. E. Gilpin, and J. D. Aber (eds), *Restoration Ecology: A Synthetic Approach to Ecological Research*. Cambridge University Press, Cambridge.

Bradshaw, A. D. 1983. The reconstruction of ecosystems. *Journal of Applied Ecology 20*, 1–17.

Bradshaw, A. D. 1984. Ecological principles and land reclamation practice. *Landscape Planning 11*, 35–48.

Bradshaw, A. D. 1989. Management problems arising from successional processes. In G. P. Buckley (ed.), *Biological Habitat Reconstruction*. Belhaven Press, London.

Bradshaw, A. 1993. Understanding the fundamentals of succession. In J. Miles and D. H. Walton (eds), *Primary Succession on Land*. Blackwell, Oxford.

Bradshaw, A. D. and Chadwick, M. J. 1980. *The Restoration of Land: The Ecology and Reclamation of Derelict and Degraded Land*. Blackwell, Oxford.

Bradshaw, A. D., Humphreys, R. N., Johnson, M. S., and Roberts, R. D. 1978. The restoration of vegetation on derelict land produced by industrial activity. In M. W. Holdgate and M. J. Woodward (eds), *The Breakdown and Restoration of Ecosystems*. Plenum, New York.

Buckley, G. P. (ed.). 1989. *Biological Habitat Reconstruction*. Belhaven Press, London.

Bunce, R. G. H. and Jenkins, N. R. 1989. Land potential for habitat reconstruction in Britain. In G. P. Buckley (ed.), *Biological Habitat Reconstruction*. Belhaven Press, London.

Department of the Environment 1994. *The Reclamation of Metalliferous Mining Sites*. Her Majesty's Stationery Office, London.

Down, G. S. and Morton, A. J. 1989. A case study of whole woodland transplanting. In G. P. Buckley (ed.), *Biological Habitat Reconstruction*. Belhaven Press, London.

Helliwell, D. R. 1989. Soil transfer as a means of moving grassland and marshland vegetation. In G. P. Buckley (ed.), *Biological Habitat Reconstruction*. Belhaven Press, London.

Jordan, W. R., Gilpin, M. E., and Aber, J. D. (eds). 1987. *Restoration Ecology: A Synthetic Approach to Ecological Research*. Cambridge University Press, Cambridge.

Lee, I. W. Y. 1985. A review of vegetative slope stabilisation. *The Journal of the Hong Kong Institution of Engineers* July, 9–22.

Marrs, R. H. and Bradshaw, A. D. 1993. Primary succession on man-made wastes: the importance of resource acquisition. In J. Miles and D. H. Walton (eds), *Primary Succession on Land*. Blackwell, Oxford.

Pywell, R. F., Bullock, J. M., Roy, D. B., Warman, L., Walker, K. J., and Rothery, P. 2003. Plant traits as predictors of performance in ecological restoration. *Journal of Applied Ecology* 40, 65–77.

8 Scales

Section 8.1

Bolger, D. T., Scott, T. A., and Rotenberry 2001. Use of corridor-like landscape structures by bird and small mammal species. *Biological Conservation* 102, 213–224.

Dunning, J. B., Danielson, B. J., and Pulliam, H. R. 1992. Ecological processes that affect populations in complex landscapes. *Oikos* 65, 169–175.

Forman, R. T. T. 1995. *Land Mosaics: The Ecology of Landscapes and Regions*. Cambridge University Press, Cambridge.

O'Neill, R. V., DeAngelis, D. L., Waide, J. B., and Allen, T. F. B. 1986. *A Hierarchical Concept of Ecosystems*. Princeton University Press, Princeton, NJ.

O'Neill, E. G., O'Neill, R. V., and Norby, R. J. 1991. Hierarchy theory as a guide to mycorrhizal research on large-scale problems. *Environmental Pollution* 73, 271–284.

Santos, T., Telleria, J. L., and Carbonell, R. 2002. Bird conservation in fragmented Mediterranean forests of Spain: effects of geographical location, habitat and landscape degradation. *Biological Conservation* 105, 113–125.

Turner, M. G., Romme, W. H., Gardner, R. H., O'Neill, R. V., and Kratz, T. K. 1993. A revised concept of landscape equilibrium: disturbance and stability on scaled landscapes. *Landscape Ecology* 8, 213–227.

Section 8.2

Archibold, O. W. 1995. *Ecology of World Vegetation*. Chapman & Hall, London.

Whittaker, R. H. 1975. *Communities and Ecosystems*, 2nd edn. MacMillan, New York.

Section 8.3

Adams, J. M., Faure, H., Faure-Denard, L., McGlade, J. M., and Woodward, F. I. 1991. Increases in terrestrial carbon storage from the last glacial maximum to the present. *Nature* 348, 711–714.

Aspinall, R. and Matthews, K. 1994. Climate change impact on distribution and abundance of wildlife species: an analytical approach using GIS. *Environmental Pollution* 83, 217–223.

Baker, J. T. and Allen, L. H. 1994. Assessment of the impact of rising carbon dioxide and other potential climate changes on vegetation. *Environmental Pollution* 83, 223–235.

Behling, H. 2002. Carbon storage increases by major forest ecosystems in tropical South America since the Last Glacial Maximum and the early Holocene. *Global and Planetary Change* 33, 107–116.

Bekkering, T. D. 1992. Using tropical forests to fix atmospheric carbon: the potential in theory and practice. *Ambio* 21, 414–419.

Bierregaard, R. O., Lovejoy, T. E., Kapos, V., dos Santos, A. A., and Hutchings, R. W. 1992. The biological dynamics of tropical rain forest fragments. *BioScience* 42, 859–866.

Clayton, K. 1995. The threat of global warming. In T. O'Riordan (ed.), *Environmental Science for Environmental Management*. Longman, Harlow

Davey, P. A., Parson, A. J., Atkinson, L., Wadge, K., and Long, S. P. 1999. Does photosynthetic acclimation to elevated CO_2 increase photosynthetic nitrogen-use efficiency? A study of three native UK grassland species in open-top chambers. *Functional Ecology* 13 (Suppl. 1), 21–28.

Dormann, C. F. and Woodin, S. J. 2002. Climate change in the Arctic: using plant functional types in a meta-analysis of field experiments. *Functional Ecology* 16, 4–17.

Houghton, J. T., Jenkins, G. J., and Ephraums, J. J. (eds). 1990. *Climate Change: The IPCC Scientific Assessment*. Cambridge University Press, Cambridge.

Innes, J. L. 1994. Climatic sensitivity of temperate forests. *Environmental Pollution* 83, 237–243.

Leemans, R. and Zuidema, G. 1995. Evaluating changes in land cover and their importance for global change. *Trends in Ecology and Evolution* 10, 76–81.

Lloyd, D. and Jenkinson, D. S. 1995. The exchange of trace gases between land and atmosphere. *Trends in Ecology and Evolution* 10, 2–4.

Lloyd, J. 1999. The CO_2 dependence of photosynthesis, plant growth responses to elevated CO_2 concentrations and their interactions with soil nutrient status, II. Temperate and boreal forest productivity and the combined effects of increasing CO_2 concentrations and increased nitrogen deposition at a global scale. *Functional Ecology 13*, 439–459.

Lloyd, J. and Farquhar, G. D. 1996. The CO_2 dependence of photosynthesis, plant growth responses to elevated atmospheric CO_2 concentrations and their interaction with soil nutrient status. 1. General principles and forest ecosystems. *Functional Ecology 10*, 4–32.

Mitchell, J. F. B., Johns, T. C., Gregory, J. M., and Tett, S. F. B. 1995. Climate response to increasing levels of greenhouse gases and sulphate aersols. *Nature 376*, 501–504.

Niklaus, P. A., Stocker, R., Korner, C. H., and Leadley, P. W. 2000. CO_2 flux estimates tend to overestimate ecosystem C sequestration at elevated CO_2. *Functional Ecology 14*, 546–559.

Norby, R. J., Gunderson, C. A., Wullschleger, S. D., O'Neill, E. G., and McCracken, M. K. 1992. Productivity and compensatory responses of yellow-poplar trees in elevated CO_2. *Nature 357*, 322–324.

Phillips, O. L., Martinez, R. V., Arroyo, L., Baker, T. R., Killeen, T., Lewis, S. L., Malhi, Y., Mendoza, A. M., Neill, D., Vargas, P. N., Alexiades, M., Cerón, C., Di Fiore, A., Erwin, T., Jardim, A., Palacios, W., Saldias, M., and Vinceti, B. 2002. Increasing dominance of large lianas in Amazonian forests. *Nature 418*, 770–774.

Pastor, J. and Post, W. M. 1988. Response of northern forests to CO_2-induced climate change. *Nature 334*, 55–58.

Sarmiento, J. L. and Orr, J. C. 1991. Three-dimensional simulations of the impact of Southern Ocean nutrient depletion on atmospheric CO_2 and ocean chemistry. *Limnology and Oceanography 36*, 1928–1950.

Sedjo, R. A. 1992. Temperate forests ecosystems in the global carbon cyle. *Ambio 21*, 274–277.

Stott, P. A., Tett, S. F. B., Jones, G. S., Allen, M. R., Mitchell, J. F. B., and Jenkins, G. J. 2000. External control of 20th century temperature by natural and anthropogenic forcings. *Science 290*, 2133–2137.

Thompson, R. D. 1992. The changing atmosphere and its impact on Planet Earth. In A. W. Mannion and S. R. Bowlby (eds), *Environmental Issues in the 1990s*. John Wiley, Chichester.

Wigley, T. M. L. and Raper, S. C. B. 1992. Implications for climate and sea level of revised IPCC emission scenarios. *Nature 357*, 293–300.

Wigley, T. M. L., Richels, R., and Edmonds, J. A. 1996. Economic and environmental choices in the stabilization of atmospheric CO_2 concentrations. *Nature 379*, 240–243.

9 Checks

Introduction

Raymo, M. E. and Ruddiman, W. F. 1992. Tectonic forcing of the late Cenozoic climate. *Nature 359*, 117–122.

Section 9.1

Hubbell, S. P. 1997. A unified theory of biogeography and relative species abundance and its application to tropical rain forests and coral reefs. *Coral Reefs 16* (Suppl.), S9–S21.

Jackson, J. B. C. 1991. Adaptation and diversity of reef corals. *BioScience 41*, 475–482.

MacArthur, R. H. and Wilson, E. O. 1967. *The Theory of Island Biogeography*. Princeton University Press, Princeton, NJ.

Section 9.2

Coope, G. R. 1995. Insect faunas in ice age environments: why so little extinction? In J. H. Lawton and R. M. May (eds), *Extinction Rates*. Oxford University Press, Oxford.

Stirling, I., Lunn, N. J., and Iacozza, J. 1999. Long-terms trends in the population ecology of polar bears in western Hudson Bay in relation to climate change. *Arctic 52*, 294–306.

Jablonski, D. 1995. Extinctions in the fossil record. In J. H. Lawton and R. M. May (eds), *Extinction Rates*. Oxford University Press, Oxford.

Jackson, J. B. C. 1995. Constancy and change of life in the sea. In J. H. Lawton and R. M. May (eds), *Extinction Rates*. Oxford University Press, Oxford.

Labandeira, C. C. and Sepkoski, J. J. 1993. Insect diversity in the fossil record. *Science 261*, 310–315.

May, R. M. 1992. How many species inhabit the Earth? *Scientific American 261*, 18–24.

Myers, N. 1993. Questions of mass extinction. *Biodiversity and Conservation 2*, 2–17.

Section 9.3

Briand, F. and Cohen, J. C. 1987. Environmental correlates of food chain length. *Science 238*, 956–960.

Chapin, F. S., Schulze, E.-D., and Mooney, H. A. 1992. Biodiversity and ecosystem processes. *Trends in Ecology and Evolution 7*, 107–108.

Collins, S. L. 1995. The measurement of stability in grasslands. *Trends in Ecology and Evolution 10*, 95–96.

May, R. M. 1986. The search for patterns in the balance of nature: advances and retreats. *Ecology 67*, 1115–1126.

McCann, K., Hastings, A., and Huxel, G. R. 1998. Weak trophic interactions and the balance of nature. *Nature 395*, 794–798.

McNaughton, S. J. 1988. Diversity and stability. *Nature 333*, 204–205.

Moore, J. C. and Hunt, H. W. 1988. Resource compartmentation and the stability of real ecosystems. *Nature 333*, 261–263.

Polis, G. A. 1998. Stability is woven by complex webs. *Nature 395*, 744–745.

Schwartz, M. W., Brigham, C. A., Hoeksema, J. D., Lyons, K. G., Mills, M. H., and van Mantgem, P. J. 2000. Linking biodiversity to ecosystem function: implications for conservation ecology. *Oecologia 122*, 297–305.

Solow, A. R. 1993. Measuring biological diversity. *Environmental Science and Technology 27*, 25–26.

Steele, J. H. 1991. Marine functional diversity. *BioScience 41*, 470–474.

Tilman, D. and Downing, J. A. 1994. Biodiversity and stability in grasslands. *Nature 367*, 3633–3635.

Walker, B. H. 1992. Biodiversity and ecological redundancy. *Conservation Biology 6*, 18–23.

Waltho, N. and Kolasa, J. 1994. Organization of instabilities in multispecies systems, a test of hierarchy theory. *Proceedings of the National Academy of Sciences of the United States of America 91*, 1682–1685.

Section 9.4

Angel, M. V. 1994. Spatial distribution of marine organisms: patterns and processes. In P. J. Edwards, R. M. May, and N. R. Webb (eds), *Large-Scale Ecology and Conservation Biology*. Blackwell, Oxford.

Auspurger, C. K. 1983. Offspring recruitment around tropical trees: changes in cohort distance with time. *Oikos 40*, 189–196.

Chown, S. L. and Gaston, K. J. 2000. Areas, cradles and museums: the latitudinal gradient in species richness. *Trends in Ecology and Evolution 15*, 311–315.

Clark, D. A. and Clark, D. B. 1984. Spacing dynamics of a tropical rain forest tree: evaluation of the Janzen-Connell model. *American Naturalist 124*, 769–788.

Cook, S. 1998. A diversity of approaches to the study of species richness. *Trends in Ecology and Evolution 13*, 340–341.

Currie, D. J. 1991. Energy and large-scale patterns of animal- and plant-species richness. *American Naturalist 137*, 27–49.

Clarke, A. 1992. Is there a latitudinal diversity cline in the sea? *Trends in Ecology and Evolution 7*, 286–287.

Rex, M. J., Stuart, C. T., Hessler, R. R., Alen, J. A., Sanders, H. L., and Wilson, G. D. F. 1993. Global-scale latitudinal patterns of species diversity in the deep sea benthos. *Nature 365*, 636–639

Ricklefs, R. E. 1990. *Ecology*, 3rd edn. Freeman, New York.

Vincent, A. and Clarke, A. 1995. Diversity in the marine environment. *Trends in Ecology and Evolution 10*, 55–56.

Section 9.5

Andren, O. and Balandreau, J. 1999. Biodiversity and soil functioning—from black box to can of worms. *Applied Soil Ecology 13*, 105–108.

Biswas, M. R. 1994. Agriculture and environment: a review, 1972–1992. *Ambio 23*, 192–197.

Cragg, R. G. and Bardgett, R. D. 2001. How changes in soil faunal diversity and composition within a trophic group influence decomposition processes. *Soil Biology & Biochemistry 33*, 2073–2081.

Crosson, P. R. and Rosenberg, N. J. 1989. Strategies for agriculture. *Scientific American 261*, 128–135.

Huston, M. 1993. Biological diversity, soils and economics. *Science 262*, 1676–1679.

Kendall, H. W. and Pimentel, D. 1994. Constraints on the expansion of the global food supply. *Ambio 23*, 198–205.

Lawler, S. P. 1993. Species richness, species composition and population dynamics of protists in experimental microcosms. *Journal of Animal Ecology 62*, 711–719.

Liiri, M., Setätä, H., Haimi, J., Pennanen, T., and Fritze, H. 2002. Soil processes are not influenced by the functional complexity of soil decomposer food webs under disturbance. *Soil Biology & Biochemistry 34*, 1009–1020.

Naeem, S., Thompson, L. J., Lawler, S. P., Lawton, J. H., and Woodfin, R. M. 1994. Declining biodiversity can alter the performance of ecosystems. *Nature 368*, 734–737.

Neilson, R., Robinson, D., Marriot, C. A., Scrimgeour, C. M., Hamilton, D., Stocking, M. 1995. Soil erosion and land degradation. In T. O'Riordan (ed.), *Environmental Science for Environmental Management*. Longman, Harlow.

Tuckwell, H. C. and Koziol, J. A. 1992. World population. *Nature 359*, 200.

Wishart, J., Boag, B., and Handley, L. L. 2002. Above-ground grazing affects floristic composition and modifies soil trophic interactions. *Soil Biology & Biochemistry 34*, 1507–1512.

INDEX

Numbers in italics refers to figures, those in bold to tables

PLATE 2.1 Blackbirds and bluebells. Photographs (a) and (b) show the European blackbird (*Turdus merula*) and bluebell (*Hyacinthoides non-scripta*); (c) and (d) their North American namesakes, *Agelaius phoeniceus* and *Phacelia tanacetifolia*.

PLATE 2.2 The marsupial thylacine is remarkably similar to placental dogs, or perhaps a blend of dog, cat, and hyaena. As one of the few major carnivores in Australia throughout the Pleistocene, it probably filled the role of all three, and shows how evolution can converge on similar adaptive solutions.

PLATE 2.3 The red wolf of North America. Classified as *Canis rufus*, the red wolf is actually a hybrid of the coyote (*Canis latrans*) and the grey wolf (*Canis lupus*).

PLATE 2.4 (a) The southern marsh orchid (*Dactylorhiza praetermissa*) and (b) the common spotted orchid (*Dactylorhiza fuchsii*). These two species naturally hybridize to produce an interpecific hybrid orchid (c).

PLATE 2.5 Apple (*Malus × domestica*), in this case, Kidd's Red Orange, a mid-season variety that is not self-fertile. The stigma of the flowers distinguish between self and non-self pollen on the basis of a protein coded by a single gene.

PLATE 3.1 Mature lowland forest—the Bialowieza National Park, Poland. This forest, on the Belorus border, is perhaps the only remaining fragment of pristine lowland forest in Europe. It is home to the only remaining population of the European bison.

PLATE 3.2 Regenerating woodland, with large numbers of small saplings.

PLATE 3.3 Placing a tracking collar on a sedated female polar bear on closed drift ice in the Barents Sea, April 1998. The adult female and her two one-year-old cubs are attended by Andrei Boltunov, Andrew Derocher, and Øystein Wiig (left to right). The white collar with its satellite transmitter lies between the two cubs.

PLATE 3.4 Elephant grazing on Acacia trees, Masai Mara, Kenya.

PLATE 3.5 Elephant damage to Acacia trees, Masai Mara, Kenya.

PLATE. 4.1 Pictured above are musk oxen (*Ovibos moschatus*) in a head to head confrontation. Horns serve several purposer. Their size and shape are primarily determined by the battles between males for a female. They are also an indication of the status of the male in the hierarchy of the hard and are recognised by both males and females.

PLATE 4.2 *Cistanche*, a close relative of the broomrape, has no chlorophyll of its own and is totally dependent on its host plant.

PLATE 5.1 The leaves of six typical sclerophyllous plants. Clockwise, from top left: holm oak (*Q. ilex*), kermes oak (*Q. coccifera*), olive (*O. europaea*), California lilac (*Ceanothus*), rosemary (*R. officinalis*) and juniper (*J. phoenicea*).

PLATE 5.2 Phrygana on Kalymnos, a Greek island off the coast of Turkey.

PLATE 5.8 Photo-montage of dune succession in the Algarve region of Portugal.
Photograph (a) shows sand being trapped by vegetation on the strandline, whilst in (b) pine trees are seen colonising the oldest dunes. Images b-e are the intermediate stages of the developing dune system.

PLATE 5.9 *Carpobrotus* invading Mediterranean sand dunes.

PLATE 5.10 The uniform pine forest in the lee of the Pilat sand dune system near Arcachon, in Les Landes of western France.

PLATE 5.11 *Eucalyptus* displacing native Mediterranean oak/pine forest.

PLATE 6.1 Well-managed meadows can be both productive and diverse. This traditional meadow in the south east of England has been managed for centuries by grazing and cutting and supports a wide range of plants including, in this particular case, green-winged orchid (*Orchis morio*).

PLATE 6.2 Monoculture: the familiar face of intensive agriculture.

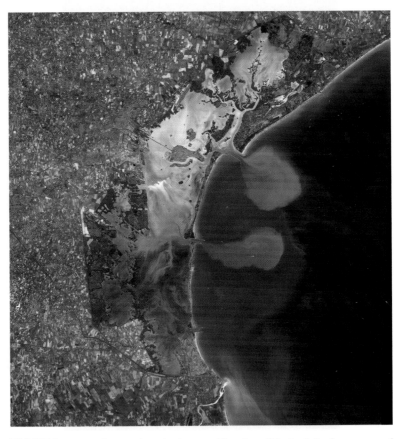

PLATE 7.3 An aerial view of Venice lagoon. The city of Venice is in the centre and is linked to the mainland by a causeway. Also clearly visible are the three inlets that connect the lagoon with Adriatic Sea.

PLATE 7.4 When viewed from space the highly eutrophic Black Sea appears darker than the oligotrophic waters of the eastern Mediterranean.

PLATE 7.5 Seabirds like this male eider (*Somateria mollisima*) are among the first casualties of marine oil-spills.

PLATE 7.6 Restoration in progress. Vegetation being used to stabilize chalk spoil produced in the construction of the Channel Tunnel that links Britain and France.

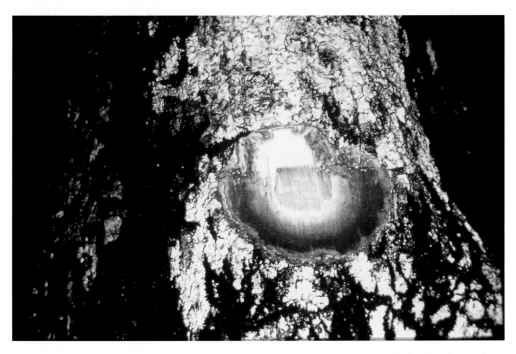

PLATE 7.7 The New Caledonian nickel tree (*Sebertia acuminata*) grows on nickel-rich soils overlying serpentine rock and secretes a blue-green nickel-rich latex.

PLATE 7.8 Heathland translocation in action. When it comes to habitat translocation, such as this patch of the West Pennine Moors in Lancashire, speed and accuracy is required to remove and replant the plant material with the minimum of damage.

PLATE 8.1 The straight lines of intensive viticulture. The monoculture of the vineyards would be regarded as the background matrix of landscape, with the ecotones represented by the sharp boundaries at patches of woodland. The roadsides or the edges of streams will be important corridors for the movement of wildlife. Languedoc, southern France.

PLATE 8.2 (a) An ecosystem stranded: abandoned boats at Muynak, formerly a fishing port on the Aral sea. (b) Wind blown dust over the Aral Sea seen by the Aqua satellite on 18 April 2003. The line shows the border between Uzbekistan in the south and Kazakhstan in the north. Maps published as late as 1991 do not show the sea divided into three distinct lobes and the central tongue of land is there seen as a relatively small island. Today, Muynak, in the south, is about 27 km from the shore.

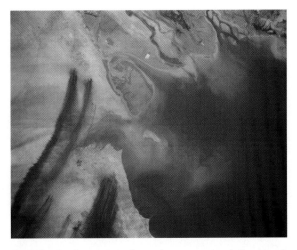

PLATE 8.3 The plumes from the Kuwaiti oil fires south of Kuwait City. This photograph, taken by the crew of the space shuttle on mission 37 is an almost vertical view. It was taken in April 1991, 2 months after the fires were ignited.

PLATE 8.4 The Teleki valley, Kenya. This montane habitat on the equator has a unique plant community reflecting the large temperature range created by the high altitude. Here giant groundsels (*Senecio* spp.) flourish with a range of adaptations to survive the very low temperatures. Different species with similar adaptations grow on the equator in the Andes and Hawaii.

PLATE 8.5 Tropical rain forest.

PLATE 8.6 East African savanna.

PLATE 8.7 Temperate broad-leaf forest.

PLATE 8.8 Prairie.

PLATE 8.9 Desert.

PLATE 8.10 Taiga.

PLATE 8.11 Tundra.

PLATE 8.12 Alpine ecosystems in a high latitude, high altitude landscape.

PLATE 8.13 Tropical forest fires in the Yucatan peninsula in Mexico, photographed on 20 April 2003 by the Aqua satellite. The red dots mark individual fires. In the past, fires from this region have caused poor air quality in areas of Texas and Oklahoma. The extent of these fires is thought to be due to the dry conditions associated with the intense El Niño event the previous winter.

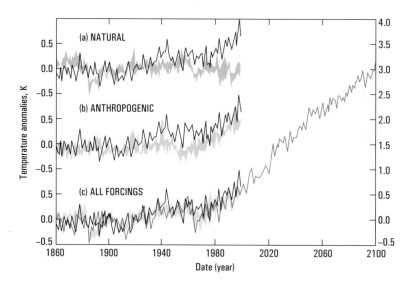

PLATE 8.14 The mean temperature of the air 1.5 m above the Earth since 1860 and its projected rise over the next 100 years. This figure, derived from a Hadley Centre GCM uses a variety of starting positions and includes the effect of sulfate aerosols. These scenarios model the effect of (a) natural, (b) anthropogenic factors and their combined effect (c). The black lines shows the observed changes since 1860 in each case and the coloured area the range of predictions from a variety of different starting positions (each of which is based on an IPCC scenario that predicts mid-range changes in greenhouse gases and set to zero for 1860). The match between the predicted and observed data is generally close. Using the left-hand scale, models of natural processes show little real effect, but human-induced warming (b) begins to be detected from 1940. Overall, and primarily due to human activity, global air temperatures are predicted to rise by 2.5 °C by 2100 (shown by the red line in (c)), a rate close to that observed over the last three decades.

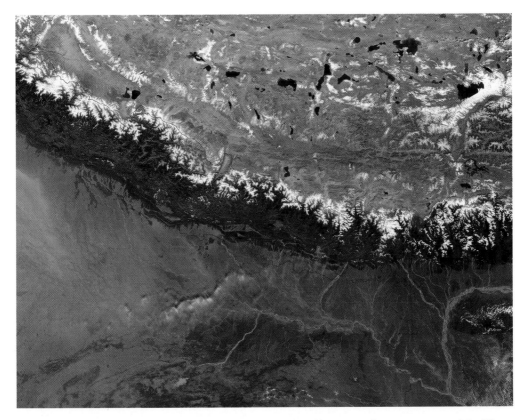

PLATE 9.1 Was the raising of the Himalaya the prime cause of human evolution? Raymo and Ruddiman propose a sequence of events that link the collision of India with Asia about 50 million years ago with the climatic changes in East Africa that gave rise to the first humans. This image shows the eastern himalayan, north of the Ganges and Brahmaputra, and encompasses the whole of Nepal. Mount Everest is to the right of centre.